AF182422

INTRODUCTION TO INTEGRATED OPTICS

CONTRIBUTORS

MICHAEL K. BARNOSKI
Hughes Research Laboratories
Malibu, California

JAMES GOELL
Bell Telephone Laboratories
Holmdel, New Jersey
(Present address: ITT, Roanoke, Virginia)

JAY H. HARRIS
University of Washington
Seattle, Washington

ERIC LEAN
IBM
T. J. Watson Research Laboratories
Yorktown Heights, New York

ROBERT MAURER
Corning Glass Works
Corning, New York

SASSON SOMEKH
Bell Telephone Laboratories
Murray Hill, New Jersey

C. L. TANG
Cornell University
Ithaca, New York

P. K. TIEN
Bell Telephone Laboratories
Holmdel, New Jersey

EDWARD WOLF
Hughes Research Laboratories
Malibu, California

AMNON YARIV
California Institute of Technology
Pasadena, California

INTRODUCTION TO INTEGRATED OPTICS

Edited by
MICHAEL K. BARNOSKI
Hughes Research Laboratories
Malibu, California

SPRINGER SCIENCE+BUSINESS MEDIA, LLC

Library of Congress Cataloging in Publication Data

Main entry under title:

Introduction to integrated optics.

"Series of lectures presented at an annual short course on integrated optics, sponsored by the University of California, Santa Barbara, in March of 1973."

Includes bibliographical references.

1. Integrated optics. 2. Optical wave guides. I. Barnoski, Michael K., 1940– ed.

TA1660.157 621.36 74-5444

ISBN 978-1-4684-2084-5 ISBN 978-1-4684-2082-1 (eBook)

DOI: 10.1007/978-1-4684-2082-1

Series of lectures presented at an annual short course on Integrated Optics, sponsored by the University of California, Santa Barbara, in March of 1973

Copyright© 1974 Springer Science+Business Media New York
Originally published by University of California in 1974
Softcover reprint of the hardcover 1st edition 1974

Santa Barbara, California

Preface

The material presented in this volume is based on a series of lectures presented at an annual short course entitled "Integrated Optics" sponsored by the University of California, Santa Barbara, in March of 1973. Each chapter has been authored by the individual lecturing on the particular subject matter contained in that chapter.

The primary intent of this book is to provide the students with a written version of the lecture material. The text treats a large variety of the basic structures and analysis techniques currently being employed in the rapidly growing area of technology termed "Integrated Optics." The volume contains both material that has previously been published in scientific journals and material which has not been published elsewhere.

Since the field of integrated optics is in a stage of rapid expansion, a complete coverage of all the experimental results is difficult. As a result the emphasis is placed on the fundamental theoretical ideas and experimental results.

The editor gratefully acknowledges the contributing authors and the institutions with which they are associated for their wholehearted cooperation in the preparation of this book. Particular thanks are extended to the Hughes Research Laboratories, a division of the Hughes Aircraft Company.

The assistance and advice of Mrs. Julie DeVaux and the help of Mrs. Anita Palmer and Mrs. Ann Kirby in the preparation of parts of the text was much appreciated. Grateful acknowledgment is also extended to the staff of the University of California, Santa Barbara, in particular, Dr. Sidney Goren, Ms. Judy Weisman, and Mr. Larry Nicklin.

The editor is particularly indebted to Dr. James F. Lotspeich for his critical reading and useful comments on the content of the three chapters written by the editor.

— M.K. Barnoski

Malibu California
December 1973

Contents

CHAPTER PAGE

Chapter 1 — GENERAL BACKGROUND REVIEW

M.K. Barnoski

Hughes Research Laboratories
3011 Malibu Canyon Road
Malibu, California 90265

1. Maxwell's Equations

Electromagnetic fields are governed by a classical set of equations developed by Maxwell. In differential form, these equations are

$$\nabla \times \overline{E} = - \frac{\partial \overline{B}}{\partial t} \tag{1}$$

$$\nabla \times \overline{H} = \frac{\partial \overline{D}}{\partial t} + \overline{J} \tag{2}$$

$$\nabla \cdot \overline{D} = \rho \tag{3}$$

$$\nabla \cdot \overline{B} = 0 \tag{4}$$

$$\nabla \cdot \overline{J} = - \frac{\partial \rho}{\partial t} \tag{5}$$

where

$\overline{E}$ = the electric field intensity

$\overline{D}$ = the electric displacement vector

$\overline{H}$ = the magnetic field intensity

$\overline{B}$ = the magnetic flux density

$\overline{J}$ = the conduction current density

ρ = the total volume charge density in the volume under consideration.

These basic equations can also be expressed in integral[1] form, that is

$$\oint_{\ell} \overline{H} \cdot d\overline{\ell} = \int_{s} \left(\frac{\partial \overline{D}}{\partial t}\right) + \overline{J} \cdot d\overline{s} \quad \text{(Ampere's law)} \tag{6}$$

$$\oint_{\ell} \overline{E} \cdot d\overline{\ell} = -\int \frac{\partial \overline{B}}{\partial t} \cdot d\overline{s} \quad \text{(Faraday's law)} \tag{7}$$

$$\int_{V} (\nabla \cdot \overline{D})\, dV = \int_{s} \overline{D} \cdot d\overline{s} = \int_{V} \rho dV \quad \text{(Gauss's law)} \tag{8}$$

$$\int_{s} \overline{B} \cdot d\overline{s} = 0 \quad . \tag{9}$$

In addition to Maxwell's field equations, involving the electric displacement vector $\overline{D}$, electric field vector $\overline{E}$, magnetic flux density $\overline{B}$, and magnetic field intensity vector $\overline{H}$, there exists a set of auxiliary equations which are dependent on the material medium in which fields being described exist. These equations are

$$\overline{D} = \tilde{\epsilon} \cdot \overline{E} \tag{10}$$

$$\overline{B} = \tilde{\mu} \cdot \overline{H} \tag{11}$$

where $\tilde{\epsilon}$ and $\tilde{\mu}$ are the material permittivity and permeability respectively, both in general being dyadics (tensors of rank 2).

2. Definitions of Various Types of Medium

The physical constants $\tilde{\epsilon}$ and $\tilde{\mu}$ are the defining constants of the material medium. If $\tilde{\epsilon}$ and $\tilde{\mu}$ are not functions of position in the material, the medium is said to be homogeneous. If, however, they have a functional dependence on position the medium is termed inhomogeneous. An anisotropic medium is one in which $\tilde{\epsilon}$ and $\tilde{\mu}$ are tensors while an isotropic medium is one in which $\tilde{\epsilon}$ and $\tilde{\mu}$ are scalar quantities. It is clear that the most general medium is inhomogeneous and anisotropic.

3. Wave Equation

Maxwell's field equations and the auxiliary equations can be manipulated so as to result in derived equations which describe the dynamics of the field vectors. For simplicity it will be assumed from this point on that the material medium is isotropic but in general inhomogeneous. The permittivity and permeability are generally complex quantities, a fact denoted by the notation $\epsilon^* = \epsilon' - i\epsilon''$ and $\mu^* = \mu' - i\mu''$ where prime and double prime denote real and imaginary parts.

The differential wave equations describing both $\overline{E}$ and $\overline{H}$ can be obtained by taking the curl of Maxwell's field equations, that is

$$\nabla \times (\nabla \times \overline{E}) = - \nabla \times \mu^* \frac{\partial \overline{H}}{\partial t} \tag{12}$$

$$\nabla \times (\nabla \times \overline{H}) = \nabla \times \epsilon^* \frac{\partial \overline{E}}{\partial t} + \nabla \times \overline{J} \ . \tag{13}$$

The following are useful vector identities[2]

$$\nabla \times (\phi\overline{A}) = \phi \nabla \times \overline{A} + \nabla\phi \times \overline{A} \tag{14}$$

$$\nabla \times \nabla \times \overline{A} = \nabla(\nabla \cdot \overline{A}) - \nabla^2\overline{A} \tag{15}$$

$$\nabla \cdot \phi\overline{A} = \phi\nabla \cdot \overline{A} + \overline{A} \cdot \nabla\phi \quad . \tag{16}$$

Utilizing these vector identity equations (12) and (13) can be reduced into the two equations

$$\nabla^2\overline{E} - \nabla\left(\frac{\rho}{\epsilon^*}\right) + \nabla\left(\frac{\nabla\epsilon^*}{\epsilon^*} \cdot \overline{E}\right) - \frac{\nabla\mu^*}{\mu^*}$$

$$x\ (\nabla \times \overline{E}) = \mu^*\epsilon^* \frac{\partial^2\overline{E}}{\partial t^2} + \mu^* \frac{\partial\overline{J}}{\partial t} \qquad (17)$$

$$\nabla^2\overline{H} + \nabla\left(\frac{\nabla\mu^*}{\mu^*} \cdot \overline{H}\right) + \frac{\nabla\epsilon^*}{\epsilon^*} \times (\nabla \times \overline{H})$$

$$= \mu^*\epsilon^* \frac{\partial^2\overline{H}}{dt^2} + \frac{\nabla\epsilon^*}{\epsilon^*} \times \overline{J} - \nabla \times \overline{J}\,. \qquad (18)$$

For most cases to be considered $\rho = 0$, $J = \sigma\overline{E}$ and μ^* is a constant. Under these conditions the above equations reduce to

$$\nabla^2\overline{E} + \nabla\left(\frac{\nabla\epsilon^*}{\epsilon^*} \cdot \overline{E}\right) = \mu^*\epsilon^* \frac{\partial^2\overline{E}}{\partial t^2} + \mu^*\sigma \frac{\partial E}{\partial t} \qquad (19)$$

$$\nabla^2\overline{H} + \frac{\nabla\epsilon^*}{\epsilon^*} \times (\nabla \times \overline{H}) = \mu^*\epsilon^* \frac{\partial^2\overline{H}}{dt^2} + \frac{\nabla\epsilon^*}{\epsilon^*}$$

$$\times\ \overline{J} - \nabla \times \overline{J}\ . \qquad (20)$$

If the material is homogeneous, further reduction is possible. The resulting equations are:

$$\nabla^2\overline{E} = \mu^*\epsilon^* \frac{\partial^2\overline{E}}{dt^2} + \mu^*\sigma \frac{\partial E}{\partial t} \qquad (21)$$

$$\nabla^2\overline{H} = \mu^*\epsilon^* \frac{\partial^2\overline{H}}{\partial t^2} + \mu^*\sigma \frac{\partial\overline{H}}{\partial t}\ . \qquad (22)$$

Without a loss in generality a time variation of the form $e^{i\omega t}$ can be assumed. The differential equations for $\overline{E}$ and $\overline{H}$ then become

$$\nabla^2 \begin{pmatrix} \overline{E} \\ \overline{H} \end{pmatrix} = -\omega^2 \left[\mu^*\left(\epsilon' - i\left[\epsilon'' + \frac{\sigma}{\omega}\right]\right)\right] \begin{pmatrix} \overline{E} \\ \overline{H} \end{pmatrix} \tag{23}$$

or

$$\nabla^2 \begin{pmatrix} \overline{E} \\ \overline{H} \end{pmatrix} = \mu^*\epsilon^* \frac{\partial^2}{\partial t^2} \begin{pmatrix} \overline{E} \\ \overline{H} \end{pmatrix} \tag{24}$$

where in this equation the complex dielectric constant ϵ^* includes dissipative effects; that is, ϵ^* is redefined to be $\epsilon^* = \epsilon' - i(\epsilon'' + \sigma/\omega)$. The vector wave equation (eq. (24)) holds for each of the scalar components of $\overline{E}$ and $\overline{H}$; that is, each component satisfies the scalar wave equation

$$\nabla^2\psi = \mu^*\epsilon^* \frac{\partial^2\psi}{\partial t^2} \quad . \tag{25}$$

The plane wave solutions of the scalar wave equation are of the form

$$\psi = \psi_o e^{i\omega t - \gamma \overline{k} \cdot \overline{r}} \tag{26}$$

where

$$\gamma = i\omega \sqrt{\mu^*\epsilon^*} = \alpha + i\beta \tag{27}$$

is the complex propagation constant and $\overline{k}$ is a unit vector in the direction of propagation. The real part of the propagation constant is the amplitude absorption coefficient and the imaginary part is the magnitude of the wave vector ($\beta\overline{k}$).

The $\overline{E}$, $\overline{H}$ and $\overline{k}$ vectors of a plane wave are all orthogonal with the ratio of the magnitude of $\overline{E}$ and $\overline{H}$ being defined as the intrinsic wave impedance

$$Z = \frac{|\overline{E}|}{|\overline{H}|} = \sqrt{\frac{\mu^*}{\epsilon^*}} \quad . \tag{28}$$

Since $\overline{E}$ and $\overline{H}$ are both orthogonal to the propagation direction $\overline{k}$, the wave is called a transverse electromagnetic wave or TEM wave.

4. Description of Other Material Parameters

Alternate parameters to ϵ^* and μ^* are often used to describe wave propagation in a material medium. Some standard parameters used are listed below

λ_o = wavelength in vacuum

λ = wavelength in material

$$n = \frac{\lambda_o}{\lambda} = \text{real refractive index} = \frac{\beta}{\frac{2\pi}{\lambda_o}}$$

$$k = \text{extinction coefficient} = \frac{\alpha\lambda}{2\pi} = \frac{\alpha}{\beta}$$

$n^* = n(1 - ik)$ = complex refractive index

$$n^* = \sqrt{\frac{\mu^*\epsilon^*}{\mu_o\,\epsilon_o}} .$$

A good review of the description of dielectrics by various sets of parameters, along with useful nomographic charts, can be found in Chapter 9 of reference 3.

5. Boundary Conditions

Maxwell's equations in integral form can be used to determine the conditions which must be satisfied by $\overline{E}$, $\overline{D}$, $\overline{H}$, and $\overline{B}$ at the boundary between two media with differing ϵ^* and/or μ^*. Since the derivation of the boundary conditions can be found in any elementary text (see, for example, Refs. 1 and 3) on electromagnetic theory the results only will be listed, these are

$$H_{1t} - H_{2t} = J_s \quad \text{or in vector form} \quad \hat{n} \times (\overline{H}_1 - \overline{H}_2) = \overline{J}_s \tag{29}$$

$$E_{1t} - E_{2t} = 0 \qquad \hat{n} \times (\overline{E}_1 - \overline{E}_2) = 0 \tag{30}$$

$$B_{1n} - B_{2n} = 0 \qquad \hat{n} \cdot \overline{B} = 0 \tag{31}$$

$$D_{1n} - D_{2n} = \rho_s \qquad \hat{n} \cdot \overline{D} = \rho_s \tag{32}$$

where t denotes the component tangent to the surface separating media (1) and (2) and n denotes the component normal to that surface. The quantities J_s and ρ_s are surface current and charge densities respectively which in the cases to be treated here are zero. The component equations can also be expressed in vector notation using the unit normal to the interface surface as shown in Fig. 1.

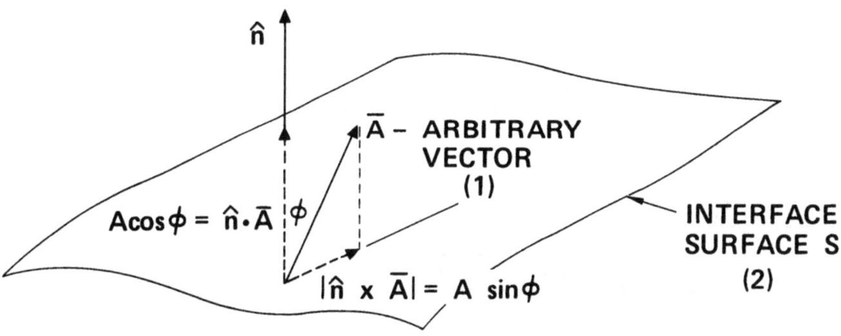

$\hat{n} \cdot \overline{A}$ = COMPONENT OF $\overline{A}$ NORMAL TO SURFACE S

$|\hat{n} \times \overline{A}|$ = COMPONENT OF $\overline{A}$ TANGENT TO SURFACE S

Fig. 1. Orientation of the normal and tangential components of an arbitrary vector $\overline{A}$ at the interface surface S.

6. Fresnel Equations

The derivation of the relationship between the incident, reflected and transmitted components of a transverse electromagnetic (TEM) wave propagating in a medium with constants ϵ_1^*, μ_1^* and which crosses an interface surface into a medium with constants ϵ_2^*, μ_2^* can also be found in any good text on electromagnetic theory. The following treatment follows that of Von Hippel (Ref. 3).

The (TEM) electric and magnetic fields for the three waves (incident, reflected, transmitted) are denoted as

<u>incident:</u>

$$\overline{E}_i = \overline{E}_o\, e^{i\omega t - \gamma_1 \overline{k}_i \cdot \overline{r}} \tag{33}$$

$$\overline{H}_i = \overline{H}_o\, e^{i\omega t - \gamma_1 \overline{k}_i \cdot \overline{r}} = \frac{1}{Z_1} (\overline{k}_i \times \overline{E}_i) \tag{34}$$

<u>reflected:</u>

$$\overline{E}_r = \overline{E}_1\, e^{i\omega t - \gamma_1 \overline{k}_r \cdot \overline{r}} \tag{35}$$

$$\overline{H}_r = \overline{H}_1\, e^{i\omega t - \gamma_1 \overline{k}_r \cdot \overline{r}} = \frac{1}{Z_1} (\overline{k}_r \times \overline{E}_r) \tag{36}$$

<u>transmitted:</u>

$$\overline{E}_t = \overline{E}_2\, e^{i\omega t - \gamma_2 \overline{k}_t \cdot \overline{r}} \tag{37}$$

$$\overline{H}_t = \overline{H}_2\, e^{i\omega t - \gamma_2 \overline{k}_t \cdot \overline{r}} = \frac{1}{Z_2} (\overline{k}_t \times \overline{E}_t) \tag{38}$$

It is customary to decompose the fields into components normal and parallel to the plane of incidence which is defined as the plane containing the incident unit propagation vector $\overline{k}$ and the unit normal $\hat{n}$ to the interface surface as shown in Fig. 2.

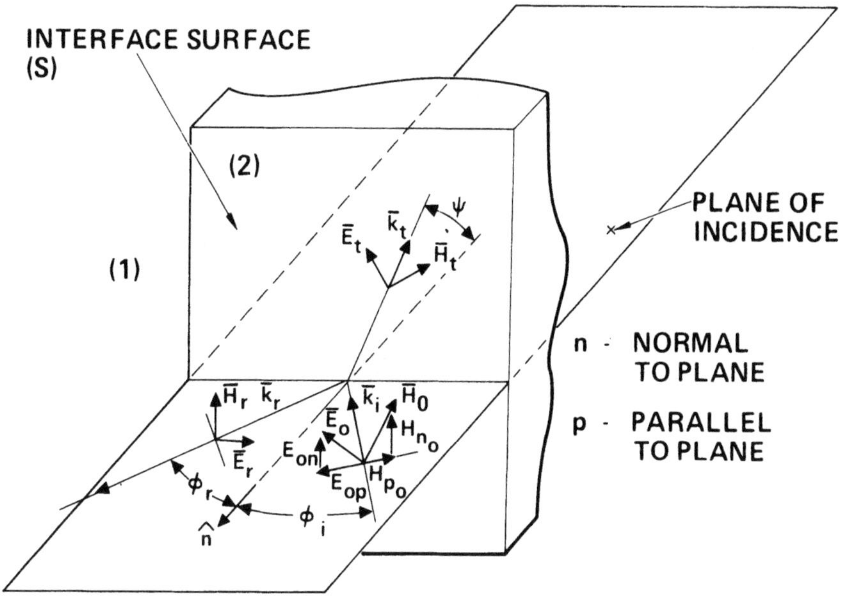

Fig. 2. Illustration shows decomposition of the electric and magnetic field quantities into components normal to and parallel to the plane of incidence.

Matching boundary conditions at the interface results in the following relationships

$$\hat{n} \times (\overline{E}_o + \overline{E}_1) = \hat{n} \times \overline{E}_2 \tag{39}$$

$$\hat{n} \times (\overline{H}_o + \overline{H}_1) = \hat{n} \times \overline{H}_2 \tag{40}$$

which can be reduced to the Fresnel equations and

$$\gamma_1 \overline{k}_i \cdot \overline{r} = \gamma_1 \overline{k}_r \cdot \overline{r} \tag{41}$$

$$\gamma_1 \overline{k}_i \cdot \overline{r} = \gamma_2 \overline{k}_t \cdot \overline{r} \tag{42}$$

which can be reduced to Snell's laws of reflection and refraction. The resulting relationships are

$$\frac{\sin \phi}{\sin \psi} = \frac{\gamma_2}{\gamma_1} = \frac{\sqrt{\mu_2^* \epsilon_2^*}}{\sqrt{\mu_1^* \epsilon_1^*}} \qquad \text{Snell's Law of Refraction} \qquad (43)$$

$$\phi_i = \phi_r \qquad \text{Snell's Law of Reflection} \quad (44)$$

Fresnel Equations (field amplitude reflection and transmission coefficients)

$$r_{E_n} = \left(\frac{E_1}{E_o}\right)_n = \frac{Z_2 \cos \phi - Z_1 \cos \psi}{Z_2 \cos \phi + Z_1 \cos \psi} = - r_{H_p}$$

$$= - \left(\frac{H_1}{H_o}\right)_p \qquad (45)$$

$$r_{E_p} = \left(\frac{E_1}{E_o}\right)_p = \frac{Z_2 \cos \psi - Z_1 \cos \phi}{Z_2 \cos \psi + Z_1 \cos \phi} = - r_{H_n}$$

$$= - \left(\frac{H_1}{H_o}\right)_n \qquad (46)$$

$$t_{E_n} = \left(\frac{E_2}{E_o}\right)_n = \frac{2Z_2 \cos \phi}{Z_2 \cos \phi + Z_1 \cos \psi} = \frac{Z_2}{Z_1} t_{H_p} \qquad (47)$$

$$t_{E_p} = \left(\frac{E_2}{E_o}\right)_p = \frac{2Z_2 \cos \phi}{Z_2 \cos \psi + Z_1 \cos \phi} = \frac{Z_2}{Z_1} t_{H_n} \qquad (48)$$

The Fresnel equations are, of course, only directly applicable when the second medium is infinitely thick.

Equations (45) through (48) relate the amplitudes of the incident, reflected and transmitted fields. The energy coefficients of reflection and transmission are related[3] to amplitude coefficients in the following fashion. The energy coefficient of reflection or reflectivity is

$$R = \frac{|E_1|^2}{|E_o|^2} = |r_E|^2 \tag{49}$$

while the energy coefficient of transmission is

$$T = \left| t_E^2 \frac{Z_1 \cos \psi}{Z_2 \cos \phi} \right| . \tag{50}$$

Conservation of energy requires that

$$R + T = 1 . \tag{51}$$

7. Special Examples

In the first example to be considered the Fresnel equations will be applied to the interface between two loss-free media, since there is no loss in $\epsilon_1^* = n_1^2$ and $\epsilon_2^* = n_2^2$. The amplitudes of the reflection coefficients for polarization normal and parallel to the plane of incidence for the separate cases of reflection when $n_2 > n_1$ and $n_2 < n_1$ are shown in Figs. 3 and 4 as functions of the angle of incidence. The Brewster angle and the critical angle for total internal reflection are defined as (here rays are always incident from (1) onto (2))

$$\phi_B = \text{Brewster's angle} = \tan^{-1}\left(\frac{n_2}{n_1}\right) \tag{52}$$

$$\phi_t = \text{critical angle} = \sin^{-1}\left(\frac{n_2}{n_1}\right) \quad n_1 > n_2 . \tag{53}$$

For the case where the wave is propagating from a dense to a rare medium ($n_1 > n_2$) it is interesting to consider what happens to the transmitted beam for angles of incidence greater than or equal to the critical angle ϕ_t. It is easy to show that for $\phi \geq \phi_t$

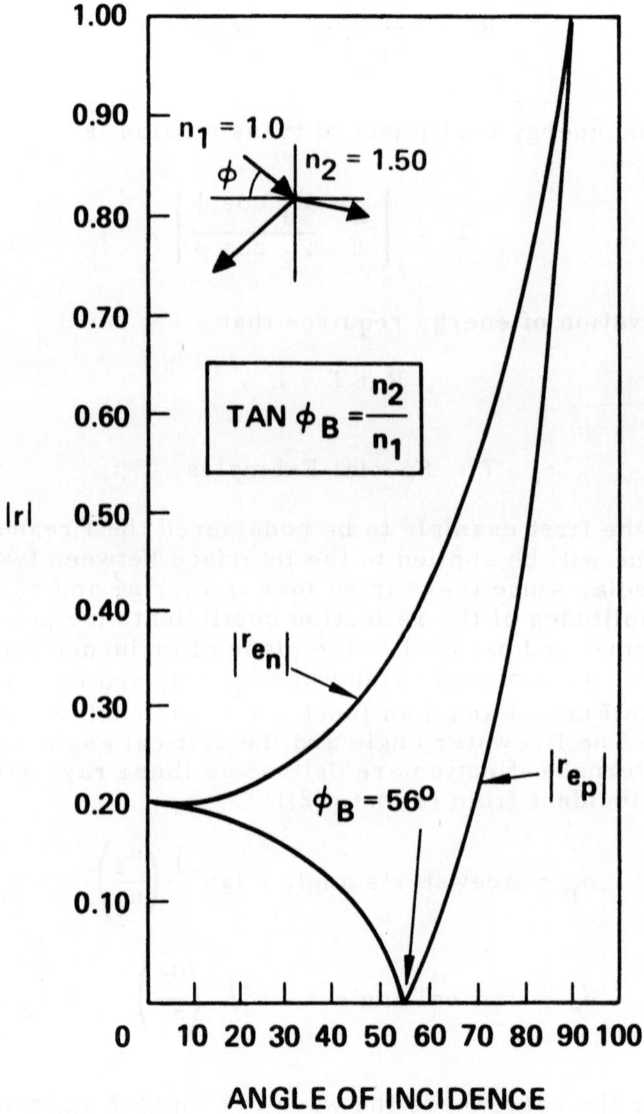

Fig. 3. Magnitude of reflection coefficient as a function of angle of incidence.

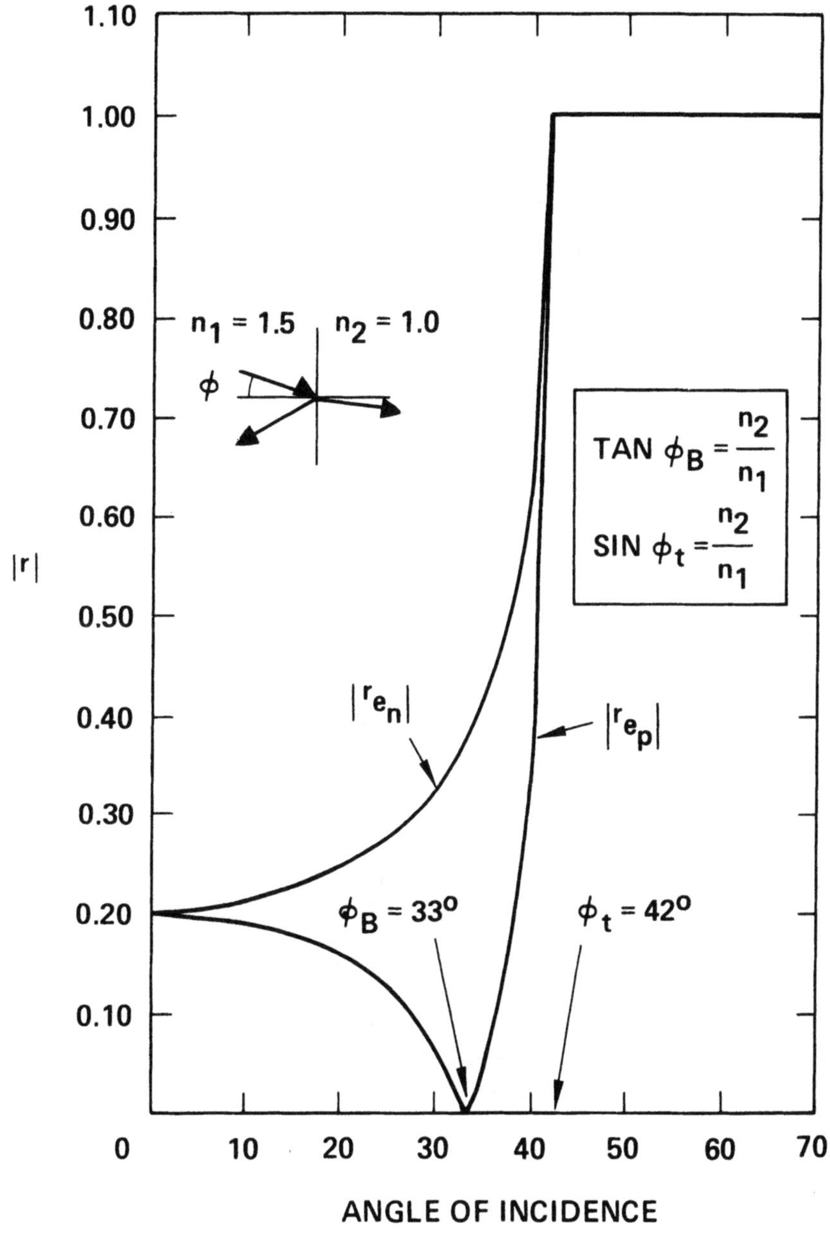

Fig. 4. Magnitude of reflection coefficient as a function of angle of incidence.

$$R_n = R_p = 1 \text{ and hence } T = 1 - R = 0 . \tag{54}$$

There is no power carried by the transmitted wave, it is all reflected.

Using the coordinate system shown in Fig. 5 the transmitted electric field component can be expressed as

$$E_t = E_2\, e^{i\omega t - \gamma_2 \overline{k}\cdot r} = E_2\, e^{i\omega t - \gamma_2 (z \cos\psi + y \sin\psi)} \tag{55}$$

where since medium (2) is loss free

$$\gamma_2 = i\,\frac{2\pi}{\lambda_2} \tag{56}$$

and Snell's law of refraction becomes

$$\frac{\sin\phi}{\sin\psi} = \frac{n_2}{n_1} = \frac{\lambda_1}{\lambda_2} . \tag{57}$$

For reasons which will soon become clear it is of interest to express $\cos\psi$ in terms of the angle of incidence ϕ. The cosine of the angle of refraction is related to the incident angle as follows

$$\cos\psi = \sqrt{1 - \sin^2\psi} = \frac{i\sqrt{\sin^2\phi - \left(\frac{n_2}{n_1}\right)^2}}{\left(\frac{n_2}{n_1}\right)} . \tag{58}$$

Substitution of these expressions for $\sin\psi$ and $\cos\psi$ into the transmitted field results in

$$E_t = E_2 e^{i\omega t}\, \underbrace{e^{-\frac{2\pi}{\lambda_1}\sqrt{\sin^2\phi - \left(\frac{n_2}{n_1}\right)^2}\, z}}_{\text{amplitude}}\; \underbrace{e^{-i\left(\frac{2\pi}{\lambda_1}\sin\phi\right) y}}_{\text{phase}} . \tag{59}$$

The planes of constant amplitude are

$$\frac{2\pi}{\lambda_1} \sqrt{\sin^2 \phi - \left(\frac{n_2}{n_1}\right)^2} \, z = \text{constant} \tag{60}$$

while the planes of constant phase are

$$\frac{2\pi}{\lambda_1} (\sin \phi) \, y = \text{constant} \quad . \tag{61}$$

The expression for the electric field (eq. (59)) is the transmitted field for the general case of any two loss-free medium.

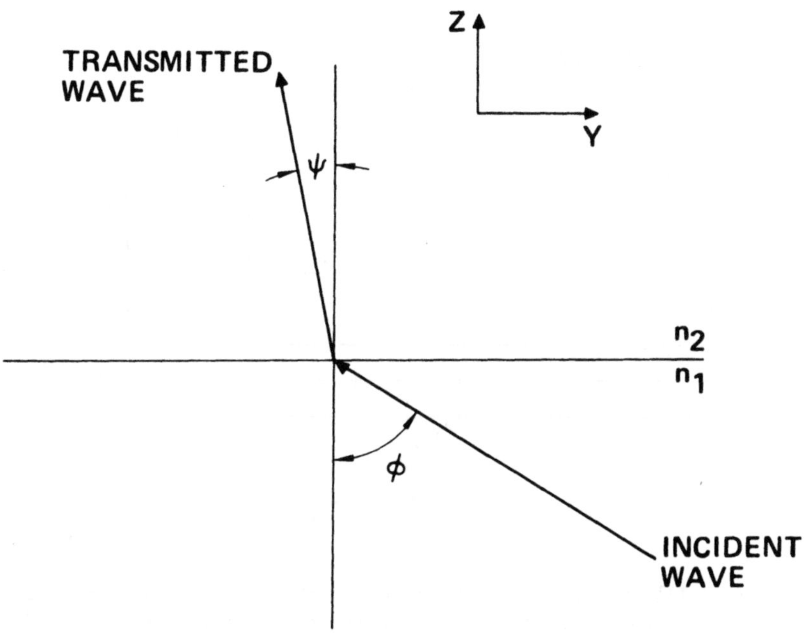

Fig. 5. Illustration shows coordinate system used in calculating transmitted field components.

When $n_1 < n_2$ then $\sqrt{\sin^2 \phi - (n_2/n_1)^2}$ is imaginary for all angles ϕ and hence the exponential containing the z variation becomes a phase term. As to be expected the transmitted field is a plane wave with a constant amplitude.

When $n_1 > n_2$ then $\sqrt{\sin^2 \phi - (n_2/n_1)^2}$ is again imaginary for angles of incidences such that $\sin \phi < n_2/n_1$. For these angles ϕ the transmitted wave is a plane wave of constant amplitude. For angles of incidence ϕ such that $\sin \phi > n_2/n_1$ the $\sqrt{\sin^2 \phi - (n_2/n_1)^2}$ is real. In this case the planes of constant amplitude are not parallel to the planes of constant phase. As shown in Fig. 6, the planes of constant phase are perpendicular to surface normal whereas the planes of constant amplitude are parallel to the surface normal. Hence for angles such that $\sin \phi > n_2/n_1$ the transmitted field is not a TEM wave. The amplitude of the transmitted field decreases exponentially from the surface with a decay constant

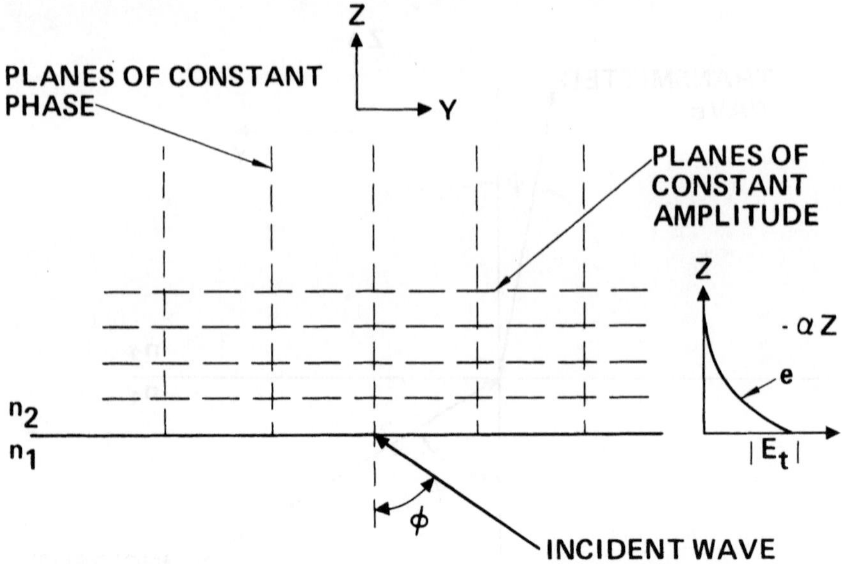

Fig. 6. Illustration shows planes of constant phase and planes of constant amplitude of the transmitted field component under the condition of total internal reflection.

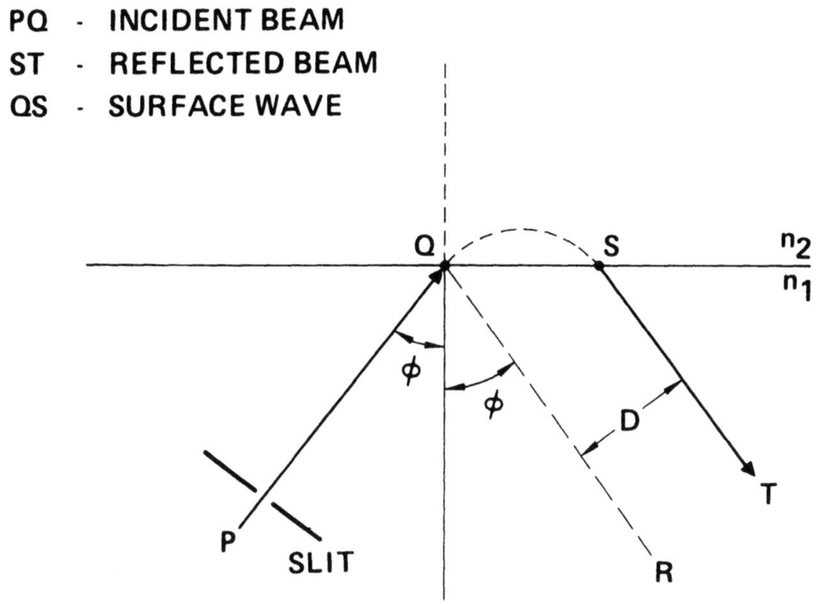

Fig. 7. Diagram illustrating the Goos-Hänchen experiment (after Von Hippel, Ref. 3).

$$\alpha = \frac{2\pi}{\lambda_1} \sqrt{\sin^2 \phi - \left(\frac{n_2}{n_1}\right)^2} \quad . \tag{62}$$

Since the planes of constant phase are perpendicular to the surface the transmitted wave is a surface wave with an amplitude which decays exponential away from the surface. The solutions presented assume that medium (2) is infinitely thick which is essentially the case if it is of finite thickness of at least several 1/e lengths; that is, if the back surface does not extend appreciably into the decaying exponential. If this is not the case the total internal reflection can be frustrated; that is, power can be made to flow in the transmitted surface wave. This is the foundation of the prism film coupler.

An optical beam incident at an interface under the condition of total internal reflection $\phi > \phi_t$ was found by F. Goos and H. Hänchen[4] not to be directly reflected at the point of incidence, but to penetrate for some distance into the material and emerge at a distance D parallel to the surface at an angle equal to the angle of reflection as shown in Fig. 7. The distance D was determined to be

$$D = 0.52\, n_2 \frac{\lambda_1}{\sqrt{\sin^2 \phi - \left(\frac{n_2}{n_1}\right)^2}} \qquad (63)$$

or

$$= 0.52\, n_2 \frac{2\pi}{\alpha} \quad . \qquad (64)$$

The transmitted field (eq. (59)) can be expressed in terms of the shift D,

$$E_t = E_2\, e^{-2\pi(0.52)\, n_2 \frac{z}{D}}\, e^{-i\left(\frac{2\pi}{\lambda_1} \sin \phi\right) y}\, e^{i\omega t} \ . \qquad (65)$$

As the angle of incidence is decreased toward the limiting angle ϕ_t the distance D increases as does the depth of penetration. In the limit when ϕ approaches ϕ_t then D approaches infinity and α approaches zero, the transmitted wave travels as a surface wave of uniform amplitude parallel to the boundary.

In the previous example both media were assumed lossless. Although this is often a valid assumption, a better approximation to the often encountered air-material interface is the more general case where medium (1) is again loss free but medium (2) is any arbitrary material. The Fresnel equations in this case are extremely complicated. The transmitted field components are not TEM. The planes of constant phase and constant amplitude are not coplanar but at an angle which depends on the direction of propagation of the refracted beam. For the interested reader a rather complete analysis of this more general case can be found in Chapter 20 of Ref. 3.

8. Separation of Longitudinal and Transverse Coordinates

For the majority of the boundary value problems dealt with it is possible to separate the variable. That is, the fields can be written as the product of a function which depends only on the coordinates transverse to the propagation direction with one which is dependent only on the longitudinal coordinate.

An arbitrary field vector can therefore be expressed as

$$\overline{F} = \overline{f}_T (\mu, v)\, e^{-\gamma z} + \overline{f}_z (\mu, v)\, e^{-\gamma z} \qquad (66)$$

where (μ, v) are arbitrary transverse coordinates, $\overline{f}_T (\mu, v)$ is a vector in the plane of the transverse coordinates, f_z is a vector parallel to the longitudinal direction of propagation z.

After some manipulation it can be shown that the transverse components of $\overline{E}$ and $\overline{H}$ can be expressed in terms of the longitudinal components E_z and H_z; that is,

$$\overline{H}_T = - i \frac{k}{Z k_c^2} (\hat{\mu}_z \times \nabla_T E_z) - \frac{\gamma}{k_c^2} \nabla_T H_z \qquad (67)$$

$$\overline{E}_T = i \frac{kZ}{k_c^2} (\hat{\mu}_z \times \nabla_T H_z) - \frac{\gamma}{k_c^2} \nabla_T E_z \qquad (68)$$

$$k_c^2 = k^2 + \gamma^2 \qquad (69)$$

$$k^2 = \omega^2 \mu^* \epsilon^* \qquad (70)$$

$$Z = \sqrt{\frac{\mu^*}{\epsilon^*}} \qquad (71)$$

TABLE I

Listing of the Various Types of Modes

Nomenclature	Longitudinal Components	Transverse Components	Propagation Constant
TEM transverse electro-magnetic	$E_z = 0$ $H_z = 0$	E_T, H_T	$\gamma = i\omega\sqrt{\mu^*\epsilon^*}$
TE transverse electric	$E_z = 0$ $H_z \neq 0$	E_T, H_T	$\gamma = (k_c^2 - \omega^2\mu^*\epsilon^*)^{1/2}$
TM transverse magnetic	$H_z = 0$ $E_z \neq 0$	E_T, H_T	$\gamma = (k_c^2 - \omega^2\mu^*\epsilon^*)^{1/2}$
HE or EH hybrid	$H_z \neq 0$ $E_z \neq 0$	E_T, H_T	$\gamma = (k_c^2 - \omega^2\mu^*\epsilon^*)^{1/2}$

where $\hat{\mu}_z$ is a unit vector in the z direction. The longitudinal components E_z and H_z must each (if they are both nonzero) be determined from the wave equations. Equations (67) to (71) are valid only for homogeneous materials.

An important point to note is that the transverse components $\overline{E}_T$ and $\overline{H}_T$ are completely determined by the longitudinal components E_z and H_z. The field components E_z and H_z are determined by the wave equation and the boundary conditions by the equation

$$\nabla_T^2 \begin{pmatrix} E_z \\ H_z \end{pmatrix} + k_c^2 \begin{pmatrix} E_z \\ H_z \end{pmatrix} = 0 \quad . \qquad (72)$$

The wave equation has therefore been reduced to Helmholtz's equation. The set of all solutions to Helmholtz equation which also satisfy the boundary conditions are defined as modes.

9. Definition of Various Types of Modes

A straightforward nomenclature, which is based on whether or not the boundary conditions require the existence of either E_z or H_z, separately or simultaneously, has been developed. A listing of the various types of modes is given in Table I. The constant k_c which appears in the propagation constant for TE, TM, and the hybrid modes depends on the boundary conditions. One-dimensional slab optical waveguides can support TE and TM modes, while circular fiber waveguides support only hybrid modes.

REFERENCES

1. R.E. Collin, Field Theory of Guided Waves, McGraw-Hill, Inc., New York (1960).

2. C.R. Wylie, Jr., Advanced Engineering Mathematics, 3rd ed., McGraw-Hill (1966).

3. A. Von Hippel, Dielectrics and Waves, John Wiley and Sons, Inc., New York (1954).

4. F. Goos and H. Hänchen, "A New and Fundamental Experiment on Total Reflection," Ann. Phys. (Leipz.), Series 6, 1 333 (1947).

Chapter 2 - FILM WAVEGUIDES AND ZIG ZAG WAVES

P. K. Tien

Bell Telephone Laboratories, Incorporated
Holmdel, New Jersey 07733

I. AN INTRODUCTION OF FILM-WAVEGUIDES-ZIGZAG WAVES

Film-waveguides are the basic structures for both the passive and active devices in integrated optical circuits. The physical principles involved in these waveguides and other related thin-film structures, coupled with those of the modern laser electronics, form the foundation of this new and exciting field of integrated optics. This lecture is to provide a physical picture of the waves in film-waveguides and of their formation into a waveguide mode. For this, we introduce the concepts of zigzag waves[1,2,3] and describe the various properties of such waves based on ray and wave-optics. After having gained some familiarity with these waves, we will show how these concepts can be used advantageously to solve the problems of thin-film optics.

An optical film-waveguide is simply a sheet of thin film deposited on a substrate, which has a refractive index, n_1, larger than that of the substrate, n_0, below and also larger than that of the air-space, n_2, above (Fig. 1). We use the subscripts 0, 1 and 2 for the substrate, film and air-space, respectively. Such a three-layered arrangement, with the highest refractive index in the middle layer, forms a lens-like medium which has a natural tendency of focusing a light beam into the film. Because of this focusing action, it is not surprising that a thin

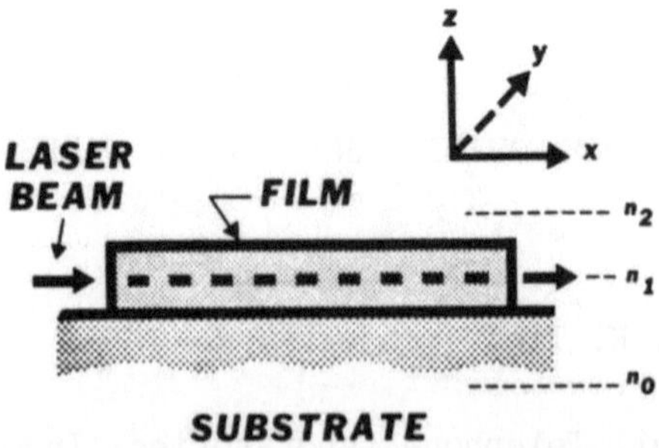

Figure 1 A film-waveguide.

film can be used as a waveguide. Consider in Fig. 2 a ray of light starting in the film and propagating toward the upper surface of the film. Since the refractive index of the film is larger than that of the air space, and if in addition, the angle between the surface normal and the direction of light wave propagation, θ_1, is larger than the critical angle ($\theta_c = \sin^{-1} n_2/n_1$), the light will be totally reflected at the upper surface. The reflected light then propagates toward the lower surface of the film. For the same reason, the light is also totally reflected at the lower surface (except that, in this case, $\theta_c = \sin^{-1} n_0/n_1$). Therefore, the light wave propagates in the film following a zigzag path and may be called a zigzag wave. Such zigzag waves involve the total reflection phenomenon in optics and have some unique properties which will be described below.

(i) <u>Dual Concept of Ray- and Wave-Optics</u>. In wave-optics, the ray of light discussed above represents a plane wave. Let the film be in the x-y plane. Imagine in Fig. 3 a plane wave which extends infinitely both in the x

Figure 2 A zigzag wave in a film-waveguide.

and y directions, is reflected back and forth between the upper and lower surfaces of the film in a zigzag manner. We will switch our discussion frequently between ray- and wave-optics. Ray-optics gives us a true sense of motion. It singles out a particular ray of light from this infinitely wide plane wave and shows how this ray zigzags through space. On the other hand, wave-optics makes us realize that there is a sea of electromagnetic fields associated with the plane wave which fills the entire space of the film. As the wave is repeatedly reflected between the two film surfaces, this plane wave folds back and forth into itself. The resultant wave in the waveguide contains thus the sum of the fields of the original wave and those of the reflected waves. It is then clear that the wave intensity can build up rapidly only if the fields of the original wave and those of the reflected waves add in phase. This phase condition produces a resonant effect, and the wave configuration which satisfies this condition manifests itself as a waveguide mode. We will show later that a film-waveguide can have many different waveguide modes.

(ii) The A and B Waves. Our dual concept of a zigzag wave based on ray- and wave-optics is reinforced by simultaneously a bookkeeping system which can be used to track the phase of an optical ray as well as the phase of a plane wave at any point along the waveguide. In most cases, ray-optics is more convenient to use. It simplifies the problem and so gives a clear physical picture. It is thus reassuring to know that precisely the same mathematics and the same results found to be valid in ray-optics apply to wave-optics as well. Our bookkeeping system starts by designating the downward leg of the zigzag wave as the A wave and the upward leg of the zigzag wave as the B wave (Fig. 2). Different A and B waves are denoted by A, A',

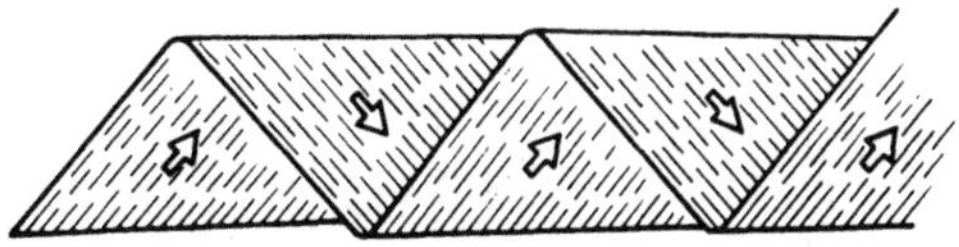

Figure 3 In wave-optics, a zigzag wave may be considered as a plane-wave which is infinitely wide in the x-direction and which is folded back and forth between the upper and lower film-surfaces in a zigzag manner.

A" ..., and B, B', B" ..., respectively. They can be expressed mathematically as

$$A \exp\{-\omega t+i\beta x-ibx\} \quad (1)$$

and

$$B \exp\{-i\omega t+i\beta x+ibz\}, \quad (2)$$

respectively. Here, A and B are the amplitudes of the waves. The film is in the x-y plane and the z-axis is along the thickness of the film. These waves can be represented by the wave vectors, $\vec{A}$ and $\vec{B}$, respectively. Since they are plane waves in the film, the wave vectors have a magnitude kn_1, where $k = \omega/c$, ω being the laser angular frequency and c being the velocity of light in vacuo. $\vec{A}$ and $\vec{B}$ have the same horizontal component, $\beta = kn_1 \sin\theta_1$. The $\vec{A}$ wave vector has a downward vertical component, $-b_1 = -kn_1\cos\theta_1$, and the $\vec{B}$ wave vector, an upward vertical component, $+b_1 = +kn_1\cos\theta_1$. We can easily see that (1) and (2) apply to both ray- and wave-optics.

(iii) <u>Electric and Magnetic Field</u>. In this lecture, for the purpose of simplification, we consider only TE waves which contain the components E_y, H_x, and H_z. Since A and B waves are plane waves, the magnetic field must be normal to both the electric field and the direction of wave propagation. It has a magnitude of n_1E_y in Gaussian units. For those who are familiar with the relations between E and H of a plane wave, we can write the following expressions immediately:

$$A_1 \text{ wave:}\quad E_y = A \exp[-i\omega t+i\beta x-ib_1z]$$
$$H_x = An_1\cos\theta_1 \exp[-i\omega t+i\beta x-ib_1z]$$
$$H_z = An_1\sin\theta_1 \exp[-i\omega t+i\beta x-ib_1z] \quad (3)$$

and

$$B_1 \text{ wave:}\quad E_y = B \exp[-i\omega t+i\beta x+ib_1z]$$
$$H_x = -Bn_1\cos\theta_1 \exp[-i\omega t+i\beta x+ib_1z]$$
$$H_z = Bn_1\sin\theta_1 \exp[-i\omega t+i\beta x+ib_1z] \quad (4)$$

The direction of the wave vector and those of the electric and magnetic fields of an A wave are illustrated in Fig. 4. The expressions shown in (3) and (4), of course satisfy the wave equation

$$\frac{\partial^2 E_y}{\partial x^2} + \frac{\partial^2 E_y}{\partial z^2} = -k^2 n_1^2 E_y . \qquad (5)$$

We assume in (5), $\partial/\partial y = 0$, since we consider a light beam which extends infinitely and uniformly in the y direction. The fields therefore vary only in the x and z directions. Substituting (3) and (4) into (5), we have

$$\beta^2 + b_1^2 = (kn_1)^2 \qquad (6)$$

which agrees with our earlier statement that

$$\beta = kn_1 \sin\theta_1$$

and

$$b_1 = kn_1 \cos\theta_1$$

are simply the horizontal and vertical components of the wave vectors $\vec{A}$ or $\vec{B}$.

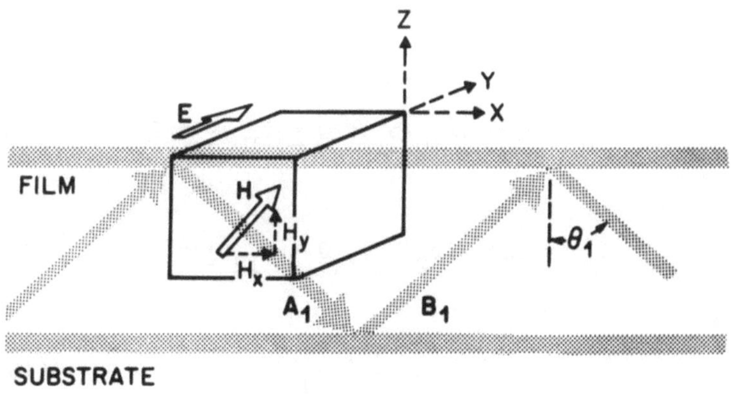

Figure 4 The downward and upward legs of a zigzag wave are denoted as the A and B waves, respectively. The figure illustrates the directions of the electric and magnetic fields in an A (TE) wave.

(iv) Waveguide Modes and the Total Reflection Phenomenon. The total reflection phenomenon[4] and its associated Goos-Haenchen shift[5] have been studied by many authors. Consider in Fig. 5, a ray starting from x_{n-1} which is reflected into B at the lower surface (at $z=-W_{10}$), and which then is reflected into A' at the upper surface (at $z=+W_{12}$). There is a complete zigzag between x_{n-1} and x_n. We choose the upper surface of the film at $z = +W_{12}$, the lower surface at $z = -W_{10}$, and the origin $z = 0$ somewhere between the upper and lower surfaces; the thickness of the film is thus $W = W_{10} + W_{12}$.

Now let us slice the film into two parts at $z = 0$. The upper part of the film contains the space $z = 0$ and $z = W_{12}$, and the lower part contains the space between $z = 0$ and $z = -W_{10}$. The interesting phenomenon in optical total reflection is that B and A waves in the lower part of the film must have the same amplitude and their combined electric field at $(z=-W_{10})$ and $(x=x_{n'})$ has a magnitude

$$|E_y|(z=-W_{10},\ x=x_{n'}) = 2A \cos \Phi_{10}. \tag{7}$$

Similarly A' and B waves in the upper part of the film must have the same amplitude and their combined electric field at $(z=W_{12})$ and $(x=x_n)$ has a magnitude

$$|E_y|(z=+W_{12},\ x=x_n) = 2B \cos \Phi_{12}. \tag{8}$$

According to Born and Wolf,[4] we have

$$\Phi_{10} = \tan^{-1}(p_0/b_1),\ 0 \le \Phi_{10} \le \pi/2; \tag{9a}$$

$$\Phi_{12} = \tan^{-1}(p_2/b_1),\ 0 \le \Phi_{12} \le \pi/2; \tag{9b}$$

also,

$$p_0^2 = \beta^2 - (kn_0)^2; \tag{10a}$$

$$p_2^2 = \beta^2 - (kn_2)^2. \tag{10b}$$

We note interestingly that in order to satisfy (7) and (8) simultaneously and for A, B, and A' to be the

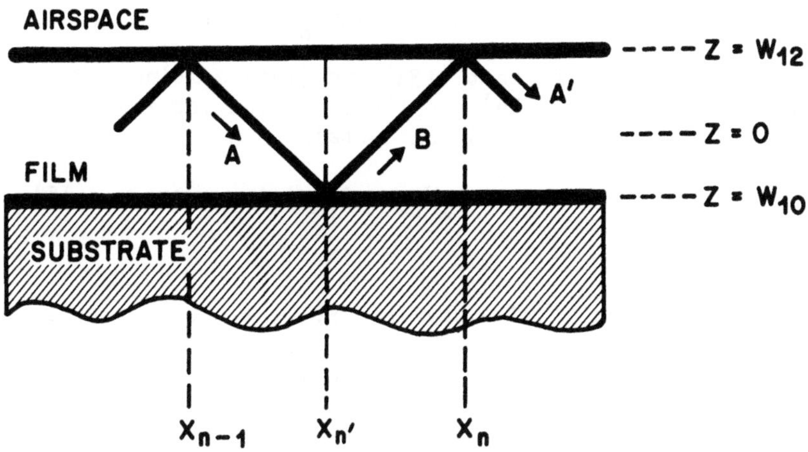

Figure 5 To study the waveguide mode, we consider a ray of light A starting from x_{n-1}. It is reflected into B at $x_{n'}$ and the B wave is then reflected into the A' wave at x_n.

members of a continuous zigzag wave in a film, we must have:

$$A = B = A' \tag{11}$$

and

$$b_1 W_{10} = \Phi_{10} + m'\pi > 0; \tag{12a}$$

$$b_1 W_{12} = \Phi_{12} + m''\pi > 0; \tag{12b}$$

Here m' and m" can be any integer 0, 1, 2, ... For then, we can write according to our recipe (1) and (2)

$$A(x_{n'}): \quad E_y = A \exp[-i\omega t + i\beta x_{n'} + i\Phi_{10} + im'\pi]; \tag{13a}$$

$$B(x_{n'}): \quad E_y = A \exp[-i\omega t + i\beta x_{n'} - i\Phi_{10} - im'\pi], \tag{13b}$$

and

$$B'(x_n): \quad E_y = A \exp[-i\omega t + i\beta x_n + i\Phi_{12} + im''\pi]; \tag{14a}$$

$$A'(x_n):\quad E_y = A\exp[-i\omega t + i\beta x_n - i\Phi_{12} - im''\pi]. \qquad (14b)$$

We can easily verify that (13a) and (13b) combine satisfying (7) and (14a) and (14b) combine satisfying (8). Combining (12a) and (12b), one obtains

$$b_1W = b_1W_{10} + b_1W_{12} = \Phi_{10} + \Phi_{12} + m\pi \qquad (15)$$

and

$$m = m' + m'' = \text{any integer } 0, 1, 2... \qquad (16)$$

Eq. (15) is the important waveguide mode equation and m is the order of the mode. For a given m and θ_1 (note that θ_1 determines b_1, p_0 and p_2 and thus Φ_{10} and Φ_{12}), the thickness of the film, W, must satisfy (15) in order for the wave in the film to form a waveguide mode. Alternatively, for a given film of thickness W, the angle θ_1 of the zigzag wave can be computed from (15).

Let us investigate (13a) to (14b) more critically. Divide (13b) by (13a), and (14b) by (14a), we have

$$B(x_{n'})/A(x_{n'}) = \exp[-2\Phi_{10}] \qquad (17)$$

$$A'(x_n)/B(x_n) = \exp[-2\Phi_{12}]. \qquad (18)$$

The zigzag wave therefore, suffers a phase change of $(-2\Phi_{10})$ because of the total reflection at the lower surface and it suffers another phase change $(-2\Phi_{12})$ because of the total reflection at the upper surface. Now consider a complete zigzag between x_{n-1} and x_n. Let the zigzag wave considered starting at $x = x_n$ of the same surface. Because of the distance $x_n - x_{n-1}$, there is a phase change, during this zigzag, $\beta(x_n - x_{n-1})$, the wave being progressing as a traveling wave with the wave number β. Vertically in the z direction, the zigzag wave traces the thickness of the film twice, once between x_{n-1} and $x_{n'}$, and another between $x_{n'}$ and x_n; the phase change associated with this is $2b_1W$. In addition, there are phase changes $-2\Phi_{10}$ and $-2\Phi_{12}$, at the upper and lower surfaces, because of total reflection. The combined phase change because of the vertical part of the wave motion is $2b_1W - 2\Phi_{10} - 2\Phi_{12}$ which is equal to $2m\pi$ according to the mode equation (15). Therefore, when a zigzag wave forms a waveguide mode, the wave progresses horizontally as a traveling wave. The total phase change due to the vertical part of the wave motion is a complete cycle $2m\pi$; the wave is resonant in the vertical direction.

So far we have considered A, B, and A' as optical rays. Returning to wave-optics, A, B and A' are plane waves; each of them extends infinitely in the x direction and are superposed one on top of the other. Based on the argument given in the last paragraph, the A' and A waves at $x = x_{n-1}$ and $x = x_n$, respectively can be expressed such as

$$A' \ (x=x_n,\ z=W_{12}): \quad E_y = A \exp[-i\omega t+i\beta x_n-ib_1W_{12}]$$

$$A \ (x=x_{n-1},\ z=W_{12}): \quad E_y = A \exp[-i\omega t+i\beta x_{n-1}-ib_1W_{12}]$$

They, in fact, belong to the same plane wave expressed by (1); the A and A' waves have then exactly the same phase at any x,z,t, and the fields of these two waves add numerically. When a zigzag wave forms a waveguide mode, the wave is reflected into itself after a zigzag. As the wave progresses it is repeatedly reflected back and forth between the two surfaces of the film. The reflected waves are in phase with the original wave and the resultant field in the waveguide is the numerical sum of the field of the original wave and those of the reflected waves. The field in the waveguide, therefore, builds up rapidly while the wave is progressing in the x direction. The waveguide mode is a resonant phenomenon.

(v) Field Distribution of a Waveguide Mode. Combining (13a) and (13b) and (14a) and (14b) and also using (3) and (4), we have at the lower film surface,

$$E_y(x_{n'}) = A(x_{n'}) + B(x_{n'})$$

$$= 2A \cos \Phi_{10} \exp[-i\omega t+i\beta x_{n'}], \qquad (19)$$

$$H_x(x_{n'}) = -n_1\cos\theta_1 [A(x_{n'})-B(x_{n'})]$$

$$= -i2An_1\cos\theta_1\sin\Phi_{10}\exp[-i\omega t+i\beta x_{n'}] \qquad (20)$$

The electric and magnetic fields at the lower surface are $2A \cos \Phi_{10}$ and $-i2An_1\cos\theta_1\sin\Phi_{10}$, respectively. Similarly, those at the upper surface are $2A \cos \Phi_{12}$ and

$i2An_1\cos\theta_1\sin\Phi_{12}$. These quantities will be useful in the later analysis. At any point inside the film, the field E_y varies as cos b_1z, can be seen by adding (3) and (4). The field distribution is plotted in Fig. 6. We see that there is at least one maximum for E_y in the waveguide. It is instructive to divide the waveguide into three layers. The top layer has a thickness equal to (Φ_{12}/b_1), so that E_y at the upper surface can be $2A\cos\Phi_{12}$. The middle layer has a thickness $(m\pi/b_1)$, where m = 0, 1, 2 ..., determines the order of the mode. Finally, the thickness of the lower layer is (Φ_{10}/b_1), in order that E_y at the lower surface be $\pm 2A\cos\Phi_{10}$. The fields in the film are sinusoidal, but they are exponential in z in the substrate and airspace such that

$$E_y = 2A\cos\Phi_{10}\exp[-i\omega t+\beta x+p_o(z+W_{10})] \qquad (21)$$

for $z < -W_{10}$, and

$$E_y = 2A\cos\Phi_{12}\exp[-i\omega t+i\beta x-p_2(z-W_{12})] \qquad (22)$$

for $z > +W_{12}$. Substituting (21) or (22) into the wave Eq. (5) and replacing n_1 in the equation by n_0 or n_2, accordingly, we again obtain (10a) and (10b) as they should be.

(vi) Effective Thickness of the Waveguide and Power Flow. It is interesting that the fields extend into the airspace within a distance of about $(1/p_2)$ above the upper surface, and that they extend within a distance of about $(1/p_0)$ below the lower surface. It is natural to consider the effective thickness of the film-waveguide as

$$W_{eff} = W + \frac{1}{p_o} + \frac{1}{p_2}. \qquad (23)$$

We shall use this thickness to calculate the power flow in the waveguide. Consider a unit distance in the y-direction. Remember that A and B are plane waves. The energy densities contained in the A and B waves are identical since A = B. It is $(1/8\pi)n_1A^2$ in Gaussian units. The wave velocity in the $\vec{A}$ or $\vec{B}$ direction is c; the wave velocity in the x-direction is thus c sin θ. The power flow in the x-direction is the product of three quantities - the combined energy density of the A and B waves, their wave velocity in the direction of power flow, and the effective

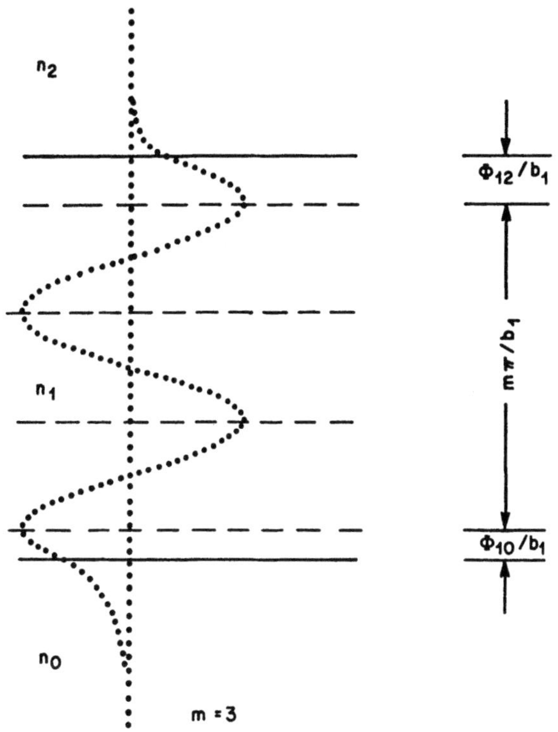

Figure 6 The electric field distribution of a TE wave in a film-waveguide.

thickness of

$$P = \frac{c}{4\pi} n_1 \sin\theta_1 A^2 \left(W + \frac{1}{p_0} + \frac{1}{p_2} \right). \qquad (24)$$

(vii) Different Waveguide Modes. It is obvious that the different waveguide modes are simply different zigzag waves. For example, a Ta_2O_5 film 0.78 μm thick deposited on a glass substrate has four waveguide modes. Figure 7 shows the zigzag waves associated with these waveguide modes. For the m = 0 mode, the rays nearly graze the surfaces of the film. As the mode order, m, increases, the angle between the ray and the surface normal becomes correspondingly smaller. For the last mode, the angle approaches the critical angle ($\theta_c = \sin^{-1}(n_0/n_1)$). The field distributions are shown in the right side of the figure. We see that the number of nodes in the field distribution is equal to the order of the mode. The angle θ_1 of each mode can be obtained

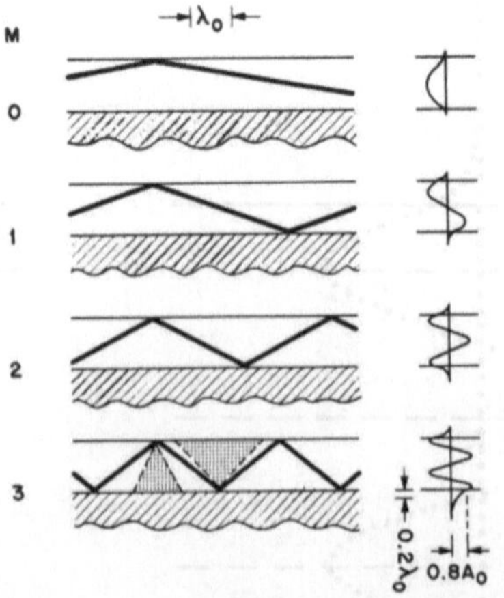

Figure 7 A Ta_2O_5 film of 0.776 microns thick propagates four waveguide modes. The electric field distributions and the zigzag waves associated with these waveguide modes are shown at the right and the left of the figure, respectively.

by solving the mode equation (15) for b_1, since $b_1 = kn_1\cos\theta_1$.

There are many advantages in using zigzag waves to solve a problem. First, the zigzag waves are simple plane waves and their field distributions can be written immediately using the recipe given by Eq. (3) and (4). Second, to solve a problem of electromagnetic waves involving several coupled media, one normally writes the expressions for the fields in each medium and then establishes, as the boundary conditions, two algebraic equations for each interface. For example, a prism-film coupler has four coupled media (prism, air-gap, film and substrate) and three interfaces resulting in a total of six simultaneous equations. The quantity of interest cannot be found until all these equations are solved. Even then, the physical meaning of the result is not clear because of the complexity of the equations. In the zigzag-wave method, however, we assume θ_1 be known, and often, the solution of the problem can be derived immediately based on the field distributions of the A and B waves. The results are then expressed in θ_1 which is assumed to satisfy the mode equation, but needs not to be solved for a particular waveguide mode. Finally, by using the concept of ray-optics, wherein a ray in the film strikes the surfaces of the film only at certain discrete points, we need only single out one of these points and study the wave behavior in the vicinity of this point in order to understand how the waves in different media are coupled with one another. It is always easier to visualize a ray at one point, instead of a sea of electromagnetic fields which vary both in time and space.

We have already shown above how the mode equation can be derived by means of some simple arguments using zigzag waves. We shall further demonstrate their advantages in an analysis of the prism-film coupler.

II. PRISM-FILM COUPLERS AND ZIGZAG WAVES

When the field of integrated optics began to open up about four years ago, considerable difficulties were encountered in focusing a laser beam into the edge of a thin film which was usually less than one micron thick. To overcome these difficulties, the prism-film coupler[3] was developed, thus making many later experiments in thin films possible. In the prism-film coupler, a prism is placed on top of a thin film leaving a small air-gap between the film and the base of the prism (Fig. 8). The spacing of the gap is on the order of a fraction of one optical wavelength. A laser beam entering into the prism is totally reflected at its base. The field distribution is sinusoidal inside the prism but has an exponential tail which extends below the base. This exponential tail is called the evanescent field. The evanescent field does not propagate in z, but can serve as a coupling by which the wave energy in the prism can be transferred into the film, and vice versa. As will be discussed later, it is a very efficient coupling method. For a uniform air-gap, about 81% of the laser beam can be coupled into the film. For a nonuniform gap, an efficiency of 100% is theoretically possible, and an efficiency of more than 90% has been achieved experimentally.

Before going into a detailed calculation, let us consider how the prism coupler works. In Fig. 9, we have a zigzag wave in the film forming a waveguide mode. The wave strikes the upper film surface at the points 1, 2, 3 and 4. In the same figure, we show a wide laser beam which enters the prism as the A_3 wave. From this A_3 wave, we pick four parallel rays which hit the prism base at the points 1', 2', 3', and 4', respectively. These points are directly opposite the points 1, 2, 3, and 4 on the other side of the gap

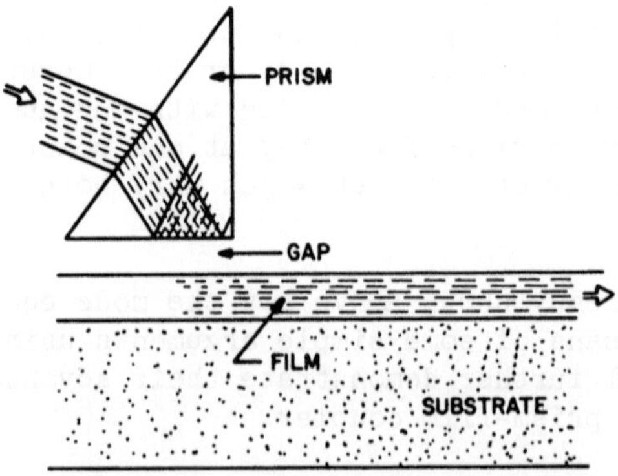

Figure 8 A prism-film coupler.

Let us draw a dotted line 1'-2"-3"-4" starting from the point 1' and normal to the laser beam in the prism. This dotted line is a wave front along which the A_3 wave has the same phase. Now consider a burst of light generated by a laser. This light reaches points 1', 2", 3" and 4" at the same time. At this time, since ray 1' has already reached the gap, it excites a wave A_1', at point 1 in the film. This wave then propagates in the film, arriving at point 2 after

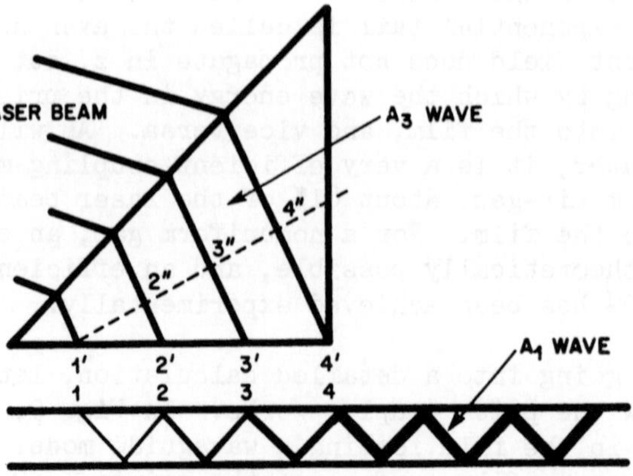

Figure 9 The wave in the prism may be considered as four parallel rays which strike the base of the prism at points 1', 2', 3' and 4'. Each ray in the prism excites a new wave in the film and these new waves add in phase such that the resultant wave in the waveguide increases rapidly in distance.

a zigzag. If the A_1' wave in the film reaches point 2 precisely at the same time when ray 2' in the prism reaches the air-gap, ray 2' will excite a new wave, A_1', which will add in phase with the A_1' wave already in the film. Consequently, the resultant wave at point 2 will have its amplitude doubled and its intensity quadrupled. Similarly, intensities of the resultant waves at point 3 and point 4 will be nine times and sixteen times that at the point 1, respectively. The field intensity of the wave in the film thus increases very rapidly in x, if the coupling between the film and the prism extends to over many zigzags.

The condition that the A_1' wave in the film and ray 2' in the prism must reach points 2 and 2', respectively, at precisely the same time is called the synchronous condition. It requires the phase delay of the path 2"-2' to be equal to that of one complete zigzag. Let the distance between 1 and 2 along x be d, and let n_3 and θ_3 be the refractive index of the prism and the angle between ray $\vec{A}_3$ and the surface normal, respectively. The phase delay between 1 and 2 for the wave in the film will then be $kn_1d \sin \theta_1$, and that between 1' and 2' for the wave in the prism will be $kn_3d\sin\theta_3$. Since the horizontal component of $\vec{A}_3$ is $\beta_3 = kn_3\sin\theta_3$, and that for A_1' is $\beta_1 = kn_1\sin\theta_1$, we have, for the synchronous condition,

$$\beta_1 = \beta_3 = \beta \tag{25}$$

or

$$kn_1\sin\theta_1 = kn_3\sin\theta_3. \tag{26}$$

For any waveguide mode, we have a value of θ_1 associated with the zigzag wave. It is always possible to adjust the direction of the incoming laser beam, and thus, the value of θ_3, so that condition (25) or (26) is met. Usually, the refractive index, n_3, of the prism is larger than that of the film, n_1. We often use a rutile prism for semiconductor films and a flint-glass prism for glass or organic films. The ability to excite any selected waveguide mode in a film is a very important property of the prism-film coupler.

The waves in the prism are designated as the A_3 and B_3 waves and those in the film as the A_1 and B_1 waves. Without the film, A_3 and B_3 are governed by the total reflection phenomenon. Similarly, without the prism, B_1 is the total reflection of A_1. By placing the prism next to the film, the total reflection phenomenon is modified and we call the resultant phenomenon frustrated total reflection. The coupling between the prism and the film extends to hundreds of optical wavelengths along the gap and thus, the coupling per wavelength is very small. It is then possible to analyze the problem by use of a perturbation method.

For the prism-film coupler, we study the power transfer between the prism and the film and are concerned with the power flow in the z-direction only. We have mentioned earlier that the power flow in the x-direction for the B_1 (or A_1) wave through a unit area of the y-z plane is $c/8\pi\ n_1 \sin\theta_1 B_1 B_1^*$. The power flow in the z-direction through a unit area of the x-y plane is thus $c/8\pi\ n_1 \cos\theta_1 B_1 B_1^*$. It is therefore convenient to normalize the amplitudes of the waves so that

$$(A_i) = (n_i \cos\theta_i)^{1/2} A_i;$$

$$(B_i) = (n_i \cos\theta_i)^{1/2} B_i; \quad i = 1 \text{ or } 3. \qquad (27)$$

The quantities $(A_i)(A_i)^*$, $(B_i)(B_i)^*$ are then proportional to the respective power flows. Consider in Fig. 10, an (A_3) wave incident on the gap. It is partially reflected as the (B_3'') wave and partially transmitted into the film as the (A_1'') wave. Similarly, a (B_1) wave incident on the gap is partially reflected as the (A_1') wave and is partially trans-

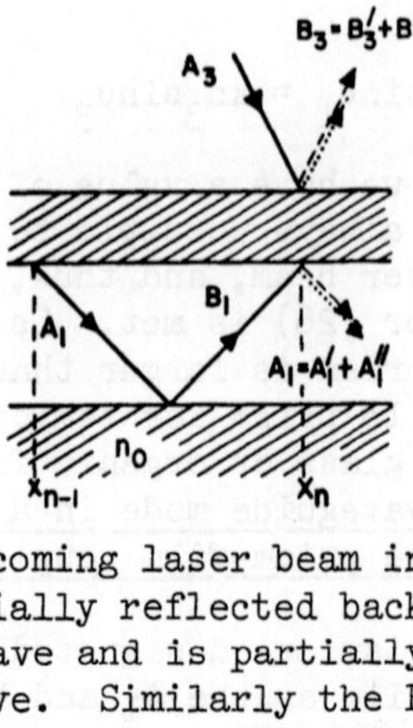

Figure 10 The incoming laser beam in the prism is the A_3 wave. It is partially reflected back at the base of the prism as the B_3'' wave and is partially transmitted into the film as the A_1'' wave. Similarly the B_1 wave in the film is partially reflected back at the upper film-surface as the A_1' wave and is partially transmitted into the prism as the B_3' wave.

mitted into the prism as the (B_3') wave. The resultant (B_3) wave in the prism is therefore the sum of (B_3') and (B_3''), and the resultant (A_1) wave in the film is the sum of (A_1') and (A_1''). Let T_3 and R_3 be some sort of transmittance and reflectance coefficients for the prism and T_1 and R_1 be those for the film. We have then

$$(B_3'') = R_3(A_3)' \quad (B_3') = T_1(B_1);$$
$$(A_1'') = T_3(A_3); \quad (A_1') = R_1(B_1); \tag{28}$$

and

$$(B_3) = (B_3'') + (B_3') = R_3(A_3) + T_1(B_1);$$
$$(A_1) = (A_1'') + (A_1') = T_3(A_3) + R_1(B_1). \tag{29}$$

For those who are familiar with network theory, it is clear that the matrix

$$\begin{pmatrix} R_3 & T_1 \\ T_3 & R_1 \end{pmatrix} \tag{30}$$

is in the form of a scattering matrix. We can thus consider the (A_3) and (B_3) waves as the incident and reflected waves for the "prism" port, and the (B_1) and (A_1) waves as the incident and the reflected waves for the "film" port. Our problem then becomes simply that of a two-port network. The scattering matrix should be symmetrical. Moreover, for a lossless network, R_1 and R_3 will be real, and T_1 and T_3 will be imaginary and numerically equal.

One advantage of the zigzag wave method is that we can write the expressions for the fields in the air-gap immediately, according to the recipe given in Section I, except that in this problem, it is more convenient to take $z = 0$ at the upper film surface, and $z = s$ at the prism base. The fields at $x = x_n$ and in the gap, $0 \leq z \leq s$, are:

$$E_{y3} = (n_3\cos\theta_3)^{-1/2} \exp[-i\omega t+i\beta x_n-p_2(s-z)]$$

$$\left\{(A_3)e^{i\Phi_{32}} + [T_1(B_1)+R_3(A_3)]e^{-i\Phi_{32}}\right\};$$

$$H_{y3} = (n_3\cos\theta_3)^{1/2} \exp[-i\omega t+i\beta x_n-p_2(s-z)]$$

$$\left\{(A_3)e^{i\Phi_{32}} - [T_1(B_1)+R_3(A_3)]e^{-i\Phi_{32}}\right\};$$

$$E_{y1} = (n_1\cos\theta_1)^{-1/2} \exp[-i\omega t+i\beta x_n-p_2 z]$$

$$\left\{T_3(A_3) + R_1(B_1)e^{-i\Phi_{12}} + B_1 e^{i\Phi_{12}}\right\};$$

$$H_{y1} = (n_1\cos\theta_1)^{1/2} \exp[-i\omega t+i\beta x_n-p_2 z]$$

$$\left\{T_3(A_3) + R_1(B_1)e^{-i\Phi_{12}} - B_1 e^{i\Phi_{12}}\right\}; \qquad (31)$$

where

$$\Phi_{12} = \tan^{-1}(p_2/b_1), \quad 0 \le \Phi_{12} \le \pi/2;$$

$$\Phi_{32} = \tan^{-1}(p_2/b_3), \quad 0 \le \Phi_{32} \le \pi/2;$$

$$b_1 = kn_1\cos\Phi_1 = \left[(kn_1)^2-\beta^2\right]^{1/2};$$

$$b_3 = kn_3\cos\theta_3 = \left[(kn_3)^2-\beta^2\right]^{1/2};$$

$$p_2 = \left[\beta^2-(kn_2)^2\right]^{1/2}. \qquad (32)$$

Some of the above quantities have already been defined in Section I. For power conservation, we obtain immediately

$$T_1T_1^* + R_1R_1^* = 1;$$

$$T_3T_3^* + R_3R_3^* = 1. \qquad (33)$$

It is evident from (31) that, if the prism and the film were decoupled we would have $T_1 = T_3 = 0$, and $R_1 = R_3 = 1$. Since T_1 and T_2 should be small quantities for a weak coupling, we are able to apply the perturbation method.

To calculate T_1, we set, in (22), $(A_3) = 0$. By taking $R_1 = 1$ for the first order approximation, we find immediately that the real component of $-E_{y1}H^*_{x1}$ is zero; therefore, there is no real power flow inside the gap in the z-direction due to the fields from the film alone. Of course, because of this, the wave is totally reflected at the film surface. Inside the gap, however, we can form a Poynting vector in the form of $R_e[-E_{y3}H^*_{x1}-E_{y1}H^*_{x3}]$ which should be equal to the power transferred into the prism, $(B_3)(B^*_3)$. By equating the above two expressions and remembering that T_1 is imaginary, we find immediately that

$$T_1 = i2e^{-p_2 s}(\sin 2\Phi_{12} \sin 2\Phi_{32})^{1/2}. \qquad (34)$$

Similarly, by setting $(B_1) = 0$, and $R_3 = 1$, we can calculate T_3. We then find that it is, indeed, equal to T_1 as predicted by the scattering matrix. The method described above for the calculation of T_1 and T_3 is similar to that used in electron tunneling. Because of this, the term "optical tunneling" has frequently been used.

Now return to Fig. 10. So far, we have only considered the fields at the point $x = x_n$. Due to the coupling, the fields vary slowly in x. We have written, in (31), the (A_1) wave at $x = x_n$ in the form

$$(A_{1,n})e^{-i\Phi_{12}} \exp[-i\omega t+i\beta x_n-ib_1 z].$$

Here, the second subscript, n, denotes the position of the ray at x_n. Similarly, we can write for the (A_1) wave at $x = x_{n-1}$

$$(A_{1,n}) = T_3(A_{3,n}) + R_1(A_{1,n-1}), \qquad (35)$$

or,

$$(A_{1,n}) - (A_{1,n-1}) = T_3(A_{3,n}) - (1-R_1)(A_{1,n-1}). \qquad (36)$$

We therefore have,

$$\frac{d(A_1)}{dn} = T_3(A_3) - (1-R_1)(A_1),$$

where n denotes the number of the zigzag. Since each zigzag occupies a distance of $2W_{eff}\tan\theta_1$ in x, we obtain

$$\frac{d(A_1)}{dx} = (2W_{eff}\tan\theta_1)\ [T_3(A_3)-(1-R_1)(A_1)]. \quad (37)$$

This is the equation of the prism-film coupler. This equation allows us to calculate the wave intensity along the coupling gap. For an air-gap of uniform spacing, s, and an input beam of constant amplitude, (A_3), the equation can be solved analytically. For a nonuniform gap and an arbitrary distribution of (A_3), the equation can only be solved by numerical computation. In (37), $T_3 = T_1$ as given in (34). From (33), $(1-R_1) = T_1T_1^*/2$. Both coefficients in (37), therefore, vary with the gap spacing. For detailed solutions of the equation, the reader is referred to the original papers.[1,2]

Let us consider now the amplitude of the (A_1) wave. When (A_1) is small, the second term in (37) may be neglected; (A_1) then increases linearly with x. This is consistent with what we have discussed in connection with Fig. 9. The amplitude of the (A_1) wave continues to increase until it saturates at some large x. At saturation, $d(A_1)/dx = 0$ and

$$(A_1) = T_3/(1-R_1) = (1/2T^*)(A_3).$$

Hence, the wave in the film cannot be increased indefinitely by simply increasing the coupling length.

Next, we will discuss the phase of the (B_3) wave. We have shown in (29) that $(B_3) = (B_3') + (B_3'')$. If (A_3) is real, (A_1) and (B_1) must be imaginary on account of the second equation in (29). Again, because T_1 is imaginary, the (B_3'') wave must be in phase and the (B_3') wave 180° out of phase with (A_3). Therefore, as x increases, we expect $(B_3) = (A_3)$ for small x, and $(B_3) = -(A_3)$ for large x. Similarly, if we vary the coupling between the prism and the film, the phase of the entire (B_3) wave will switch by 180° from the case of undercoupling to that of overcoupling. As can be seen, the zigzag wave method does provide a clear physical picture of the waves in the coupler. Furthermore, it enables us to solve the problem by considering only the fields in the coupling gap. We thus by-pass the necessity of solving the mode equation.

In addition to the prism-film couplers, grating couplers[6,7] and tapered-film couplers[8] are also frequently used in integrated optics. Recently, a method of forming light-guiding interconnections has been developed by the author and co-workers.[9] Being limited in space, we will not discuss them. All these can, however, be analyzed by the zigzag wave method. The readers are referred to the original papers for details.

III. MATERIALS FOR FILM-WAVEGUIDES AND THEIR LOSSES

When integrated optics was begun, it was thought that any transparent film would be an excellent optical waveguide. However, it turned out that a seemingly perfect film could easily have a loss as large as tens of decibels per centimeter. This large loss was later found to be due mostly to surface scattering. We can understand this by again considering zigzag waves. Figure 11 shows a zigzag wave in a film-waveguide. Because the film is very thin, a light wave traveling one centimeter in the x-direction will typically go through about 1000 zigzags. In each zigzag, the light wave will strike the upper and lower surfaces of the film once. If the surfaces are not perfectly smooth, each bounce at the surface will scatter some of the light. Scattering losses can become very large in 1000 zigzags. For example, for the film considered in Fig. 7, to limit the loss of the waveguide to within 1 db/cm the scattering per reflection must be less than 1 part in 10,000. This should be compared with the scattering loss of 1 part in 1000 per reflection for the better mirrors used in gas lasers.

To study the materials used for film-waveguides, it is important to differentiate the absorption loss in the volume of the film from the scattering loss at the surfaces. Usually, the scattering loss increases rapidly with the order of the waveguide mode whereas the volume loss stays relatively constant. A simple theory[1] for the differing losses of different waveguide modes has been worked out by the author based on the Rayleigh criterion. For details,

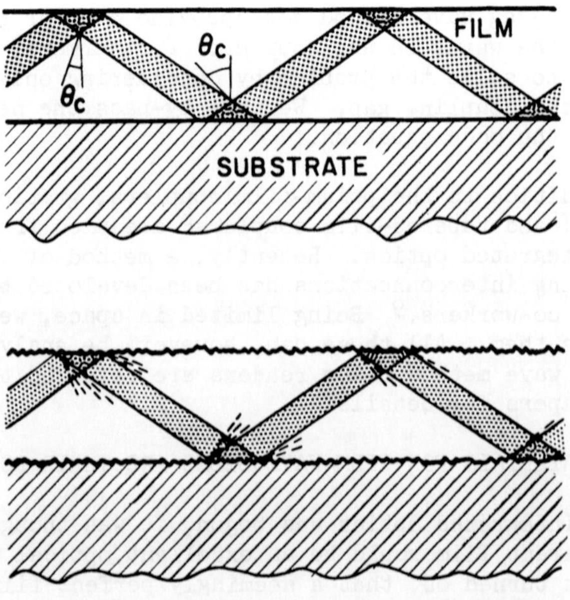

Figure 11 A typical zigzag wave bounces back and forth between two film surfaces about 1000 times before the wave advances 1 cm in the x direction. If the surfaces of the film are not smooth as illustrated in the lower figure, the light will be rapidly scattered out of the film, and thus, the major loss in the film-waveguide is caused by surface scattering.

the readers are referred to the original paper.[1] The solid curve in Fig. 12 shows the sum of the volume and surface losses for four waveguide modes in a Ta_2O_5 film. In this figure, the horizontal axis is the β/k's of the waveguide modes and the vertical axis is the loss of cm^{-1} for the left scale and in dB/cm for the right scale. An He-Ne laser of wavelength .6328 μm was used in the measurement. By extending this solid curve to $\beta/k = n_1$, where n_1 is the refractive index of the film, we obtain the volume absorption loss of the film. For all practical purposes, this volume loss can be considered as roughly the same for all the waveguide modes. The remaining loss, which is shown between the solid and the dotted curves in this figure, is then the loss due only to surface scattering. This loss

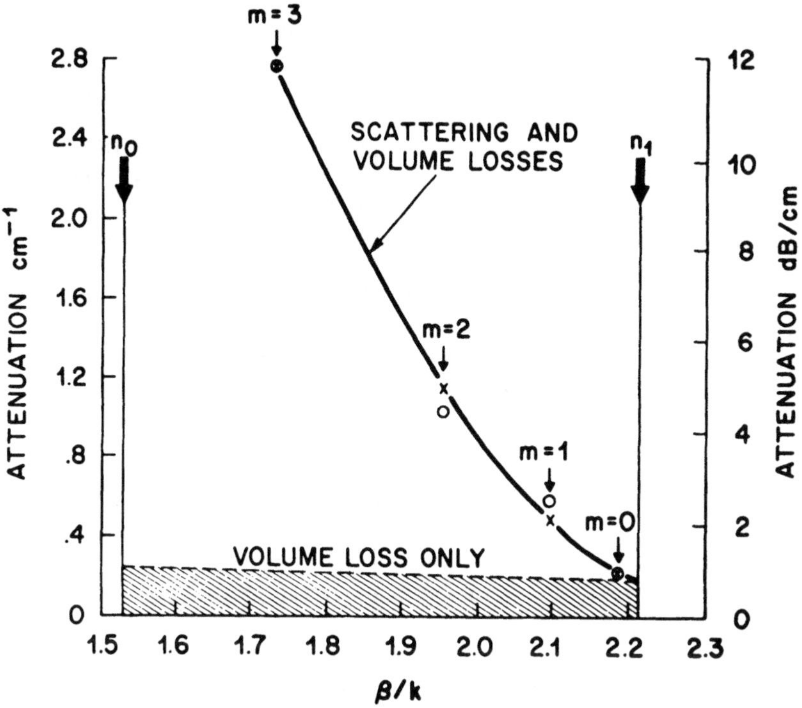

Figure 12 The figure shows how the volume absorption loss and the surface scattering loss of a film-waveguide can be separated by simple measurements.

varies from almost nothing for the m = 0 mode to about 11 dB/cm for the m = 3 mode. It is important to point out that the scattering loss varies inversely with the thickness of the film. A loss of 1 dB/cm for a film 3 μm thick is equivalent to 10 dB/cm for a film 0.3 μm thick. Therefore, one must specify the order of the waveguide mode and the thickness of the film when quoting losses.

Because of the large concentration of light intensity achieved by a wave propagating in thin films, several attempts[1] have been made since 1969 to study nonlinear optics in film-waveguides. The first semiconductor films chosen for this purpose were ZnO polycrystalline films.[1] These films were developed previously to be used in ultrasonic transducers. The films were deposited on glass substrates in an argon-oxygen atmosphere at a substrate temperature of 400°C. The particular film which will be discussed below was 1.5 μm thick. X-ray diffraction showed a well-oriented c-axis normal to the film. The average grain size of the crystallites as observed in electron micrographs, was on the order of 0.5 μm. The surface of the film had a peak-to-peak roughness as large as 1000 Å. Consequently, the film had a scattering loss larger than 50 dB/cm. Later, the film was carefully polished by lapping its surface with chromium oxide dispersed in water. After this treatment, the electron micrographs showed a reduction of the peak-to-peak variation of the surface to less than 100 Å - a reduction which was also accompanied by a reduction in loss to less than 20 dB/cm. Although these films were much too lossy to be useful, these experiments were important in the sense that they demonstrated, for the first time, the importance of surface smoothness in optical waveguides. Moreover, these experiments showed an improvement in the surface condition by polishing - a process which was frequently adopted later for epitaxial films. The refractive indices of the ZnO films are 1.9732 and 2.0428 at 0.6328 and 0.4880 μm laser wavelengths, respectively.

Recently, single-crystal epitaxial ZnO films have been grown on sapphire substrates by Hammer and his co-workers.[10] They used a method of close-spaced chemical vapor transport. In this method, a substrate was suspended close to a ZnO powder source which was heated to 800 - 900°C. Because of a mismatch in the lattice structures of the film and the substrate, the film surfaces grown are rough and consequently must be polished. For a film 10 μm thick, a loss of 4 dB/cm was reported for the m = 0 mode at 0.6328 μm laser wavelength.

For refractive indices larger than 2.0, the best semiconductor films available today, for the visible spectrum, are the Ta_2O_5 amorphous films developed by Hensler and his co-workers.[11] Ta_2O_5 has a very large energy gap (4.6 eV) and thus, neglecting serious band tailing effects, light of wavelengths as short as 3000 Å can be propagated. These films are formed by first depositing β-tantalum on glass substrates. This process involves the sputtering of high-purity tantalum in an argon atmosphere. The deposited β-tantalum is then heated in pure oxygen at 500°C for about 20 hours to form Ta_2O_5. The temperature and the time of oxidation are critical for forming low-loss waveguides. Too high a temperature or a prolonged oxidation time would make films of a polycrystalline nature, whereas the reverse would not completely convert the β-tantalum into Ta_2O_5. For films 0.8 μm thick, losses of 1 dB/cm, 2.5 dB/cm and 4.1 dB/cm have been reported at 0.6328, 0.5145 and 0.4880 μm laser wavelengths. At these wavelengths, the refractive indices of the films are 2.2136, 2.2423 and 2.2767, respectively.

To produce refractive indices between 1.455 and 1.980 at 0.546 μm laser wavelength, silicon oxynitride (SiON) films on fused silica substrates were developed by Rand and Standley.[12] The SiON films are glassy, amorphous and stable silicon-oxygen-nitrogen polymers. The refractive index of the material can be varied by varying the oxygen-nitrogen composition. The energy gap also varies with the composition but is always larger than 6 eV. SiON films are formed by the process of chemical vapor deposition. A conventional rf-heated silica tube reactor can be used for this purpose. The deposition is made at 850°C from an ambient, typically consisting of 0.2-0.5% nitric oxide, 0.02-0.07% silane, and the remainder, nitrogen. The substrates used are disks of Vitereosil or Spectrosil, 25 mm in diameter and 1 mm in thickness, supported in the reactor by a graphite susceptor. The SiON composition is controlled by the NO/SiH_4 concentration ratio. The authors discussed, in detail, methods of treating and cleaning the substrate surfaces. A loss of less than 0.4 dB/cm was reported for a film 0.5 μm thick having a refractive index of 1.50 at 0.6328 μm wavelength. However, films with higher refractive indices are not so successful as the surfaces craze easily.

If refractive indices of less than 1.6 are desired, several different types of films may be used. These include (1) sputtered glass films, (2) gas-discharge polymerized organosilicon films, and (3) various solution-deposited films such as photoresists, polyurethane, and other epoxy resins. We will discuss them separately below.

The glass films were first reported by Goell and Standley[13] in 1969. Their films were formed by rf-sputtering Corning 7059 glass on substrates consisting of ordinary laboratory glass slides. The refractive index 7059 glass has a refractive index 1.53, because of a change in composition during the sputtering process, the refractive index of the sputtered film is 1.6. Films sputtered in an oxygen atmosphere (of about 0.3 μm in thickness) have a loss of less than 1 dB/cm for the m = 0 waveguide mode at

0.6328 μm wavelength. Films sputtered in an argon atmosphere have relatively larger losses.

Recently, barium silicate glass films were investigated by Goell.[14] For the cathode, he used a hot pressed disk prepared from a mixture of barium carbonate and silica. The films were sputtered in oxygen at a pressure of 1.2×10^{-2} Torr using a 13.56 MHz rf-generator. The power level was about 25 W/cm^2 for a typical sputtering rate of 0.6 μm/hour. A loss of 0.6 dB/cm at 0.6328 μm laser wavelength was reported for films with thicknesses greater than 1 μm. The refractive index of the film could be varied from 1.48 to 1.62 by varying the content of BaO in the cathode from 0 to 40 weight percent.

In 1970, films prepared by a novel rf-discharge polymerization process using organic chemical monomers were reported by Tien and Smolinsky.[2,15] Films made from vinyltrimethylsilane (VTMS) and hexamethyldisiloxane (HMDS) monomers were studied exclusively. The films were deposited on Corning 744 Pyrex substrates having a refractive index of 1.4704. The refractive index of the films can be varied from 1.480 to 1.530 by using mixtures containing VTMS and HMDS in different proportions. A vacuum chamber 20 cm in diameter was used in the polymerization process. The substrate rested on an active cathode which was maintained at 25°C by circulating water. The active and the ground electrodes consisted respectively of brass and steel plates, 15 cm in diameter, placed 4 cm apart from one another. The monomers and the argon were introduced into the vacuum chamber through separate variable leak valves. The chamber was typically evacuated initially to a pressure of less than 2×10^{-6} Torr. Monomers and argon were metered into the chamber to pressures of 0.3 Torr and 0.1 Torr, respectively. A discharge was then established by coupling a 200 W, 13.56 MHz rf-generator to the electrodes. Polymer formation began as soon as a discharge was established. The VTMS film grew at the rate of .2 μm/minute while HMDS films grew at half the rate. Losses as low as 0.1 dB/cm at 0.6328 laser wavelength were reported for these films (2.5 μm thick). The refractive indices of the VTMS films are 1.5279, 1.5370 and 1.5398 at 0.6328, 0.5145 and 0.4880 μm laser wavelengths. Those of the HMDS films are 1.4880, 1.4960 and 1.4996, respectively.

Photoresist films were reported by Ostrowsky and Jaques[16] and by Weber et al.[17] in 1971. Ostrowsky et al., used Shipley Az 1350 (n=1.618) deposited on ordinary glass slides (n=1.5.2). These films were very lossy. To reduce the losses, Weber et al., used Kodak KPR, their material being processed to remove a photosensitizer which absorbed visible light. Losses of their films were on the order of 2 dB/cm at 0.6328 μm for films 2 μm thick. These films were prepared by flooding a clean microscope slide with the photoresist solution. The slide was then raised to a vertical position for about 30 seconds, so as to allow the excess solution to drip off. The resulting film was air-dried in a horizontal position for one hour. Finally, the film was baked for 5 minutes at 95°C. The advantage of the photoresist films is that curved or straight three-dimensional optical waveguides can be written directly on the film by exposing it first to ultraviolet light and subsequently developing it in toluene.

Polyurethane is also widely used for solution-deposited films. Ulrich et al.,[18] used the brands 9653-1 and LX500. They are marketed as two-component protective coatings. The two components are first mixed and applied to the glass slide. After evaporation of the solvent, the resulting polymer film is polymerized by baking at 65°C for about an hour. For thinner films, the mixture can be diluted by a thinner (66-C-20) before its application to the substrate. The refractive index of the film is 1.56 at 0.6328 μm wavelength, with a loss on the order of 1 dB/cm (for films 0.8 μm thick). Other two-component solutions, such as Epoxy Araldite and Lead-Silica 112, have also been used by Ulrich et al.,[18] with excellent results.

Being limited in space, we will not discuss here optical waveguides made by ion-bombarded fused quartz,[19,20] single-crystal epitaxial layers,[21,22] semiconductor films formed by solid-state diffusion,[23,24] or light-guiding layers involving nematic liquid crystal.[25]

ACKNOWLEDGMENT

The author would like to acknowledge his secretary, Mrs. L. Kohutich, for her help in preparing this chapter.

REFERENCES

1. P. K. Tien, Appl. Opt. 10, 2395 (1971).

2. P. K. Tien and R. Ulrich, J. Opt. Soc. Am. 60, 1325 (1970).

3. P. K. Tien, R. Ulrich and R. J. Martin, Appl. Phys. Letters 14, 291 (1969).

4. M. Born and E. Wolf, "Principles of Optics" (Pergamon, New York, 1970) p. 49, Eq. (60).

5. H. K. V. Lotsch, J. Opt. Soc. Am. 58, 551 (1968).

6. M. L. Dakss, L. Kuhn, P. F. Heidrich, and B. A. Scott, Appl. Phys. Letters 16, 523 (1970).

7. H. Kogelnik and T. Sosnowski, Bell Syst. Tech. J. 49, 1602 (1970).

8. P. K. Tien and R. J. Martin, Appl. Phys. Letters 18, 398 (1971).

9. P. K. Tien, R. J. Martin and G. Smolinsky, "Formation of Light-Guiding Interconnections in an Integrated Optical Circuit by Composite Tapered-Film Coupling" to be published in Applied Optics.

10. J. M. Hammer, D. J. Channin, M. T. Duffy and J. P. Wittke, Appl. Phys. Letters 21, 358 (1972).

11. D. H. Hensler, J. D. Cuthbert, R. J. Martin and P. K. Tien, Appl. Opt. 10, 1037 (1971).

12. M. J. Rand and R. D. Standley, Appl. Opt. 11, 2482 (1972).

13. J. E. Goell and R. D. Standley, Bell Syst. Tech. J., 48, 3445 (1969).

14. J. E. Goell, Appl. Opt., 12, 737 (1973).

15. P. K. Tien, G. Smolinsky and R. J. Martin, Appl. Opt. 11, 637 (1972).

16. D. B. Ostrowsky and A. Jaques, Appl. Phys. Letters 18, 556 (1971).

17. H. P. Weber, R. Ulrich, E. A. Chandross and W. J. Tomlinson, Appl. Phys. Letters 20, 143 (1972).

18. R. Ulrich and H. Weber, Appl. Opt. 11, 428 (1972).

19. E. R. Schineller, R. P. Flam, and D. W. Wilmot, J. Opt. Soc. Am. 58, 1171 (1968).

20. R. D. Standley, W. M. Gibson and J. D. Rogers, Appl. Opt. 11, 1313 (1972).

21. P. K. Tien, R. J. Martin, S. L. Blank, S. H. Wemple, and L. J. Varnerin, Appl. Phys. Letters 21, 207 (1972).

22. P. K. Tien, R. J. Martin, R. Wolfe, R. C. LeCraw and S. L. Blank, Appl. Phys. Letters 21, 394 (1972).

23. H. F. Taylor, W. G. Martin, D. B. Hall and V. N. Smiley, Appl. Phys. Letters 21, 95 (1972).

24. I. P. Kaminow and J. R. Carruthers, Appl. Phys. Letters 22, 326 (1973).

25. D. J. Channin, Appl. Phys. Letters 22, 365 (1973).

Chapter 3 — ONE-DIMENSIONAL CONFINEMENT

M. K. Barnoski

Hughes Research Laboratories
3011 Malibu Canyon Road
Malibu, California 90265

1. Guided Modes of a Slab Waveguide

Consider the dielectric slab waveguide[1] of thickness t illustrated in Fig. 1. The waveguide is assumed to be infinitely thick in the y dimension, therefore, $\partial/\partial y = 0$. For TE modes the longitudinal components of the electric field E_z vanishes. Substitution of $E_z = 0$ and $\partial/\partial y = 0$ into Equations 67 and 68 of Chapter I relating the transverse and longitudinal components results in the following field equations.

$$E_x = -i \frac{kZ}{k_c^2} \frac{\partial H_z}{\partial y} = 0 \tag{1}$$

$$H_y = -\frac{\gamma}{k_c^2} \frac{\partial H_z}{\partial y} = 0 \tag{2}$$

[1]The material in this section follows, in part, an analysis developed by D. Hall for his doctorate dissertation at California Institute of Technology.

$$E_y = i \frac{kZ}{k_c^2} \frac{\partial H_z}{\partial x} \tag{3}$$

$$H_x = - \frac{\gamma}{k_c^2} \frac{\partial H_z}{\partial x} \tag{4}$$

As can be seen, the transverse electric (TE) modes have only field components E_y, H_x, and H_z.

The boundary conditions for the slab waveguide are

$$E_y^{(1)}\Big|_{x=0} = E_y^{(2)}\Big|_{x=0} \tag{5}$$

$$H_z^{(1)}\Big|_{x=0} = H_z^{(2)}\Big|_{x=0} \tag{6}$$

$$E_y^{(2)}\Big|_{x=t} = E_y^{(3)}\Big|_{x=t} \tag{7}$$

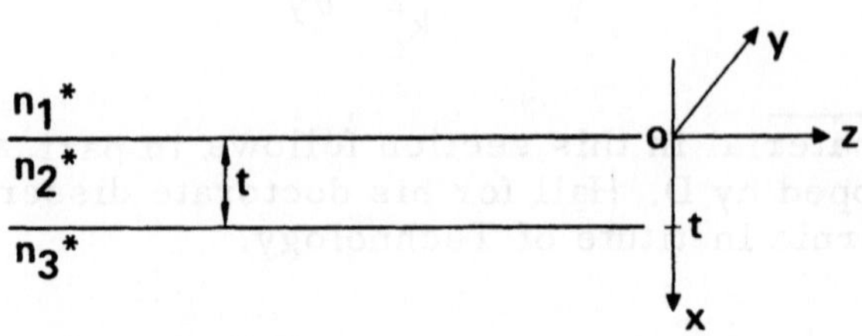

Fig. 1. The coordinate system used in describing an asymmetric waveguide.

$$H_z^{(2)}\Big|_{x=t} = H_z^{(3)}\Big|_{x=t} \quad . \tag{8}$$

Assume solutions of the form

$$E_y = A\, e^{+qx} \qquad x < 0 \tag{9}$$

$$E_y = B\, e^{ihx} + c\, e^{-ihx} \qquad 0 \leq x \leq t \tag{10}$$

$$E_y = D\, e^{-p(x-t)} \qquad x \geq t \tag{11}$$

where A, B, C, D, and q, h, p are constants to be determined.

The boundary conditions on E_y require

$$A = B + C \tag{12}$$

$$D = B\, e^{iht} + Ce^{-iht} \tag{13}$$

In addition, E_y must satisfy the wave equation. This results in the condition that

$$\frac{\partial}{\partial x} E_y = -i\omega\mu^* H_z \tag{14}$$

which, coupled with the boundary conditions on H_z, results in the additional equations

$$qA = ihB - ihC \tag{15}$$

$$-pD = ihB\, e^{iht} - ihC\, e^{-iht} \quad . \tag{16}$$

These are four linear, homogeneous equations to determine the four unknown constants A, B, C, and D.

The characteristic equation for TE modes can be found by setting the determinate of equations 12, 13, 15, and 16 equal to zero. The characteristic determinate for TE modes is therefore

$$\mathrm{Det}_{TE} = \begin{vmatrix} -q & ih & -ih & 0 \\ -1 & 1 & 1 & 0 \\ 0 & ihe^{iht} & -ihe^{-iht} & +p \\ 0 & e^{iht} & e^{-iht} & -1 \end{vmatrix} = 0 \tag{17}$$

Requiring that the determinate be zero results in the equation

$$ht = \tan^{-1}(p/h) + \tan^{-1}(q/h) + m\pi \tag{18}$$

where $m = 0, 1, 2, 3, \ldots$ is the mode quantum number. The appearance of an integral multiple of π on the right hand side of Eq. 19 results from the fact that the tan is periodic with a period of π. Thus, for a given waveguide thickness there are multiple solutions (modes) to the characteristic equation. The modes are distinguished by their mode number m.

Substituting the expressions for E_y into the wave equation results in the additional relationship

$$q^2 + \left(\frac{2\pi}{\lambda_o}\right)^2 n_1^{*2} = -\gamma^2 \tag{19}$$

$$-h^2 + \left(\frac{2\pi}{\lambda_o}\right)^2 n_2^2 = -\gamma^2 \tag{20}$$

$$p^2 + \left(\frac{2\pi}{\lambda_o}\right)^2 n_3^2 = -\gamma^2 \quad . \tag{21}$$

Eliminating the propagation constant γ from these three equations results in relationships between p, h, and q.

The complete set of equations that determines the TE mode solutions is therefore

$$q^2 + h^2 = \left(\frac{2\pi}{\lambda_o}\right)^2 \left(n_2^{*2} - n_1^{*2}\right) \tag{22}$$

$$p^2 + h^2 = \left(\frac{2\pi}{\lambda_o}\right)^2 \left(n_2^{*2} - n_3^{*2}\right) \tag{23}$$

$$ht = \tan^{-1}(p/h) + \tan^{-1}(q/h) + m\pi \quad . \tag{24}$$

Equations that determine the TM mode solutions can be derived in a similar fashion.

For TM modes the only nonzero field components are E_x, E_z, H_y.

Again assume solutions of the form

$$H_y = A\, e^{+qx} \qquad x < 0 \tag{25}$$

$$H_y = B\, e^{ihx} + C\, e^{-ihx} \qquad 0 < x < t \tag{26}$$

$$H_y = D\, e^{-p(x-t)} \qquad x > t \quad . \tag{27}$$

The characteristic equation and the relationships between the constants q, h, and p are obtained in the same fashion as was done for the TE modes. The resulting equations governing the modes are

$$q^2 + h^2 = \left(\frac{2\pi}{\lambda_o}\right)^2 \left(n_2^{*2} - n_1^{*2}\right) \tag{28}$$

$$p^2 + h^2 = \left(\frac{2\pi}{\lambda_o}\right)^2 \left(n_2^{*2} - n_3^{*2}\right) \tag{29}$$

$$ht = \tan^{-1}\left[\frac{q}{h}\left(\frac{n_2^*}{n_1^*}\right)^2\right] + \tan^{-1}\left[\frac{p}{h}\left(\frac{n_2^*}{n_3^*}\right)^2\right] + m\pi \quad . \tag{30}$$

As can be seen, the only difference between these equations and those for the TE modes (Eqs. 22 to 24) is the appearance of the ratio of the indices squared in the arguments of the inverse tangents.

The field components now can be expressed in terms of the arbitrary amplitude constant A and constants p, q, and h. For TE modes the field components are

$$E_y = A\,e^{qx} \qquad x < 0 \tag{31}$$

$$E_y = A(q/h)\,(\sin hx + (h/q)\cos hx) \qquad 0 \le x \le t \tag{32}$$

$$E_y = A(q/h)\,(\sin ht + (h/q)\cos ht)\,e^{-p(x-t)} \qquad x \ge t \tag{33}$$

$$H_z = +\frac{i}{\omega\mu^*}\left(\frac{\partial E_y}{\partial x}\right) \tag{34}$$

$$H_x = +\frac{i\gamma}{\omega\mu^*}\,E_y \quad . \tag{35}$$

The field components for TM modes are

$$H_y = Ae^{qx} \tag{36}$$

$$H_y = A(q/h)\left(\frac{n_2^*}{n_1^*}\right)^2\left\{\sin hx + \frac{h}{q}\left(\frac{n_1^*}{n_2^*}\right)^2\cos hx\right\} \quad 0 \le x \le t \tag{37}$$

$$H_y = A(q/h)\left(\frac{n_2^*}{n_1^*}\right)^2\left\{\sin ht + \frac{h}{q}\left(\frac{n_1^*}{n_2^*}\right)^2\cos ht\right\} e^{-p(x-t)} \quad x \ge t \tag{38}$$

$$E_z = + \frac{1}{i\omega n^{*2} \epsilon_o} \frac{\partial H_y}{\partial x} \tag{39}$$

$$E_x = \frac{\gamma}{i\omega n^{*2} \epsilon_o} H_y \tag{40}$$

2. Graphical Solution of the Governing Transcendental Equations

The governing equations for TE and TM mode propagation in the slab waveguide have been shown to be

$$q^2 + h^2 = \left(\frac{2\pi}{\lambda_o}\right)^2 \left(n_2^{*2} - n_1^{*2}\right) \tag{41}$$

$$p^2 + h^2 = \left(\frac{2\pi}{\lambda_o}\right)^2 \left(n_2^{*2} - n_3^{*2}\right) \tag{42}$$

Both for TE and TM modes

$$ht = \tan^{-1}(q/h) + \tan^{-1}(p/h) + m\pi \quad \text{(TE modes)} \tag{43}$$

$$ht = \tan^{-1}\left(\left[\frac{n_2^*}{n_1^*}\right]^2 \frac{q}{h}\right) + \tan^{-1}\left(\left[\frac{n_2^*}{n_3^*}\right]^2 \frac{p}{h}\right) + m\pi \quad \text{(TM modes)} \tag{44}$$

Two separate cases will now be treated. The first will be the very asymmetric waveguide. This will be followed by a treatment of the symmetric guide.

An asymmetric waveguide is one for which the refractive indices n_1^*, n_2^*, and n_3^* satisfy the requirements that $n_2^* >> n_1^*$ and $n_2^* \cong n_3^*$. Then

$$q^2 + h^2 = \left(\frac{2\pi}{\lambda_o}\right)^2 \left(n_2^{*2} - n_1^{*2}\right) = R_q^2 \tag{45}$$

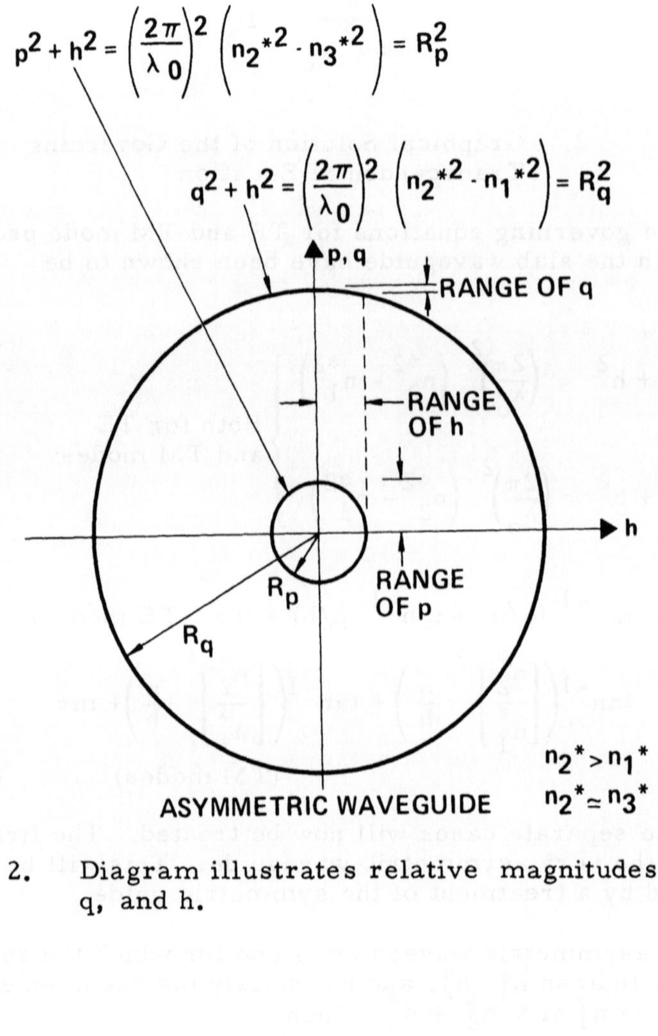

Fig. 2. Diagram illustrates relative magnitudes of p, q, and h.

$$p^2 + h^2 = \left(\frac{2\pi}{\lambda_o}\right)^2 \left(n_2^{*2} - n_3^{*2}\right) = R_p^2 \tag{46}$$

are the equations of circles of drastically different radii R_q and R_p as shown in Fig. 2. As a result, $q >> p$ and $q >> h$ for all h. It also follows that $q/h >> 1$ and $(n_2^*/n_1^*)^2 \, q/h >> 1$.

Now consider $\theta = \tan^{-1} x$ or $x = \tan\theta$. When $x >> 1$, θ takes on the value of $\pi/2$. Therefore, since q/h and $(n_2^*/n_1^*)^2 \, q/h >> 1$ it follows that Eqs. 43 and 44 reduce to

$$ht = \left(m + \frac{1}{2}\right)\pi + \tan^{-1}\left(\frac{p}{h}\right) \qquad \text{TE} \tag{47}$$

$$ht = \left(m + \frac{1}{2}\right)\pi + \tan^{-1}\left\{\left(\frac{n_2^*}{n_3^*}\right)^2 \frac{p}{h}\right\} \qquad \text{TM} \tag{48}$$

Since $n_2^* \cong n_3^*$ the governing equations for both TE and TM modes of asymmetric waveguides are

$$(qt)^2 = \left(\frac{2\pi t}{\lambda_o}\right)^2 \left(n_2^2 - n_1^2\right) \tag{49}$$

$$(pt)^2 + (ht)^2 = \left(\frac{2\pi t}{\lambda_o}\right)^2 \left(n_2^{*2} - n_3^{*2}\right) \tag{50}$$

$$pt = -ht \cot ht = ht \tan(ht - \pi/2) \; . \tag{51}$$

Within the approximations made here the TE and TM modes of a very asymmetric guide are degenerate.

The transcendental equations (Eqs. 50 and 51) governing the solutions of an asymmetric waveguide can be solved graphically as is shown in Fig. 3. For simplicity, it is assumed that all three media are lossless so that the refractive indices n_1, n_2, and n_3 are all real. In addition, since the interest here is in the bound modes the exponential constants p and q must be restricted to real positive values. If Eqs. 50 and 51 are plotted in the pt-ht

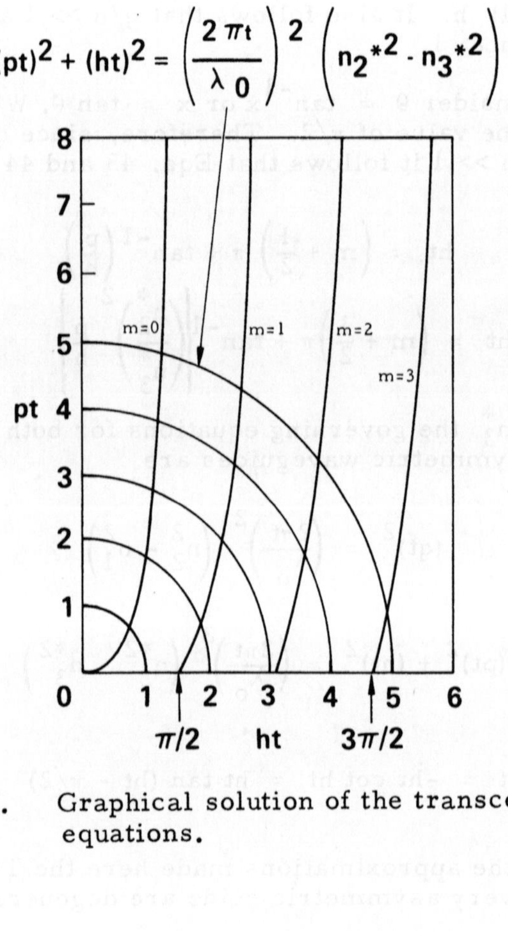

Fig. 3. Graphical solution of the transcendental equations.

plane as is done in Fig. 3, simultaneous solution occurs where the two curves intersect. Since p must be positive only the odd values of m, that is, m = 1, 3, 5, ... are valid for the asymmetric case. The others are for the symmetric waveguide to be dealt with later. The actual values of p and h for a given solution (mode) are determined by the circle radius $R = (2\pi t/\lambda_o)\sqrt{n_2^2 - n_3^2}$. For the mth mode the circle radius must be within the limits

$$(2m_{as}+1)\frac{\pi}{2} \leqq \left(\frac{2\pi t}{\lambda_o}\right)\left(n_2^2 - n_3^2\right)^{1/2} \leqq (m_{as}+1)\pi \quad (52)$$

where m_{as} = 1, 3, 5, ... It can be seen that asymmetric dielectric waveguides have a low frequency (long wavelength) cutoff below which energy does not propagate. The cutoff condition for the index change $\Delta n = n_2 - n_3$ for the mth mode is

$$\Delta n = \frac{(2m-1)^2}{32\, n_3}\left(\frac{\lambda_o}{t}\right)^2 \quad \text{for } m = 1, 2, 3, \ldots \quad (53)$$

For the same wavelength to thickness ratio the index change required for the second mode to propagate is nine times that of the lowest order mode. The number of modes that propagate are finite and are dependent on the controllable parameters n_2, n_3, λ_o, and t. It should be noted that these bound modes do not form a mathematically complete set.

A symmetric waveguide is a slab guide of index n_2 buried in a cladding material of index n_1 that is $n_1 = n_3$ in Fig. 1. Equations 41 and 42 reveal that when $n_1 = n_3$ then p = q. Substitution of p = q and $n_1 = n_3$ into equations 43 and 44 results in

$$hd = \tan^{-1}\left(\frac{p}{h}\right) + m\frac{\pi}{2} \quad \text{(TE modes)} \quad (54)$$

$$hd = \tan^{-1}\left(\left[\frac{n_2}{n_1}\right]^2 \frac{p}{h}\right) + m\frac{\pi}{2} \quad \text{(TM modes)} \quad (55)$$

where for convenience the substitution t = 2d has also been made. The governing equations for the symmetric waveguide of thickness t = 2d are therefore

$$(pd)^2 + (hd)^2 = \left(\frac{2\pi d}{\lambda_o}\right)^2 \sqrt{n_2^2 - n_1^2} \qquad \text{TE and TM} \quad (56)$$

$$pd = \begin{cases} hd \tan hd & \text{even TE modes } (m_s = 0, 2, 4, 6, \ldots) \\ -hd \cot hd & \text{odd TE modes } (m_s = 1, 3, 5, \ldots) \end{cases} \quad (57)$$

$$pd = \begin{cases} hd\left(\frac{n_1}{n_2}\right)^2 \tan hd & \text{even TM modes } (m_s = 0, 2, 4, 6 \ldots) \\ -hd\left(\frac{n_1}{n_2}\right)^2 \cot hd & \text{odd TM modes } (m_s = 1, 3, 5, \ldots) \end{cases} \quad (58)$$

As with the asymmetric waveguide the transcendental equations (Eqs. 56 to 58) can be solved graphically as is shown in Fig. 3. For the symmetric waveguide all values of $m_s = 0, 1, 2, 3, \ldots$ both even and odd are allowed solutions.

To help clarify the relationships between the propagating modes of an asymmetric and symmetric guide graphically, solutions for the symmetric and asymmetric cases are shown in Fig. 4. The asymmetric waveguide propagates only the odd modes of a symmetric guide of twice its thickness. As stated above, the complete set of graphical solutions for both the symmetric and asymmetric case can therefore be obtained from Fig. 3.

To obtain the field distribution for the symmetric guide it is convenient to shift the coordinate system to the center of the slab. Let $x' = x - t/2 = x - d$ in the diagram shown in Fig. 1. Substitution of x' for x in the field equations for the TE and TM modes (Eqs. 32 to 41) and utilization of the characteristic Eqs. 57 and 59 results in the following field expressions for a symmetric guide of thickness t = 2d.

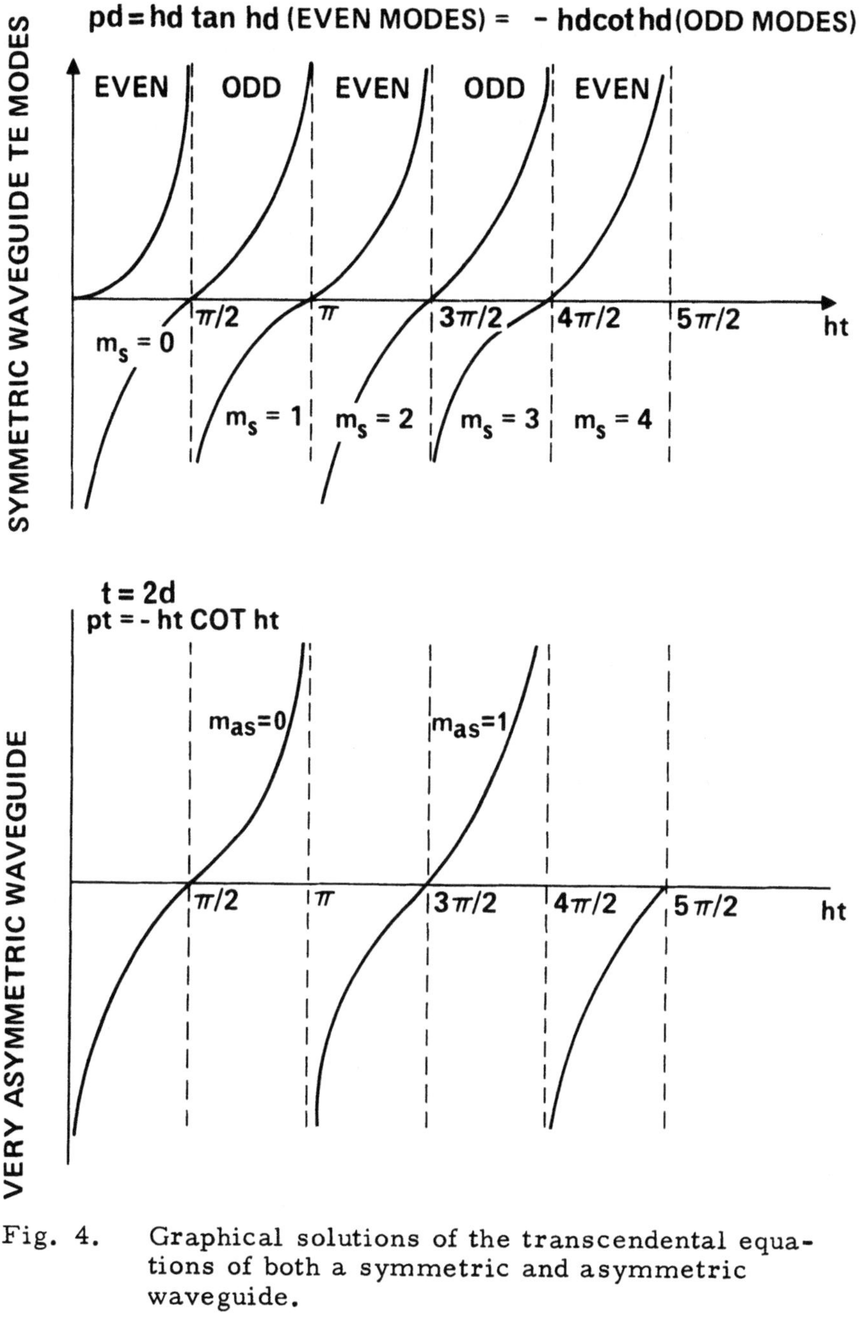

Fig. 4. Graphical solutions of the transcendental equations of both a symmetric and asymmetric waveguide.

$$E_y = \begin{cases} A \sin hx' & \text{odd TE modes } (m_s = 1, 3, 5\ldots) \\ & \qquad |x'| \leqq 2d \\ A \cos hx' & \text{even TE modes } (m_s = 0, 2, 4\ldots) \end{cases} \tag{59}$$

$$H_y = \begin{cases} A \sin hx' & \text{odd TM modes } (m_s = 1, 3, 5, \ldots) \\ & \qquad |x'| \leqq 2d \\ A \cos hx' & \text{even TM modes } (m_s = 0, 2, 4, \ldots) \end{cases} \tag{60}$$

3. Dispersion in Thin Films

The dispersion equations (ω versus β) for both TE and TM modes of asymmetric and symmetric waveguides have been determined in terms of the constants h, q, and p (Eqs. 43 and 44). The constants h, q, and p are related to the frequency and propagation constant (Eqs. 20, 21, 22). For simplicity it will be assumed that all three media are loss free so that n_1^*, n_2^*, n_3^* are real quantities. In this case $\gamma = i\beta$ and Eqs. 19, 20, 21 become

$$q^2 = \beta^2 - k_1^2 \tag{61}$$

$$p^2 = \beta^2 - k_3^2 \tag{62}$$

$$h^2 = k_2^2 - \beta^2 \tag{63}$$

where

$$k_i = \frac{2\pi}{\lambda_o} n_i = \frac{\omega n_i}{c}$$

Substitution of these expressions for p, q, and h into the dispersion equations (Eqs. 44, 45) results in the expressions

TE modes

$$t(k_2^2-\beta^2)^{1/2} = \tan^{-1}\left[\frac{(\beta^2-k_1^2)^{1/2}}{(k_2^2-\beta^2)^{1/2}}\right] + \tan^{-1}\left[\frac{(\beta^2-k_3^2)^{1/2}}{(k_3^2-\beta^2)^{1/2}}\right] + m\pi \tag{64}$$

TM modes

$$t\left(k_2^2 - \beta^2\right)^{1/2} = \tan^{-1}\left\{\left(\frac{n_2}{n_1}\right)^2\left[\frac{(\beta^2 - k_1^2)^{1/2}}{(k_2^2 - \beta^2)^{1/2}}\right]\right\} + \tan^{-1}\left\{\left(\frac{n_2}{n_3}\right)^2\left[\frac{(\beta^2 - k_3^2)^{1/2}}{(k_2^2 - \beta^2)^{1/2}}\right]\right\} + m\pi \tag{65}$$

for the TE and TM modes of an asymmetric waveguide.

The dispersion equations for the even and odd modes of a symmetric waveguide can be derived from the above expressions. The resulting equations for the symmetric guide are

Even TE modes

$$(\beta^2 - k_1^2)^{1/2} = (k_2^2 - \beta^2)^{1/2} \tan (k_2^2 - \beta^2)^{1/2} d \tag{66}$$

Odd TE modes

$$(\beta^2 - k_1^2)^{1/2} = -(k_2^2 - \beta^2)^{1/2} \cot (k_2^2 - \beta^2)^{1/2} d \tag{67}$$

Even TM modes

$$(\beta^2-k_1^2)^{1/2} = (k_2^2-\beta^2)^{1/2}\left(\frac{n_1}{n_2}\right)^2 \tan (k_2^2 - \beta^2)^{1/2} d \tag{68}$$

Odd TM modes

$$(\beta^2 - k_1^2)^{1/2} = -(k_2^2 - \beta^2)^{1/2} \left(\frac{n_1}{n_2}\right)^2 \cot (k_2^2 - \beta^2)^{1/2} \; d \quad (69)$$

Since loss-free media only are being considered, the constants p, q, and h are real for the bound waveguide modes. This places restrictions on the allowed values of ω and β for a given mode. Since p, q, and h must be real it is clear from the previous equations (Eqs. 61, 62, and 63) that

$$\beta > \frac{\omega}{c} n_1 \quad (70)$$

$$\beta > \frac{\omega}{c} n_3 \quad (71)$$

and

$$\beta < \frac{\omega}{c} n_2 \quad (72)$$

For the case when $n_1 < n_3 < n_2$ the region on the ω-β diagram where confined mode propagation occurs is indicated by the shading in Fig. 5. In physical terms this restriction requires that the phase velocity of the modes must be no greater than the phase velocity of a plane wave traveling in the substrate material of index n_3 and must be no less than the velocity of a plane wave traveling in the guiding layer whose index is n_2.

Plots of computer solutions obtained for an asymmetric waveguide whose parameters are

$n_1 = 1.0$ $t = 3.34$ microns

$n_2 = 2.23$

$n_3 = 1.393$

are shown in Fig. 6. The plots were made neglecting any natural dispersion in the materials. The dispersion curves for the first ten TE modes are shown. The dispersion curve for the lowest order TE mode of a symmetric

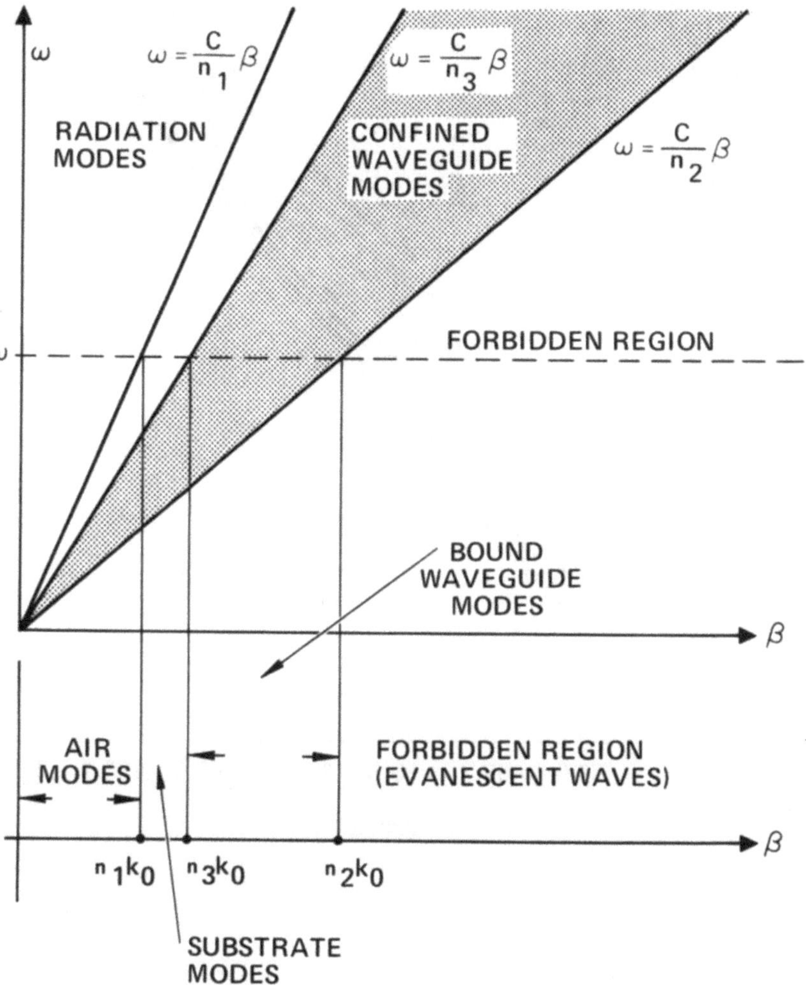

Fig. 5. Illustration shows the region on the ω-β diagram where confined modes can exist.

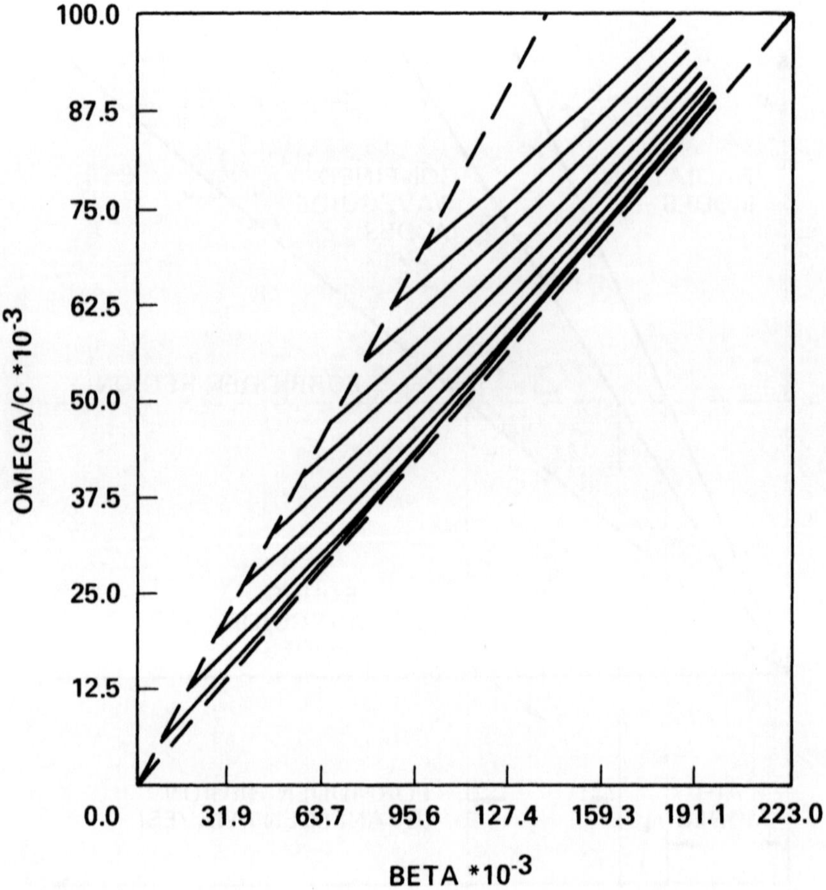

Fig. 6. Dispersion curves of ten modes of an asymmetric waveguide 3.34 μm thick with $n_1 = 1$, $n_2 = 2.23$, and $n_3 = 1.393$. The curves neglect any natural dispersion.

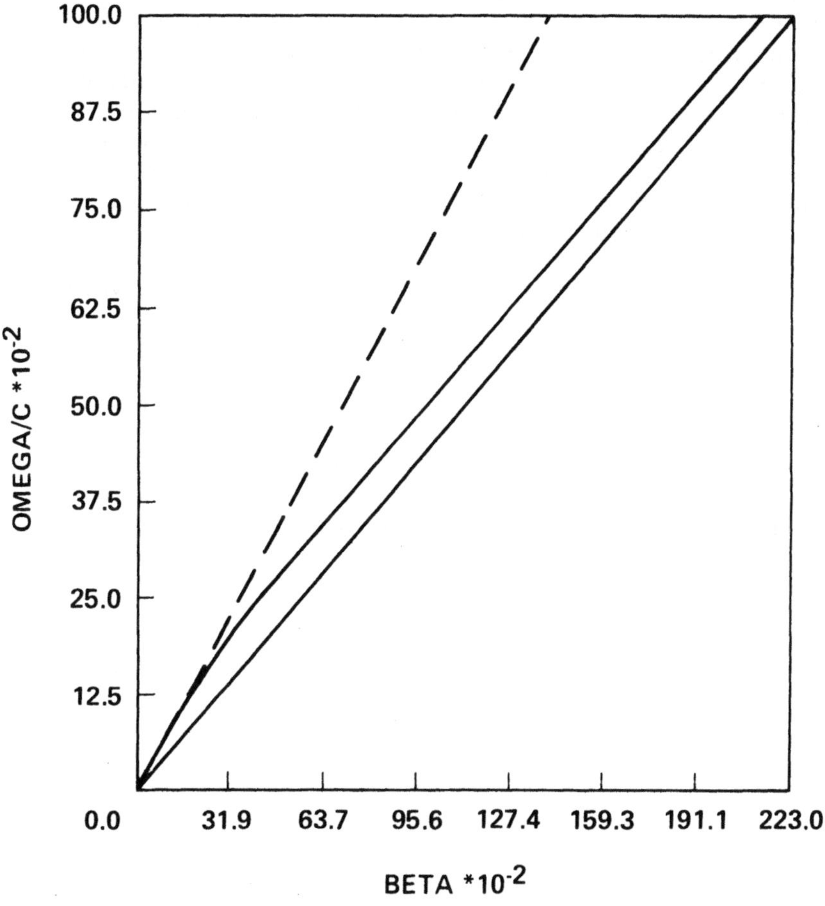

Fig. 7. Dispersion curve of the lowest order mode of a symmetric waveguide 3.34 μm thick with $n_1 = n_3 = 1.393$ and $n_2 = 2.23$. The curve neglects natural dispersion.

waveguide with parameters $n_1 = n_3 = 1.393$, $n_2 = 2.23$, and $t = 3.34$ μm is shown in Fig. 7.

The shape of the dispersion curve for a given mode can be understood physically as follows. As the frequency increases the depth of penetration of the exponential tails of the fields into regions (1) and (3) decreases, that is, p and q increase. As a result, the fields are confined more and more to region n_2. In the limit as $\omega \rightarrow \infty$ the fields are completely confined to the waveguide layer and propagate as plane waves in a medium of index n_2. The slope of the lower boundary is the velocity of light line in the guiding layer.

Near the upper boundary, whose slope is the velocity of light in the substrate of index n_3, the value of p becomes small and in the limit when (ω/c) approaches k_3 the constant p approaches zero. In this limit the field tail extends infinitely into the substrate material.

Beyond the limit lines $\omega/c = \beta/n_3$ and $\omega/c = \beta/n_2$ there exists a continuous spectrum of modes called radiation modes. In the region beyond the lower limit line the propagation constant β is imaginary and hence these modes are termed evanescent.

Chapter 4 - RECTANGULAR DIELECTRIC WAVEGUIDES

J. E. Goell

Bell Laboratories
Crawford Hill Laboratory
Holmdel, N. J. 07733

ABSTRACT

The properties of lossless rectangular dielectric waveguides will be presented. A review of the relevant analysis techniques will be given. Marcatili's rectangular harmonic analysis which yields closed form results for many integrated optics applications and the circular harmonic numerical approach will be described. Both dispersion curves and simulations of mode patterns will be presented.

I. INTRODUCTION

Dielectric waveguides are expected to be building blocks of integrated optical circuits. These waveguides can serve not only as transmission media to confine and direct optical signals, but also as the basis for components such as filters and directional couplers.[1] Thus, it is important to have a thorough knowledge of the properties of their modes. In many integrated optics applications it is expected that waveguides will consist of a rectangular or a near rectangular dielectric core imbedded in a dielectric medium of slightly lower refractive index.

Lossless rectangular dielectric waveguides have been analyzed by a number of workers including Marcatili,[2] Goell,[3] Schlosser,[4] Bartling,[5] and Shaw.[6] In Marcatili's analysis fields with sinusoidal variation in the core are matched to exponentially decaying fields in the external medium. In each region a single mode is used. The results of this method are obtained in a relatively simple form for numerical evaluation. In Goell's analysis the radial variation of the longitudinal electric and magnetic fields of the modes was represented by a sum of Bessel functions inside the waveguide core and by a sum of modified Bessel functions outside the waveguide core. Computer solutions were found by matching the fields along the perimeter of the core. Schlosser performed an analysis using a sum of rectangular harmonics, and has computed propagation constants for the case of a dielectric rod in air. This case is of interest for microwave applications, but is of a higher index difference than is anticipated for integrated optics applications. With his approach the number of terms that must be included increases as cut-off is approached for some of the modes, while for Goell's analysis it does not. On the other hand Schlosser's approach is better suited than Goell's to the analysis of guides of large aspect ratio. Bartling has solved the problem in terms of a Green's dyadic, but has not calculated specific cases.

Our purpose here is to examine the properties of rectangular dielectric waveguides to form a basis for experimental work. Marcatili's analysis, which gives analytical results in a relatively simple form will be treated first. Following this Goell's analysis and results will be described to give insight into the cases where the simpler approach cannot be employed. Finally numerical results obtained from the two methods will be compared.

II. MARCATILI'S ANALYSIS

Consider a rectangular dielectric waveguide of the geometry shown in Figure 1 consisting of a core of refractive index n_1 surrounded by media of indices n_2, n_3, n_4, and n_5. The indices of the corner (shaded) regions are unspecified and in the analysis the fields in these regions are not considered. This is justified for well guided modes. For this case when

$$\frac{n_1}{n_{\substack{2\\3\\4\\5}}} - 1 \ll 1$$

the energy in these regions is negligible. The range over which this approximation is valid will become apparent shortly.

Under this approximation two sets of modes can be found to satisfy the boundary conditions; one designated E^x_{pq} with the magnetic field assumed to be y directed and the second designated E^y_{pq} with the magnetic field assumed to be x directed. The subscripts p and q specify the number of field extrema along the x and y directions, respectively.

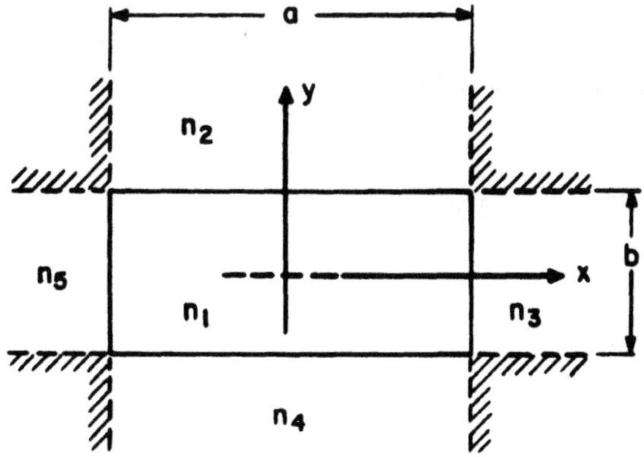

Fig. 1 - Idealized rectangular waveguide geometry.

Approximate solutions to Maxwell's equations have been found for the E^y_{pq} modes by assuming an x-directed magnetic field of the form

$$H_{x\nu} = \exp(-ik_z z + i\omega t) \begin{cases} M_1 \cos(k_x x+\alpha)\cos(k_y y+\beta) & \nu=1 \\ M_2 \cos(k_x x+\alpha)\exp(-ik_{y2} y) & \nu=2 \\ M_3 \cos(k_y y+\beta)\exp(-ik_{x3} x) & \nu=3 \\ M_4 \cos(k_x x+\alpha)\exp(ik_{y4} y) & \nu=4 \\ M_5 \cos(k_y y+\beta)\exp(ik_{x5} x) & \nu=5 \end{cases} \tag{1}$$

where the numerical subscript ν represents the region of application. From Maxwell's equations the other significant field quantity, $E_{y\nu}$ is given by

$$E_{y\nu} = \frac{k^2 n_\nu^2 - k_{y\nu}^2}{\omega \varepsilon n_\nu^2 k_z} H_{x\nu} \tag{2}$$

The other field quantities E_x, H_y, E_z, and H_z are small and need not be considered.

Eigenvalue relations are found by matching field components at the boundary using the familiar relationship

$$k_\nu^2 = k_{x\nu}^2 + k_{y\nu}^2 + k_{z\nu}^2 \tag{3}$$

where

$$k_\nu = \frac{2\pi}{\lambda_0} n_\nu$$

and λ_0 is the free space wavelength under the assumption that

$$k_{x1} = k_{x2} = k_{x4} = k_x$$

$$k_{y1} = k_{y3} = k_{y5} = k_y \qquad (4)$$

Assuming the waves impinge on the surfaces of the core at grazing incidence so that

$$\begin{matrix} k_x \\ k_y \end{matrix} << k_z$$

gives

$$k_y b = q\pi - \tan^{-1} \frac{n_2^2}{n_1^2} k_y \eta_2 - \tan^{-1} \frac{n_4^2}{n_1^2} k_y \eta_4$$

$$k_x a = p\pi - \tan^{-1} k_x \xi_3 - \tan^{-1} k_x \xi_5 \qquad (5)$$

where

$$\eta_{\substack{2\\4}} = \frac{1}{|k_{y_{\substack{2\\4}}}|} = \frac{1}{\left[\left(\frac{\pi}{A_{\substack{2\\4}}}\right)^2 - k_y^2\right]^{1/2}}$$

$$\xi_{\substack{3\\5}} = \frac{1}{|k_{x_{\substack{3\\5}}}|} = \frac{1}{\left[\left(\frac{\pi}{A_{\substack{3\\5}}}\right)^2 - k_x^2\right]^{1/2}} \qquad (6)$$

and

$$A_{2,3,4,5} = \frac{\pi}{\left(k_1^2 - k_{2,3,4,5}^2\right)^{1/2}} = \frac{\lambda}{2\left(n_1^2 - n_{2,3,4,5}^2\right)^{1/2}} \qquad (7)$$

The longitudinal propagation constant is given by

$$k_z = \sqrt{k_1^2 - k_x^2 - k_y^2} \qquad (8)$$

For well guided modes

$$\left[\frac{k_x A_{\substack{3\\5}}}{\pi}\right]^2 << 1, \left[\frac{k_y A_{\substack{2\\4}}}{\pi}\right]^2 << 1 \qquad (9)$$

When these relations hold the propagation constants can be put in the following closed form:

$$k_x = \frac{p\pi}{a}\left(1 + \frac{A_3 + A_5}{\pi a}\right)^{-1} \qquad (10)$$

$$k_y = \frac{q\pi}{b}\left(1 + \frac{n_2^2 A_2 + n_4^2 A_4}{\pi n_1^2 b}\right)^{-1}$$

where the A's are given by Eq. 7 and

$$\xi_{\substack{3\\5}} = \frac{A_{\substack{3\\5}}}{\pi}\left[1 - \left[\frac{pA_{\substack{3\\5}}}{a} \frac{1}{1 + \frac{A_3 + A_5}{\pi a}}\right]^2\right]^{-\frac{1}{2}}$$

$$n_{\frac{2}{4}} = \frac{A_{\frac{2}{4}}}{\pi}\left[1 - \left[\frac{\frac{qA_{\frac{2}{4}}}{b}}{} \frac{1}{1 + \frac{n_2^2A_2+n_4^2A_4}{\pi n_1^2 b}}\right]^2\right]^{-\frac{1}{2}} \tag{11}$$

The longitudinal propagation constant is again given by Eq. 8.

Corresponding relationships for the E^y_{pq} modes can be found by interchanging x and y, a and b, n_2 and n_3 and, n_4 and n_5.

III. CIRCULAR HARMONIC ANALYSIS

A. Analysis

The waveguides to be considered here consist of a rectangular core of dielectric constant ε_1 surrounded by an infinite medium of dielectric constant ε_0. Both media will be assumed to be isotropic, and have the permeability of free space, μ_0.* The coordinate systems (rectangular and cylindrical) and rod dimension are shown in Figure 2. The direction of propagation is in the +z direction (towards the observer).

---

* It can be assumed, without loss of generality, that the outer medium is free space.

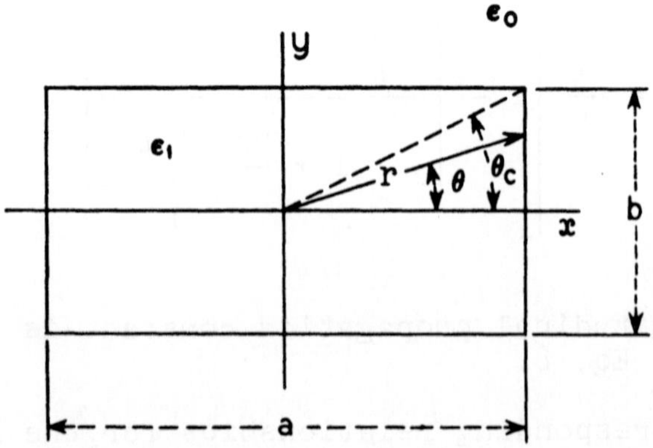

Fig. 2 - Dimensions and Coordinate system.

In cylindrical coordinates the field solutions of Maxwell's equations take the form of Bessel functions and modified Bessel functions multiplied by trigometric functions, and their derivatives. In order for propagation to take place in the z direction, the field solutions must be Bessel functions in the core and modified Bessel functions outside. Since Bessel functions of the second kind have a pole at the origin and modified Bessel functions of the first kind a pole at infinity, the radial variation of the fields will be assumed to be a sum of Bessel functions of the first kind and their derivatives inside the core and a sum of modified Bessel functions and their derivatives outside the core.

In cylindrical coordinates, the z components of the electric and magnetic fields are given by:[7]

$$E_{z1} = \sum_{n=0}^{\infty} a_n J_n(hr)\sin(n\theta+\phi_n)\exp\{i(k_z z-\omega t)\} \tag{12a}$$

and

$$H_{z1} = \sum_{n=0}^{\infty} b_n J_n(hr)\sin(n\theta+\psi_n)\exp\{i(k_z z-\omega t)\} \tag{12b}$$

inside the core, and by

$$E_{z0} = \sum_{n=0}^{\infty} c_n K_n(pr)\sin(n\theta+\phi_n)\exp\{i(k_z z-\omega t)\} \tag{12c}$$

$$H_{z0} = \sum_{n=0}^{\infty} d_n K_n(pr)\sin(n\theta+\psi_n)\exp\{i(k_z z-\omega t) \tag{12d}$$

outside the core, where ω is the radian frequency and k_z the longitudinal propagation constant. The transverse propagation constants are given by

$$h = \sqrt{k_1^2 - k_z^2}$$

$$p = \sqrt{k_z^2 - k_0^2} \tag{13}$$

where

$$k_1 = \omega\sqrt{\mu_0 \varepsilon_1}$$

and

$$k_0 = \omega\sqrt{\mu_0 \varepsilon_0} \ .$$

The terms J_n and K_n are the nth order Bessel functions and modified Bessel functions, respectively, and ψ_n and ϕ_n are arbitrary phase angles.

The transverse components of the fields are given by[7]

$$E_r = \frac{ik_z}{k^2 - k_z^2}\left[\frac{\partial E_z}{\partial r} + \left(\frac{\mu_0\omega}{k_z r}\right)\frac{\partial H_z}{\partial \theta}\right]$$

$$E_\theta = \frac{ik_z}{k^2 - k_z^2}\left[\frac{1}{r}\frac{\partial E_z}{\partial \theta} - \left(\frac{\mu_0\omega}{k_z}\right)\frac{\partial H_z}{\partial r}\right]$$

$$H_r = \frac{ik_z}{k^2 - k_z^2}\left[-\left(\frac{k^2}{\mu_0\omega k_z r}\right)\frac{\partial E_z}{\partial \theta} + \frac{\partial H_z}{\partial r}\right]$$

$$H_\theta = \frac{ik_z}{k^2 - k_z^2}\left[\left(\frac{k^2}{\mu_0\omega k_z}\right)\frac{\partial E_z}{\partial r} + \frac{1}{r}\frac{\partial H_z}{\partial \theta}\right], \quad (14)$$

where k can be either k_1 or k_0.

Finally, the component of the transverse electric field tangent to the rectangular core is given by

$$E_t = \pm(E_r \sin\theta + E_\theta \cos\theta) \qquad \begin{array}{l} -\theta_c < \theta < \theta_c \\ \pi - \theta_c < \theta < \pi + \theta_c \end{array} \quad (15)$$

or

$$E_t = \pm(-E_r \cos\theta + E_\theta \sin\theta), \qquad \begin{array}{l} \theta_c < \theta < \pi - \theta_c \\ \pi + \theta_c < \theta < -\theta_c \end{array}$$

where θ_c is the angle which a radial line to the corner in the first quadrant makes with the x axis. Similar expressions exist for the transverse tangential magnetic field.

Solutions were found from the field equations by matching the transverse and longitudinal tangential electric and magnetic fields along the boundary of the core. Eqs. 12 and 15 were put in matrix form by fixing the number of harmonics and matching the fields at discrete points on the boundary. Symmetry was taken into account to minimize computation.

Numerous test runs were made in order to obtain an estimate of the accuracy of the computed results. It was found that the propagation constants converged rapidly for aspect ratios up to 2 and it is believed that all of the data are accurate to 1 percent except for the results of calculations using even harmonics for aspect ratios other than unity which are believed to be accurate to better than two percent.

The electric and magnetic fields were calculated for representative cases by arbitrarily picking one of the field coefficients and then solving for the others by matrix manipulation[8] on the computer.

Several approaches were used to obtain information that could be used to derive the field patterns. These included computation of the field components along radial cuts of the waveguide cross section, computer generated isoclines giving the direction of the electric field, and computer generated mode pictures.

In order to draw pictures of mode patterns, the power density was calculated at each of the points to be plotted. The square root of the power density was then normalized to the square root of the peak power density and quantized into 21 levels. The figure shown in Figure 3 was then plotted on a cathode ray tube, centered for each point, starting at 1 and going to the point

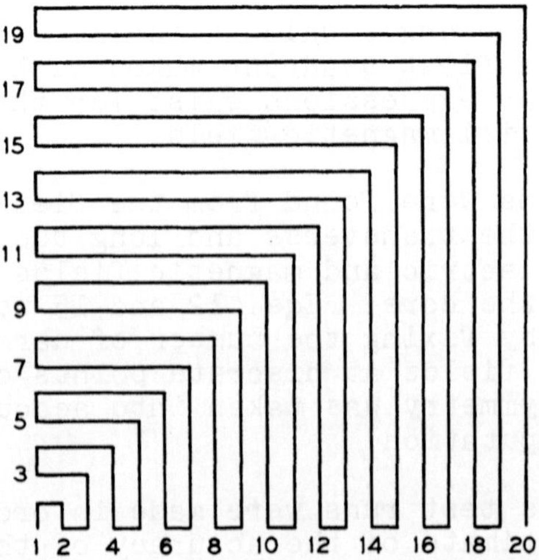

Fig. 3 - Intensity picture figure.

corresponding to the appropriate quantized level (except at the points where the normalized power was zero where no plotting was done). The size of the cathode ray tube spot is approximately equal to the line spacing in the figure so the plotted figures are filled in. Therefore, the light passed by these figures is approximately equal to the power density to be represented. Starting with the single quadrant pictures, complete pictures were generated by making quadruple exposures of the microfilm.

Unlike metallic waveguide, the field patterns of dielectric waveguide are sensitive to refractive index difference, wavelength, and aspect ratio. This complicates the problem of finding a reasonably descriptive mode designation scheme.

For rectangular metallic waveguide, the accepted approach is to designate the modes as TE (or H) and TM (or E), and to specify the number of field maxima in the x and y directions with a double subscript. When there is no variation the subscript 0 is used.

Since the rectangular dielectric waveguide modes, in general, are neither pure TE nor pure TM, a different scheme must be used. The scheme adopted is based on the fact that in the limit, for large aspect ratio, short wavelength, and small refractive index difference, the transverse electric field will be primarily parallel to one of the transverse axes. Modes will be designated as E^y_{mn} if in the limit their electric field is parallel to the y axis and as E^x_{mn} if in the limit their electric field is parallel to the x axis. The m and n subscript will be used to designate the number of maxima in the x and y directions respectively.*

The arguments of the Bessel and modified Bessel functions are given by

$$hr = \sqrt{k_1^2 - k_z^2}\, r$$

and

$$pr = \sqrt{k_z^2 - k_0^2}\, r ,$$

respectively. The first argument can be put in the form

$$hr = \sqrt{k_1^2 - k_0^2 - p^2}\, r . \qquad (16)$$

Letting

$$P^2 = \frac{(k_z/k_0)^2 - 1}{n_r^2 - 1} \qquad (17)$$

and

---
* This scheme agrees with that used by Marcatili.

$$R = rk_0 \sqrt{n_r^2 - 1} \,, \tag{18}$$

where

$$n_r = k_1/k_0 \tag{19}$$

is the index of refraction of the core relative to the outer medium, gives

$$pr = PR \tag{20}$$

and

$$hr = \sqrt{1-P^2}\; R \tag{21}$$

The curves of the propagation constant that will be given in Section IV will be drawn in terms of P^2 and B, where

$$B = \frac{2b}{\lambda_0} \sqrt{n_r^2 - 1} \tag{22}$$

and

$$\lambda_0 = \frac{2\pi}{k_0}$$

Since for fixed aspect ratio, R is proportional to B, the use of P^2 and B as plotting variables eliminates the explicit dependence of the Bessel and modified Bessel function arguments on the refractive indices of the media.

The normalized propagation constant, P^2, has two additional properties which make its use convenient. First, its range of variation is on the interval (0,1). Second, for $n \approx 1$,

$$P^2 \approx \frac{k_z/k_0 - 1}{\Delta n_r} , \tag{23}$$

where

$$\Delta n_r = n_r - 1 ,$$

so for small Δn_r, P^2 is proportional to $k_z - k_0$. The latter property is the reason that P^2 rather than P was used as a plotting variable.

B. Computed Results

1. Mode Configurations. The effect of the number of harmonics used in the field patterns is of some interest. This question has not been explored in great detail; however, a few comparisons of intensity pictures for different numbers of circular harmonics were made. In general, it was found that five harmonics were sufficient to give a good representation of the modes that will be presented. An example of this is given in Figure 4, a comparison of the E^y_{11} mode

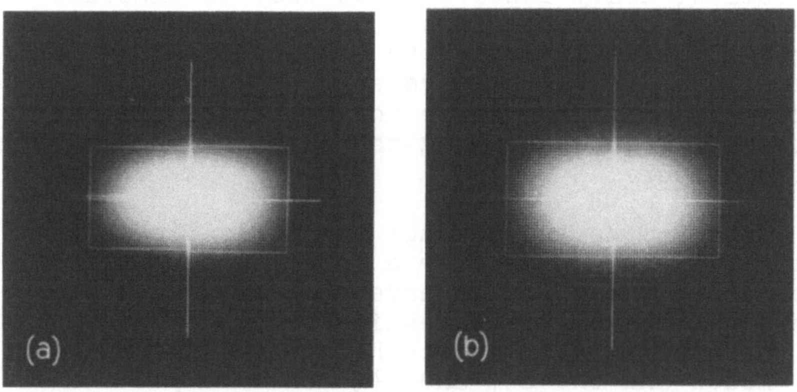

Fig. 4 - Intensity Pictures for the E^y_{11} Mode for $a/b=2$; $\beta=2$, and $\Delta n_r=.01$; a. for 5 Harmonics; b. for 9 Harmonics.

intensity patterns for 5 and 9 harmonics. For the results which follow, 5 circular harmonics were used.

Figure 5 shows intensity pictures for the first 6 modes for unity aspect ratio, $B=3.$, and an index difference of 1%. Similar data is presented in Figure 6 for an aspect ratio of two and $B=2$. In both cases, the plots are arranged in ascending order of cutoff frequency. All of the pictures are for E^y_{mn} modes. These pictures are virtually indistinguishable from the corresponding E^x_{mn} modes so both sets will not be presented. In general, for small index difference the E^y_{mn} and E^x_{mn} can be considered to be near duals, that is, to have identical field patterns except that the electric and magnetic fields are interchanged.

The field distribution patterns for the modes of Figures 5 and 6 are more complicated than those for rectangular metallic waveguide since they extend beyond the waveguide boundary and in general, their shape is dependent on the various waveguide parameters. The E^x_{11} and E^y_{11} modes have the simplest field patterns. The electric and magnetic field orientations for the E^y_{11} mode are shown in Figure 7. If this figure and the ones to follow, heavy lines will be used to indicate the fields in the regions of high intensity and light lines to indicate the continuation of the field lines in the transverse plane. Only E^y_{mn} modes will be shown since the E^x_{mn} modes can be obtained by interchanging the electric and magnetic field vectors.

Figures 8a and 8b show the field lines for the E^y_{21} and E^y_{12} modes for a large aspect ratio. (For $a/b \to \infty$ the fields have the appearance of rectangular metallic waveguide modes.) However, as the aspect ratio approaches unity, the E^y_{12} and E^x_{21} modes and the E^y_{21} and E^x_{12} modes couple and shift to the patterns shown in Figures 9a and 9b. Most of the change takes place with the aspect ratio close to unity.

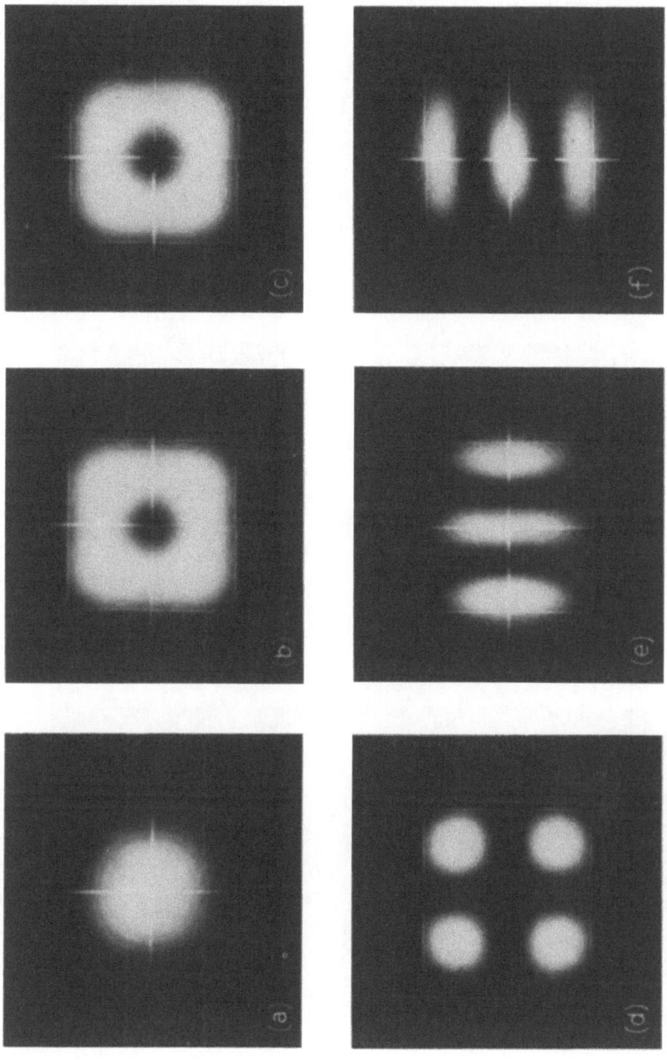

Fig. 5 - Intensity Pictures for Some E^y_{mn} Modes with Unity Aspect Ratio, $B=3$, and $\Delta n_r=.01$.

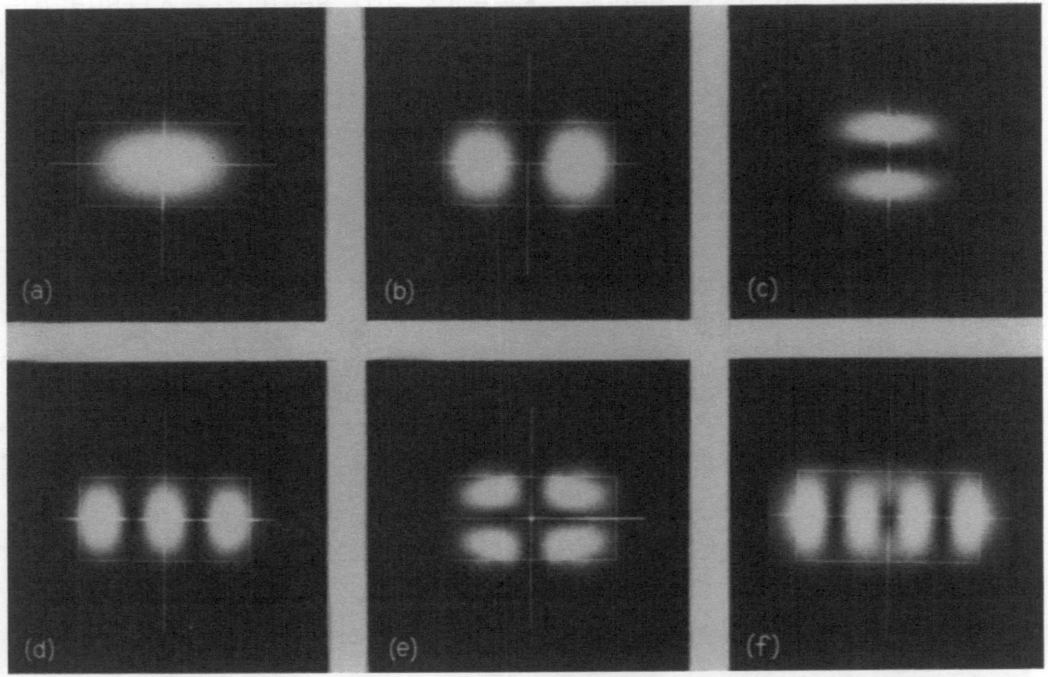

Fig. 6 - Intensity Picture for Some E^y_{mn} Modes with a/b=2, B=2, and Δn_r=.01.

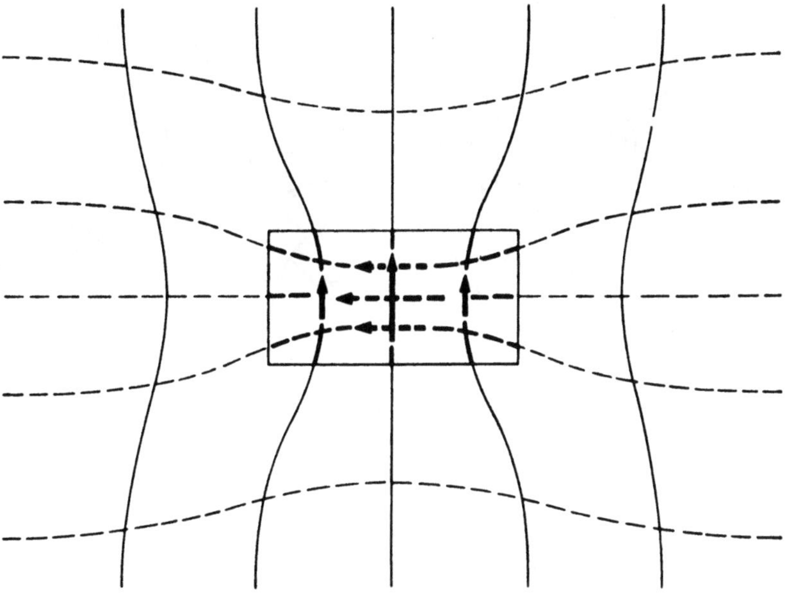

Fig. 7 - Field Configuration of the E^y_{11} Mode.

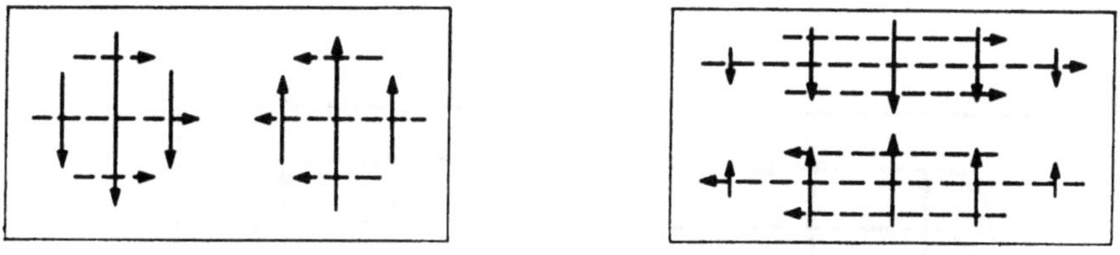

(a) (b)

Fig. 8 - Field configurations for the (a) E^y_{21} and (b) E^y_{12} modes far from cutoff.

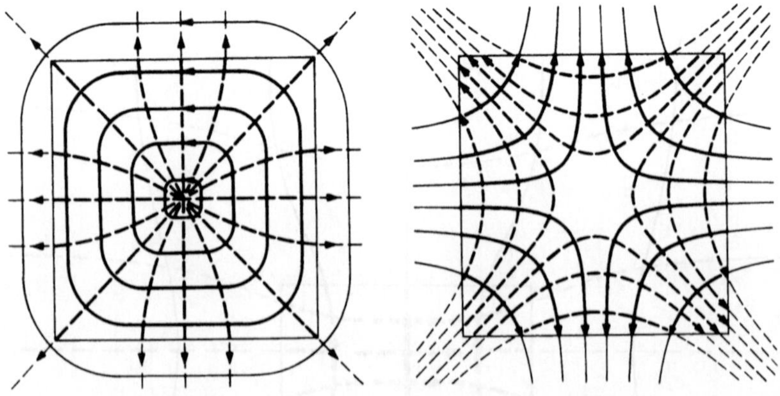

Fig. 9 - Field configurations for the square (a) E^y_{21} and (b) E^y_{12} modes.

The field configuration for the E^y_{22} mode is shown in Figure 10, the E^y_{31} mode in Figure 11, and the E^y_{13} mode in Figure 12. The field pattern of these modes do not change drastically with aspect ratio.

Figure 13a shows an intensity picture of the E^y_{32} mode and Figure 13b its field pattern for unity aspect ratio. The field pattern inside the core is similar to a sum of the TE_{23} and TE_{32} of

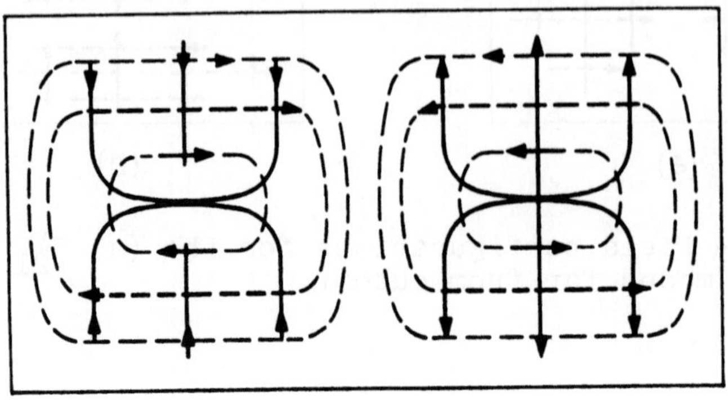

Fig. 10 - Field Configuration of the E^y_{22} Mode.

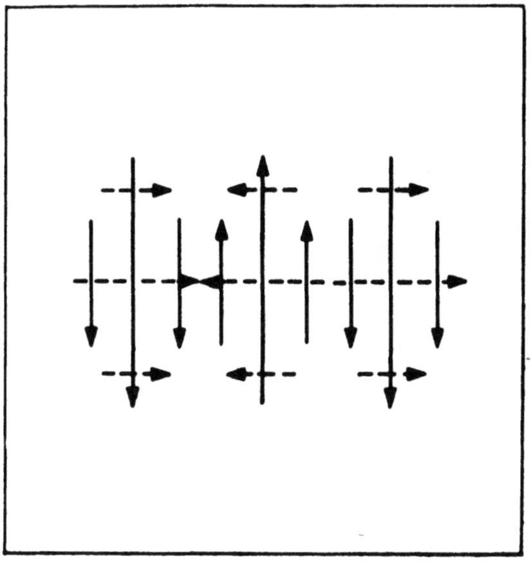

Fig. 11 - Field Configuration of the E^y_{31} Mode.

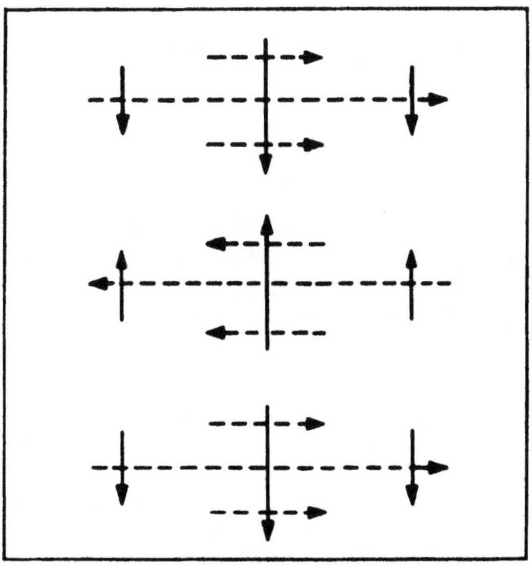

Fig. 12 - Field Configuration of the E^y_{13} Mode.

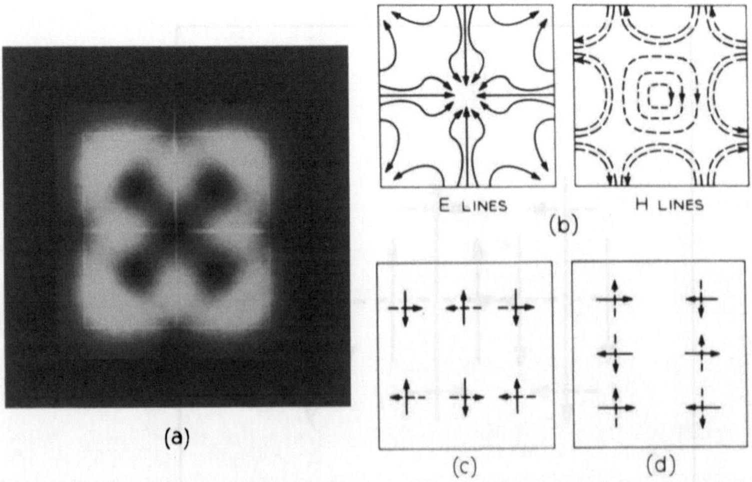

Fig. 13 - The E^y_{32} Mode for Unity Aspect Ratio.

metallic waveguide, shown in Figures 13c and 13d respectively. Figure 13a demonstrates that the circular-harmonic analysis can generate complex field patterns with a relatively small number of harmonics.

The variation of the intensity distribution with P^2 is shown in Figures 14 and 15 for the E^y_{11} and E^y_{21} modes, respectively. As one would expect, for small values of P^2 the radial extent of both modes increases very rapidly as P^2 decreases. It is of significance, however, that most of the energy is contained within the waveguide core, even for relatively small values of P^2. Thus Marcatili's assumption that very little energy propagates in the region of the corners is valid over a wide range.

2. <u>Propagation Curves</u>. This section will be devoted to a discussion of computed propagation curves. In all cases, the normalized waveguide height B, as given in Eq. (22), will be plotted on the horizontal axis and the normalized propagation constant, P^2, given in Eq. (17), along the vertical axis.

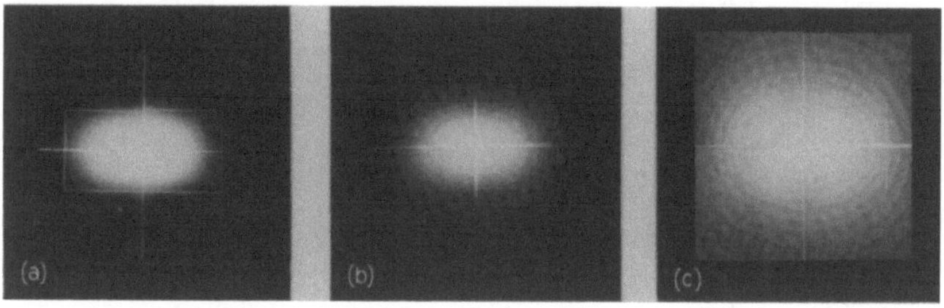

Fig. 14 - Intensity Pictures of the E^y_{11} Mode for Several Values of P^2.

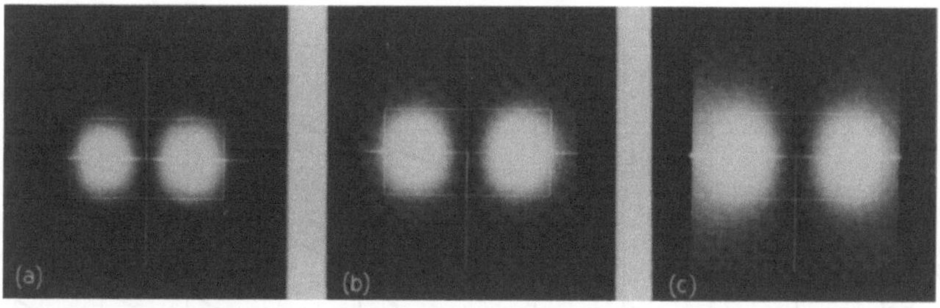

Fig. 15 - Intensity Pictures of the E^y_{21} Mode for Several Values of P^2.

The case of vanishing index difference is shown in Figure 16 for an aspect ratio of one. The first sixteen modes are shown. For this case the following six degenerate groups exist:

$$E^y_{11} , E^x_{11}$$
$$E^y_{12} , E^x_{12} , E^y_{21} , E^x_{21}$$
$$E^y_{31} , E^x_{13}$$
$$E^x_{31} , E^y_{13}$$
$$E^x_{22} , E^y_{22}$$
$$E^y_{32} , E^y_{23} , E^y_{23} , E^x_{23} .$$

In addition, the E^y_{13} and the E^x_{31} modes are almost degenerate except near cutoff. The splitting of these mode can be accounted for by the differences of the field patterns shown in Figures 11 and 12. Since the E^x_{31} mode reversals occur along the direction of the electric field lines, the electric

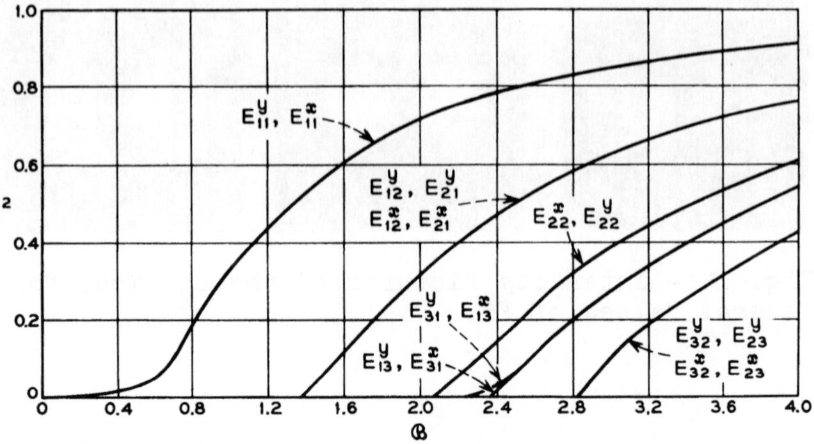

Fig. 16 - Propagation Curves for the First 16 Modes for Unity Aspect Ratio and $\Delta n_r \to 0$.

field for this mode must have a larger longitudinal field component than for the E^y_{31} mode.

All degeneracies, except the E^y_{mn}-E^x_{mn} degeneracies are broken by a change in the aspect ratio as demonstrated in Figure 17 which is drawn for the first 12 modes of a waveguide of aspect ratio 2. One interesting feature of this curve is the mode crossing of the E^y_{31} and E^y_{12} modes. Crossings of this type, which cannot occur in metallic waveguide, are possible because the field functions are frequency dependent. Qualitatively, it can be explained by noting that field reversals must take place in the core, so the central lobe of the E^y_{31} mode is constrained more than any of the E^y_{12} mode lobes as cutoff is approached. Far from cutoff, however, all fields are well constrained and the E^y_{31} mode has a larger propagation constant than the E^y_{12} mode, as it does for the similar metallic waveguide mode with an aspect ratio of 2.

The effect of finite index difference on the modes can be observed by comparing Figure 16 with Figure 18, which is for unity aspect ratio and a 50% index difference. The curves for modes whose field lines reverse direction across the origin are no longer degenerate, but those whose field lines do not reverse still are. For all degeneracies to be split, there must exist a finite index difference as well as an aspect ratio other than unity. One such case is illustrated in Figure 19.

The effect of index difference on the degenerate principal modes for unity aspect ratio is examined in Figure 20. The curve shows both a low and high index difference limit. In the range of interest for optical circuits, 0% - 10%, the vanishing difference curve is an excellent approximation. The greatest changes occur in the 10% - 1,000% range, which is the range of interest for some microwave problems.

Figure 21 presents the computed results for the effect of index changes on the principal modes

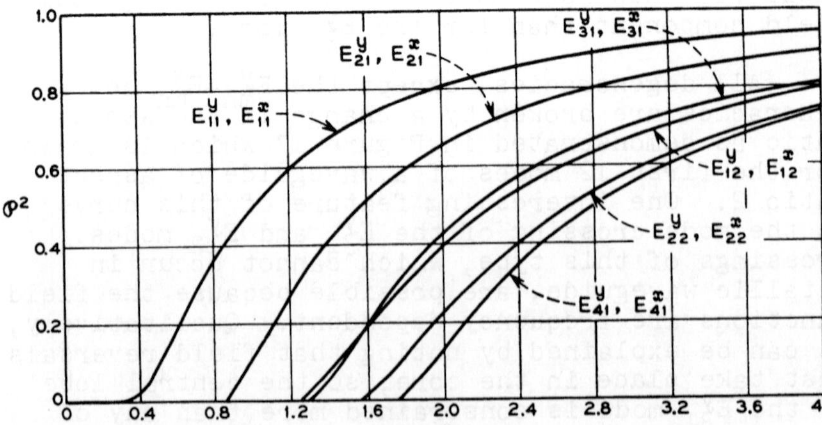

Fig. 17 - Propagation Curves for the First 12 Modes for a/b=2 and $\Delta n_r \to 0$.

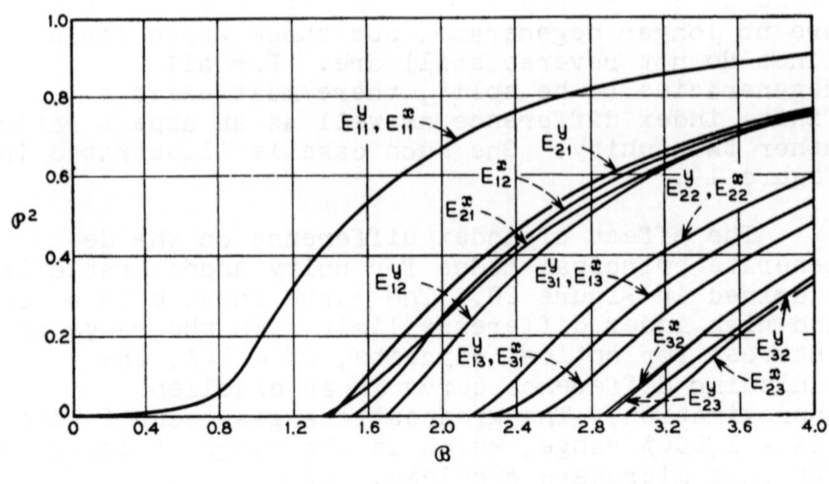

Fig. 18 - Propagation Curves for the First 16 Modes for Unity Aspect Ratio and $\Delta n_r = .5$.

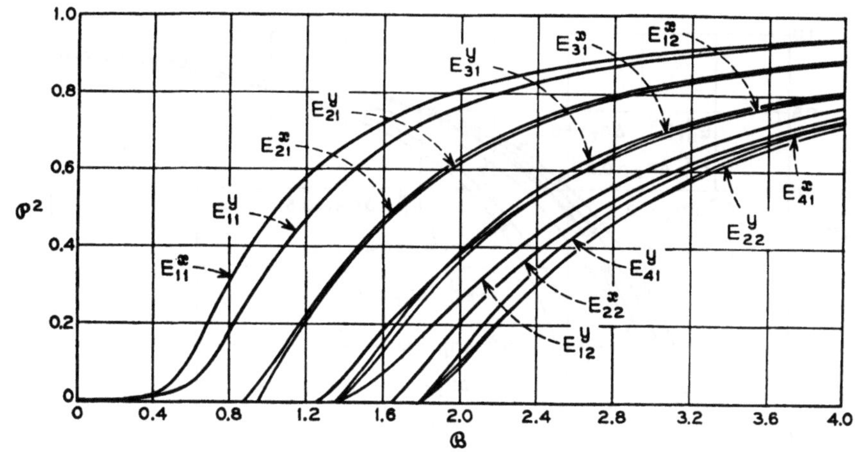

Fig. 19 - Propagation Curves for the First 12 Modes for a/b=2 and Δn_r=.5.

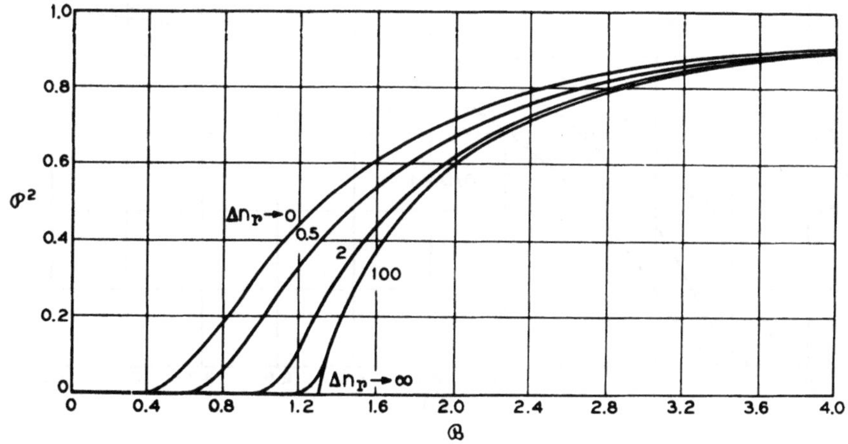

Fig. 20 - E^y_{11} and E^x_{11} mode propagation curves for several values of Δn_r with unity aspect ratio.

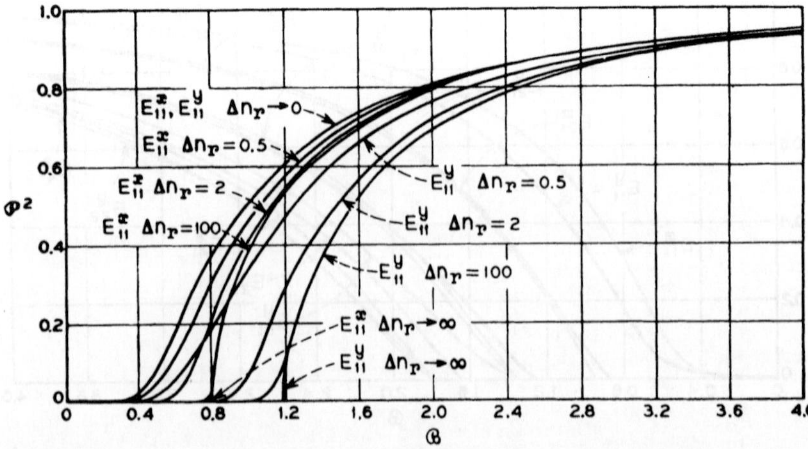

Fig. 21 - E_{11}^y and E_{11}^x mode propagation curves for several values of Δn_r with a/b = 2.

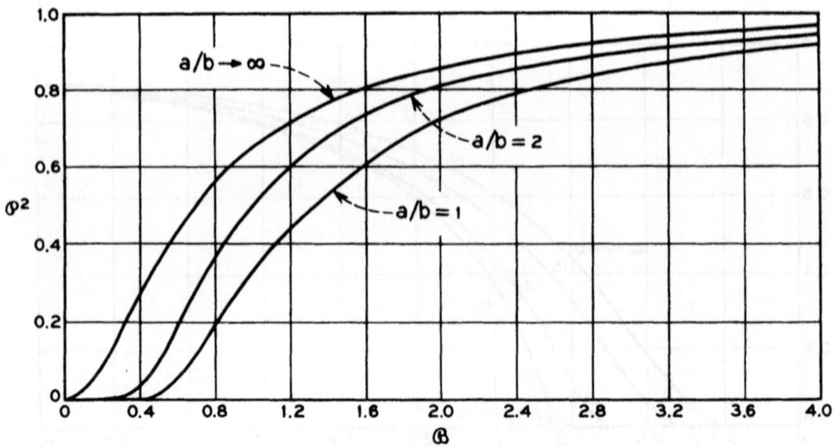

Fig. 22 - E_{11}^y and E_{11}^x mode propagation curves for several values of a/b with $\Delta n_r \to 0$.

for an aspect ratio of 2. The effect is much stronger on the E^y_{11} mode than the E^x_{11} mode. In fact, the effect on the E^x_{11} mode is comparatively small, except near cutoff.

The effect of aspect ratio on the principal modes is demonstrated for vanishing index difference in Figure 22. The curve for infinite aspect ratio was obtained from the exact analysis of the slab case.[9]

IV. COMPARISON OF METHODS

Figure 23 gives solutions from Marcatili's transcendental equations and closed form equations,

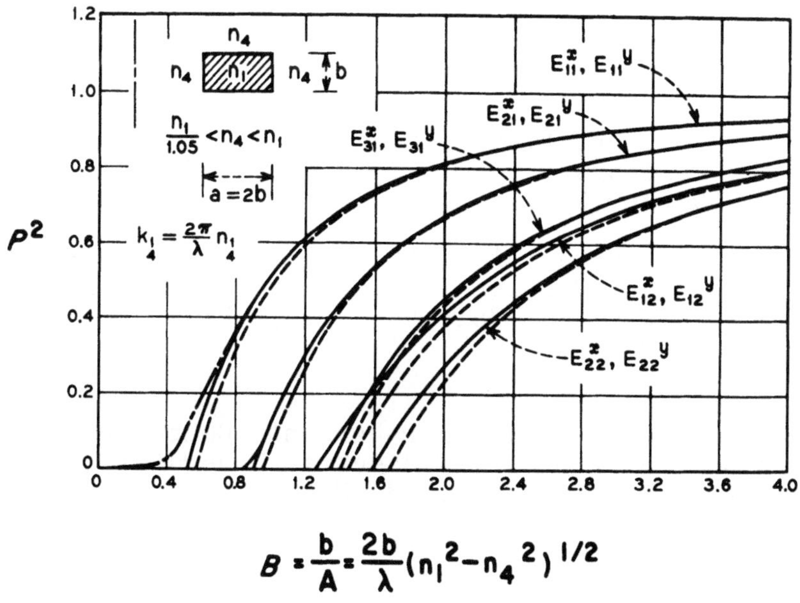

Fig. 23 - Propagation Constant for Several Modes of Rectangular Dielectric Waveguide. Transcendental Solution ______; Closed Form Solutions _ _ _ _; Goell's Computer Solutions ____ . _____.

and circular harmonic solutions for an aspect ratio of 2 with small index difference.

Comparison of the results shows that the two methods give values of the normalized propagation constant, P^2, which are within a few percent for $P^2 > .5$. Thus for P^2 in this range Marcatili's method is to be preferred since the calculations required are much simpler.

For small P^2, however, the fields begin to violate the assumption that the energy in the corner regions is negligible so the accuracy of Marcatili's analysis decreases, especially for the lowest order mode where the analytic approach predicts cutoff. Thus, for this region a more complete field analysis such as the circular harmonic method must be employed.

REFERENCES

1. S. E. Miller, "Integrated Optics," An Introduction, Bell Syst. Tech. J., Vol. 48, No. 7, Sept. 1969, pp. 2059-2070.

2. E. A. J. Marcatili, "Dielectric Waveguide and Directional Coupler for Integrated Optics," Bell Syst. Tech. J., Vol. 48, No. 7, Sept. 1969, pp. 2071-2102.

3. J. E. Goell, A Circular-Harmonic Computer Analysis of Rectangular Dielectric Waveguides, Bell Syst. Tech. J., Vol. 48, No. 7, Sept. 1969, pp. 2133-2160.

4. W. Schlosser, and H. G. Unger, "Partially Filled Waveguides and Surface Waveguides of Rectangular Cross-Section," Advances in Microwaves, New York, Academic Press, 1966, pp. 319-387.

5. J. Q. Bartling, Propagation of Electromagnetic Wave in an Infinite Rectangular Dielectric Waveguide, Journal of the Franklin Institute, Vol. 287, No. 5, May 1969, pp. 389-407.

6. C. B. Shaw, B. T. French, and C. Warner III, "Further Research on Optical Transmission Lines," Scientific Report No. 2, Contract AF449 (638)-1504 AD 625 501, Autonetics Report No. C7-929/501, pp. 13-44.

7. E. Snitzer, "Cylindrical Dielectric Waveguide Modes," Journal of the Optical Society of America, Vol. 51, No. 5, May 1961, pp. 491-498.

8. System/360 Scientific Subroutine Package, White Plains, N.Y. IBM, H20-0205-2, pp. 179-182.

9. R. E. Collin, "Field Theory of Guided Waves," New York, McGraw-Hill, 1960, pp. 480-495.

CHAPTER 5 - LOSS MECHANISMS IN DIELECTRIC WAVEGUIDES

J. E. Goell

Bell Laboratories
Crawford Hill Laboratory
Holmdel, N. J. 07733

ABSTRACT

Scattering loss due to edge roughness introduced during processing can be expected to be a significant, and possibly even a limiting factor in the performance of integrated optical circuits employing two-dimensional confinement, and radiation loss due to bends can be expected to place a lower limit on the size of multifunction circuits and also to influence performance. In this lecture the nature of these losses is discussed. The results of analyses of scattering loss due to edge imperfections and bend loss are summarized and the limitations these losses place on circuits discussed, including the effect of the latter on resonator Q.

I. INTRODUCTION

Attenuation in dielectric waveguides can be caused by absorption or scattering in the core or cladding, an irregular interface between the core and the cladding, and intentional bends. For integrated optical circuits two sources of loss are of particular interest. These are, scattering loss due to irregularities in the core-cladding interface and radiation loss due to bends. The former is significant because it places stringent requirements on the fabrication process and the latter because it places a lower limit on the size of circuits. In this lecture the nature of these losses will be summarized.

II. RADIATION LOSS

Irregularities in the boundaries of dielectric waveguides can couple energy from propagating modes into other propagating modes or into radiation modes. The latter effect is significant for two reasons. First, unintentional irregularities due to processing imperfections can cause attenuation in waveguides and reduce the Q of filters. Second, because a periodic irregularity can be used to couple light both into and out of a dielectric waveguide. The latter effect will be described briefly here and in more detail in a later chapter.

To date, the analysis of losses due to edge irregularity has been limited to the slab and circular geometries. This section will deal with the former following the work of Marcuse[1] to give insight and a rough indication of the magnitude of the losses that can be expected in optical circuits with two dimensional confinement.

Figure 1 shows the geometry to be considered. Propagation is in the z-direction and the structure is assumed to be uniform in the y-direction. The wall fluctuations f(z) are given by

$$f(z) = \frac{b}{2} + a \sin \beta_w z$$

$$h(z) = \frac{b}{2} - a \sin (\beta_w z + \alpha)$$

The index of refraction in the central region is given by n_g and outside by 1. The results to be given apply to the modes with y-directed electric fields. Results for an x-directed electric field can be found in a recent work by Marcuse.[2]

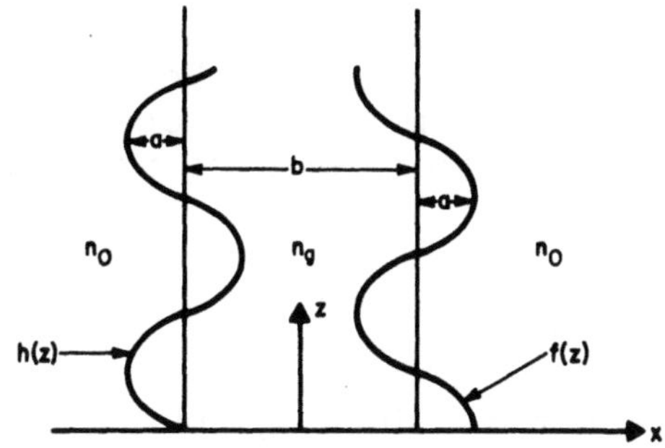

Fig. 1 - Slab geometry

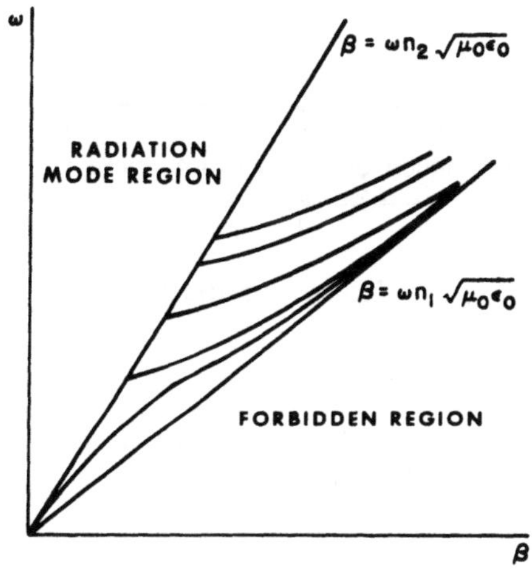

Fig. 2 - ω vs β for a symmetric slab waveguide

The modal properties for an ideal slab guide are reviewed in Figure 2. β is the propagation constant and $k = \omega\sqrt{\mu_0 \varepsilon_0}$. At any frequency there is a continuum of radiation modes and a finite number of propagating modes. For a small perturbation of the wall a mode of propagation constant β, can only couple to a mode of propagation constant β_2 if

$$\beta_2 = \beta \pm \beta_w \tag{1}$$

This relationship is shown graphically in Figure 3.

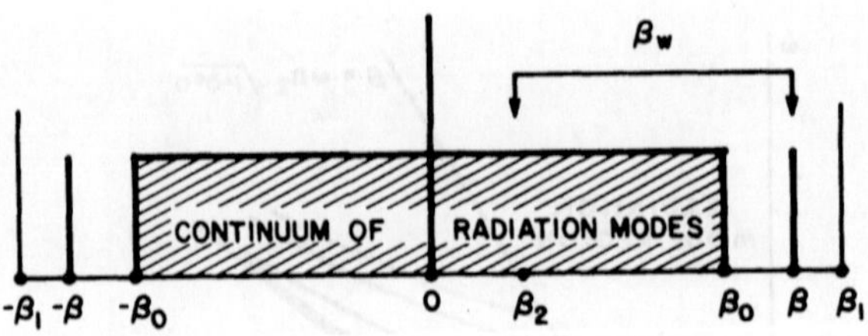

Fig. 3 - Mode coupling diagram

Assuming a weak perturbation of the boundary, the field in a slab with a boundary irregularity of the form shown in Figure 1 can be put in the form

$$E = \sum C_n(z)E_n + \sum \int_0^\infty g(\rho,z)E(\rho)d\rho \tag{2}$$

where E_n is the normalized field function of each of the propagating modes and $C_n(z)$ is the magnitude of the mode, which because of the coupling is a function of z. The normalized field distribution of the continuum is given by $g(\rho,z)$ and its magnitude by $E(\rho)$. The second summation takes account of the presence of both even and odd modes.

For the case where the coupling is between two propagating modes, designated as 0 and 1, the electric field can be written as

$$E = C_0(z)E_0 + C_1(z)E_1 \tag{3}$$

and if at z=0 all of the energy is in the 0th mode, then

$$C_0 = \cos \kappa_{01} z$$

$$C_1 = \left(\frac{\kappa_{01}^*}{\kappa_{01}}\right)^{\frac{1}{2}} \sin \kappa_{01} z \tag{4}$$

where the coupling coefficient κ_{01} is given by

$$\kappa_{01} = \frac{k^2 a}{2}(n_g^2 - 1)\frac{\cos \kappa_0 \frac{b}{2} \cos \kappa_1 \frac{b}{2} \cos \frac{\alpha}{2}}{\left[\left(\beta_0 \frac{b}{2} + \frac{\beta_0}{\gamma_0}\right)\left(\beta_1 \frac{b}{2} + \frac{\beta_1}{\gamma_1}\right)\right]^{\frac{1}{2}}} e^{i\frac{\alpha}{2}} \tag{5}$$

for even mode coupling and by

$$\kappa_{01} = \frac{k^2 a}{2}(n_g^2 - 1)\frac{\cos \frac{\kappa_0 b}{2} \sin \frac{\kappa_1 b}{2} \cos \frac{\alpha}{2}}{\left[\left(\frac{\beta_0 b}{2} + \frac{\beta_0}{\gamma_0}\right)\left(\frac{\beta_1 b}{2} + \frac{\beta_1}{\gamma_1}\right)\right]^{\frac{1}{2}}} e^{i\frac{\alpha}{2}} \tag{6}$$

for odd mode coupling, the propagation constants κ_0, κ_1, γ_0, and γ_1 are given by

$$\kappa_0 = \left(n_g^2 k^2 - \beta_0^2\right)^{\frac{1}{2}}$$

$$\kappa_1 = \left(n_g^2 k^2 - \beta_1^2\right)^{\frac{1}{2}}$$

$$\gamma_0 = \left(\beta_0^2 - k^2\right)^{\frac{1}{2}}$$

$$\gamma_1 = \left(\beta_1^2 - K^2\right)^{\frac{1}{2}}$$

The values of the propagation constants can be found from the analysis given in the chapter on slab waveguides.

From Eq. 4 it is evident that power is conserved, that is

$$|C_0|^2 + |C_1|^2 = 1 \qquad (7)$$

and that complete power transfer occurs in a distance given by

$$D = \frac{\pi}{2|\kappa_{01}|}$$

For the case of coupling of a mode with propagation constant β to the continuum the attenuation coefficient is given by

$$\alpha_t^2 = \frac{a^2k^4}{4}(n_g^2 - 1)^2 \frac{\cos^2 \frac{\kappa_0 b}{2}}{\frac{\beta_0 b}{2} + \frac{\beta_0}{\gamma_0}} \left\{ \frac{\rho \cos^2 \frac{\sigma b}{2} \cos^2 \frac{\alpha}{2}}{\rho^2 \cos^2 \frac{\sigma b}{2} + \sigma^2 \sin^2 \frac{\sigma b}{2}} \right.$$

$$\left. + \frac{\rho \sin^2 \frac{\sigma b}{2} \sin^2 \frac{\alpha}{2}}{\rho^2 \sin^2 \frac{\sigma b}{2} + \sigma^2 \cos^2 \frac{\sigma b}{2}} \right\}. \tag{8}$$

where

$$\sigma = \left(n_g^2 k^2 - \beta^2\right)^{\frac{1}{2}}$$

$$\rho = \left(k^2 - \beta^2\right)^{\frac{1}{2}}$$

In this case energy is lost continuously to the continuum in contrast to the previous case where it was engaged in a lossless cyclical exchange between two modes.

By reciprocity, if light can be coupled from a waveguide to free space by a periodic wall variation, then the reverse must also be true. This has been used to advantage in the grating coupler, which is analyzed in detail in a subsequent chapter.

In practice one would expect the waveguide walls to have random fluctuations due to processing defects. The radiation loss due to slab irregularities can be found from the preceding results. The case where the distance from the guide axis to one of the walls fluctuates will be described. If the distance to both walls fluctuates and the fluctuations are uncorrelated, the attenuations add. The loss expression is found by rederiving Eq. 8 for a variation of a single wall (can be done by summing the $\alpha=0$ and $\alpha=\pi$ cases) and then integrating the product of the result and the spectral distribution function of the wall irregularity. The latter is found from the statistics of the wall fluctuation.

For a wall roughness correlation function given by

$$R(u) = A^2 \exp\left(-\frac{|u|}{B}\right) \tag{9}$$

where A is the rms value of the wall deviation and B the correlation length, the attenuation is given by

$$\alpha_t = \frac{A^2 k^4}{4\pi B}(n_g^2 - 1)^2 \int_{-\beta_e}^{\beta_e} \left[\frac{\rho \cos^2 \frac{\kappa_0 b}{2}}{\left[(\beta_0-\beta)^2 + \frac{1}{B^2}\right]\left(\frac{\beta_0 b}{2} + \frac{\beta_0}{\gamma_0}\right)} \left(\frac{\cos^2 \frac{\sigma b}{2}}{\rho^2 \cos^2 \frac{\sigma b}{2} + \sigma^2 \sin^2 \frac{\sigma b}{2}} + \frac{\sin^2 \frac{\sigma b}{2}}{\rho^2 \sin^2 \frac{\sigma b}{2} + \sigma^2 \cos^2 \frac{\sigma b}{2}} \right) \right] d\beta. \tag{10}$$

where β_e is the value of β at the edge of the continuum.

Typical results obtained by numerical integration of the Eq. (10) are shown for $n_g = 1.01$ and for $k_b = 16$ and for $n_g = 1.5$ $k_b = 2.6$ in Figures 4 and 5 respectively. The results demonstrate that a loss maxima exists corresponding roughly to coupling into the center of the

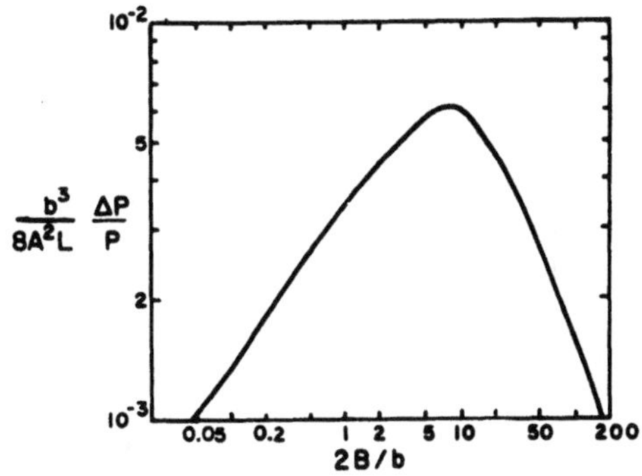

Fig. 4 - Attenuation due to radiation loss $(b^3/16A^2)(\Delta P/P)$ as a function of the normalized correlation length (2B/b) for n_1/n_2 = 1.01, kb = 16.0 and n_2 = 1 (single mode operation)

radiation region of Figure 3. For a gradual fluctuation in the wall or for an extremely short one, the loss can be very small.

It is interesting to examine what the maximum loss for a dielectric waveguide might be. Table 1 gives the peak loss, correlation length for peak loss, and attenuation coefficient for several values of guide width for a guide of index 1.515 surrounded by a medium of index 1.5 with 500Å rms roughness at .6328μ. The loss coefficient would vary as the mean-square wall roughness. It is evident from these results that there will be stringent requirements on wall smoothness for integrated optical circuits.

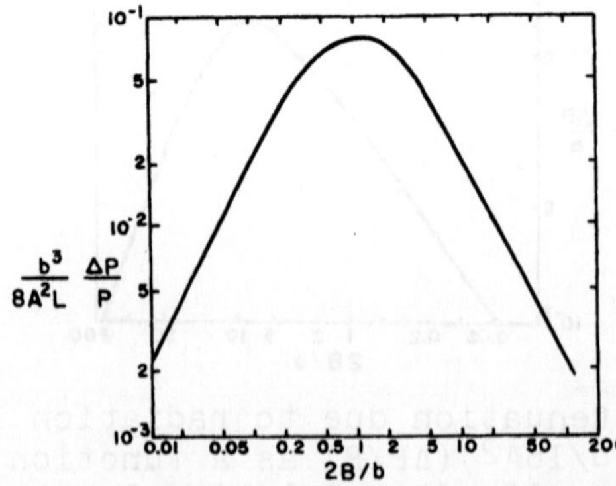

Fig. 5 - Attenuation due to radiation loss $(b^3/16A^2)(\Delta P/P)$ as a function of the normalized correlation length (2B/b) for n_1/n_2 = 1.5 kb = 2.6 and n_2 = 1. (Single mode operation)

TABLE I

Wall Loss
for
Δ = .01
n_{2_0} = 1.5, λ = .6328μ
500Å rms Edge Roughness

Guide Width μ	Correlation Length μ	Loss dB/cm
1.08	4.2	4.1
2.02	3.8	1.2
3.1	3.3	.5

III. BEND LOSS

The minumum allowable radius of curvature of dielectric waveguides is limited by bend losses. Bend losses arise from two effects. First, energy radiates from a bend to the surrounding media. The effect is physically justified by the fact that lossless propagation would imply energy propagation above the speed of light at a distance sufficiently far from the center of curvature of the guide. Second, the mode shape of a guide varies with the radius of curvature. Thus, mode coupling loss can be anticipated at an abrupt transition from a straight to a curved guide.

In this section radiation loss will first be treated using the velocity approach described by Marcatili and Miller.[3] Following this the results of Marcatili's modal analysis[4] of radiation and coupling loss will be summarized.

A. Velocity Approach

The phase velocity of fields in a curved dielectric waveguide is proportional to the distance from the center of curvature. If a straight waveguide mode is assumed to be propagating in a bend with a curved propagation axis with radius of curvature R and a propagation constant k_z, as shown in Figure 6, the radius at which the phase velocity exceeds the velocity of light in the cladding will be given by

$$x_r = \frac{k_z - k_0}{k_0} R$$

where k_0 is the propagation constant in the outer medium. In an infinite medium energy remains collimated for a distance

$$Z_c = \frac{a^2}{2\lambda_g}$$

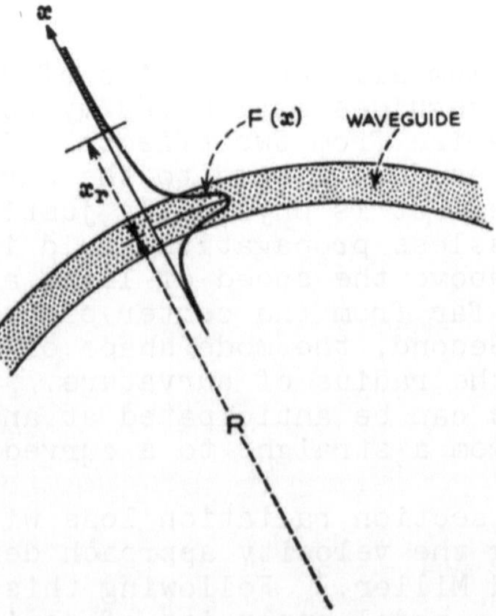

Fig. 6 - Curved waveguide

where a is the transverse field width and λ_g is the wavelength. The attenuation constant, α, is related to the loss by the relation

$$\alpha = \frac{1}{2P_0} \frac{dP}{dz}$$

where P_0 is the propagating power. Then defining E_ℓ as the energy beyond x_r and E_t as the total energy carried by the guide gives

$$\alpha = \frac{1}{2Z_c} \frac{E_\ell}{E_t} \qquad (11)$$

For a slab waveguide of width b with transverse field variation

$$F(x) = \cos k_x x \qquad -\frac{b}{2} \leq x \leq \frac{b}{2}$$
$$F(x) = \cos\left(\frac{k_x b}{2}\right) e^{\frac{-\left(|x| - \frac{b}{2}\right)}{\xi}} \qquad |x| \geq \frac{b}{2} \tag{12}$$

$$\mathcal{E}_\ell = \frac{\xi}{2} \cos^2 \frac{k_x b}{2} \, e^{\frac{-2(x_r - \frac{b}{2})}{\xi}} \tag{13}$$

$$\mathcal{E}_t = \frac{b}{2} + \frac{1}{2k_x} \sin k_x b + \xi \cos^2\left(\frac{k_x b}{2}\right) \tag{14}$$

$$Z_c = \frac{\left[b + 2\xi \cos\left(\frac{k_x b}{2}\right)\right]^2}{2\lambda_0} \tag{15}$$

where λ_0 is the wavelength in the surrounding medium and ξ is the reciprocal of the transverse decay coefficient outside the guide (see chapters on slab and rectangular guides). Since Z_c and $\mathcal{E}_t$ are independent of the radius of curvature α can be put in the form

$$\alpha = c_1 \exp(-c_2 R) \tag{16}$$

Thus α varies as an exponential of the radius of curvature.

The significance of this result cannot be over emphasized. The loss is a very fast function of radius of curvature. Thus, some value of radius of curvature exists where the loss changes almost abruptly from prohibitively large to negligibly small. Some results of the preceding analysis are shown in Table II. The loss as a function of R for case 2 is shown in Figure 7. For this case, the attenuation coefficient increases by two orders of magnitude for a 40 percent decrease in R.

TABLE II

Values for c_1 and c_2

Case	Waveguide Index of Refraction	Slab Width t (μm)	Surrounding Index of Refraction	c_1 (Nepers Per Meter)	c_2 (Meters^{-1})	R for $\alpha_r = 1$ Neper/m
1	1.5	0.198	1.0	2.57×10^6	3.47×10^6	4.25 μm
2	1.5	1.04	1.485	1.04×10^5	1.46×10^4	0.79 mm
3	1.5	1.18	1.4985	5.4×10^3	81.4	0.106 m

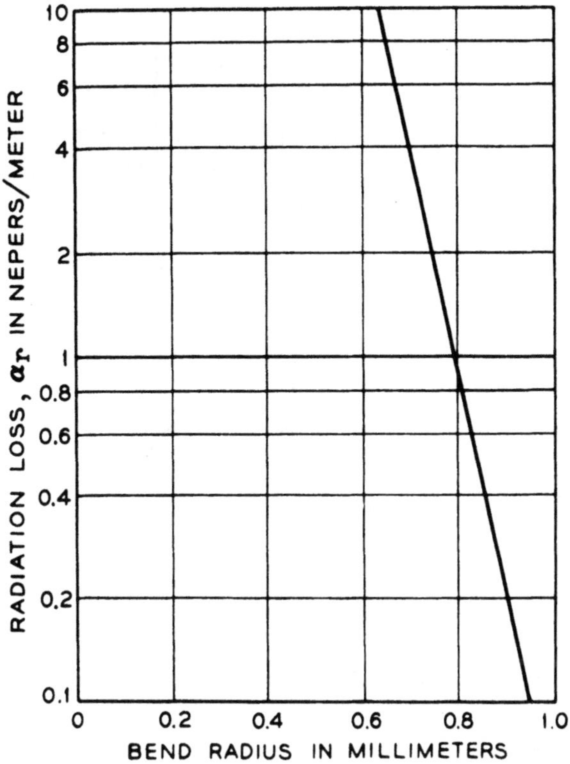

Fig. 7 - Radiation loss versus bend radius

B. Modal Analysis

Marcatili has analyzed curved guides using a similar approach to the one he has used for straight guides.[4] The geometry for which the analysis applies is shown in Figure 8. The following is based on his work.

The effect of guide curvature is to change the real part of the propagation constant, introduce an imaginary part to the propagation constant corresponding to loss, and alter the field distribution. For a guide of infinite height with $n_3 = n_5$, for the E^x_{11} mode it can be shown that the portion of the power coupled to higher order modes at the junction between a

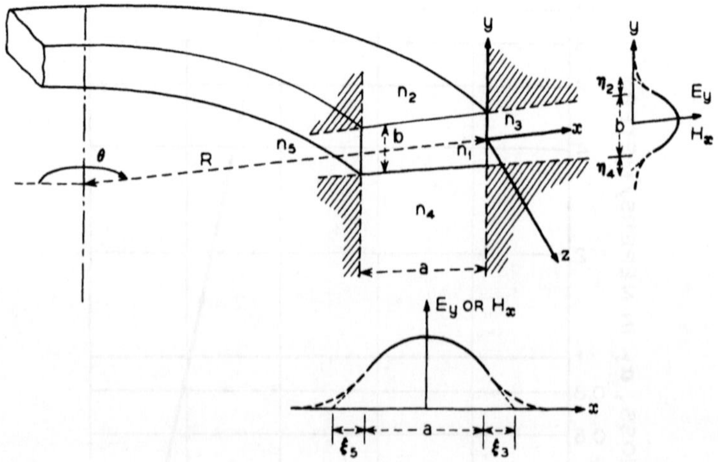

Fig. 8 - Curved dielectric waveguide

straight and curved section is given by

$$c = \frac{1}{2k_{x0}a}\left(\frac{\pi a}{A}\right)^3 \frac{1}{\mathcal{R}} \tag{17}$$

where

$$A = \frac{\lambda}{2(n_1^2 - n_3^2)} \tag{18}$$

and k_{x0} is the propagation constant of a straight waveguide. Defining a normalized radius

$$\mathcal{R} = \frac{2\pi^3 R}{k_{z0}^2 A^3} = 2\,\frac{k_1^3}{k_{z0}^2}\left(1 - \frac{n_3^2}{n_1^2}\right)^{3/2} R \tag{19}$$

where k_{z0} is the z propagation constant for a straight guide gives for the attenuation radius of curvature product

$$\alpha_c R = \frac{1}{2}\left(1 - \frac{n_3^2}{n_1^2}\right)^{-\frac{1}{2}} \left(\frac{n_3 k_{x0} a}{n_1}\right)^2 \left(\frac{A}{\pi a}\right)^3 \left[1 - \left(\frac{k_{z0}A}{\pi}\right)^2\right]^{\frac{1}{2}}$$

$$\cdot \frac{R \exp\left\{-\frac{R}{3}\left[1 - \left(\frac{k_{z0}A}{\pi}\right)^2\left(1 + \frac{2c}{ak_{x0}}\right)^2\right]^{\frac{3}{2}}\right\}}{1 - \left(1 - \frac{n_3^4}{n_1^4}\right)\left(\frac{k_{x0}A}{\pi}\right)^2 + 2\,\frac{n_3^2 A}{n_1^2 a}\left[1 - \left(\frac{k_{x0}A}{\pi}\right)^2\right]^{-\frac{1}{2}}}$$

Using this approach it is found that

$$\alpha_c = c_1 e^{-c_2 R} \tag{21}$$

if

$$\frac{2c}{ak_{x0}} \ll 1 \tag{22}$$

For the case where

$$\frac{a}{A} \gg 1 \text{ and } c \cong 1 \tag{23}$$

$$\alpha_c R = \frac{n_3^2}{n_1^2}\left[1 - \left(\frac{n_3}{n_1}\right)^2\right]^{-\frac{1}{2}} \exp\left\{-\frac{R}{3}\left[1 - \left(\frac{9\pi}{2R}\right)^{\frac{2}{3}} + \frac{4n_3^2}{n_1^2 R}\right]^{\frac{3}{2}}\right\}$$

The previous equations apply to the E^x_{11} mode. For the E^y_{11} mode in all of these formulas the refractive index ratio is replaced by unity, except for the case where it is subtracted from unity. A normalized family of curves of $\alpha_c R$ versus A for various values of R are shown in Figure 9 for the two limiting cases. An estimate for in-between cases can be made from this graph.

Li has obtained a number of interesting results from the preceding analysis.[5] These results will now be presented.

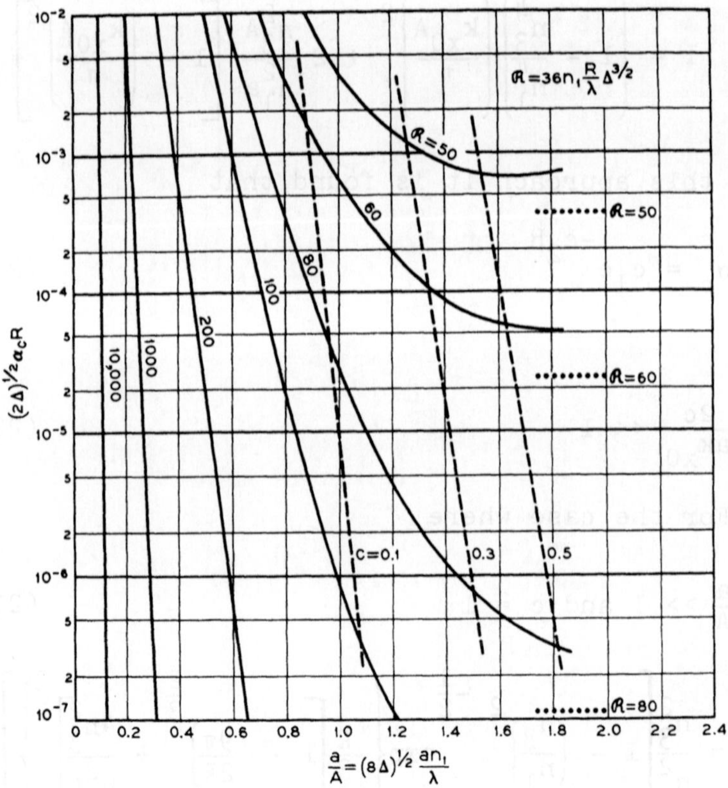

Fig. 9 - Attenuation per radian for E^x_{11} and E^y_{11} modes if $n_1/n_3 = 1 + \Delta$ and $\Delta << 1$

A plot of bend loss for a 90 degree bend versus the radius of curvature is shown in Figure 10 for the case where A is greater than 2. The factor $[1 - (k_y/k_z)^2]$ takes finite height into account. Generally it is close to 1. Table III shows results of interest for integrated optical circuit applications at .6328 microns. These results demonstrate that the loss varies rapidly with the radius of curvature and index difference.

It is of interest to attempt to estimate the minimum allowable radius of curvature that can be employed with pillbox resonator since such a circuit might have applications as a filter in future integrated optical circuits. The Q is defined as

$$Q = \frac{2\pi}{\text{Power loss per guide wavelength}} = \frac{1}{Q_t} + \frac{1}{Q_c}$$

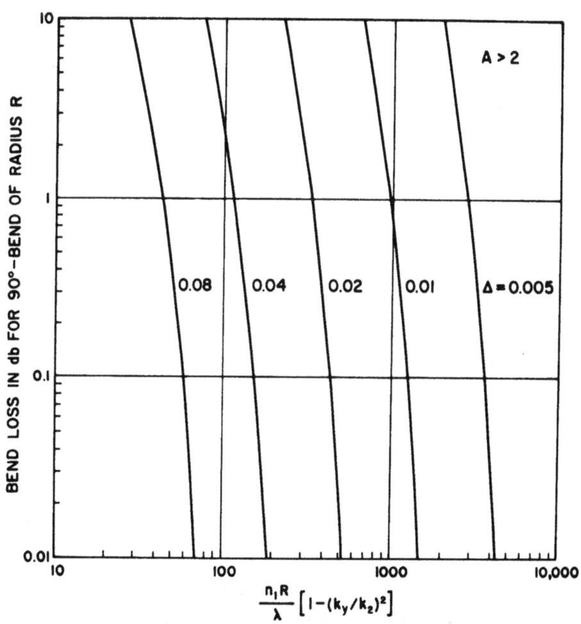

Fig. 10 - Bend loss for a 90° bend vs radius of curvature

TABLE III

Bend Loss

$n_2 = 1.5,\ \lambda = .6328\mu$

Height Factor $= 1 - (k_y/k_2)^2 = .9$

Bend Loss (90°) dB	Δ	Radius of Curvature (mm)
.1	.005	1.7
.1	.01	.59
.1	.02	.20
1.0	.005	1.3
1.0	.01	.45
1.0	.02	.16

where

$$Q_t = \frac{\pi n_1}{\lambda \alpha_t}$$

$$Q_c = \frac{\pi n_1}{\lambda \alpha_c}$$

The constant α_t is the attenuation coefficient of the straight guide and α_c is the additional attenuation coefficient due to curvature. A normalized curve of Q_c versus R is shown in Figure 11. If a straight guide attenuation of

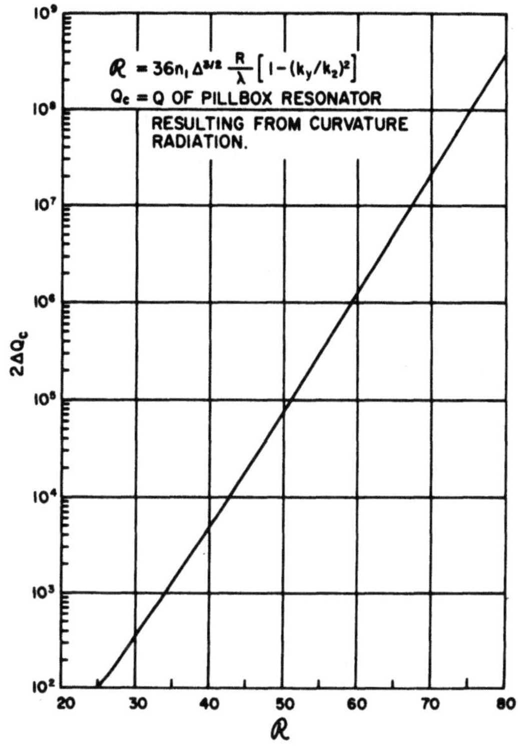

Fig. 11 - Q_c versus R for a ring resonator

.1 dB/cm is assumed, then if $n_1 = 1.5$, $\Delta = .01$, at 1 micron the Q due to the straight guide loss would be approximately 4×10^6 and the bandwidth approximately 150 MHz. The radius of curvature which gives $Q_c = Q_t$ is 1.27 millimeters. Thus, if we make the diameter of the resonator somewhat greater than 2.54 millimeters, the bend loss would have an insignificant effect on the Q of the resonator.

REFERENCES

1. D. Marcuse, "Mode Conversion Caused by Surface Imperfections of a Dielectric Slab Waveguide," Bell Syst. Tech. J., Vol. 48, No. 10, pp. 3187-3217, Dec. 1969.

2. D. Marcuse, "Coupling Coefficient for Imperfect Asymmetric Slab Waveguides," Bell Syst. Tech. J., Vol. 52, No. 1, pp. 63-82, Feb. 1972.

3. E. A. J. Marcatili, S. E. Miller, "Improved Relations Describing Directional Control in Electromagnetic Waveguidance," Bell Syst. Tech. J., Vol. 48, No. 7, pp. 2161-2188, Sept. 1969.

4. E. A. J. Marcatili, "Bends in Optical Dielectric Guides," Bell Syst. Tech. J., Vol. 48, pp. 2103-2132, Sept. 1969.

5. T. Li, "Passive Integrated Optical Circuits," Topical Meeting on Integrated Optics Guided Waves, Materials, and Device, Las Vegas, Feb. 7-10, 1972, Opt. Soc. of Am., Washington D.C.

Chapter 6 - THIN-FILM WAVEGUIDE FABRICATION AND TESTING CONSIDERATIONS

J. E. Goell

Bell Laboratories
Crawford Hill Laboratory
Holmdel, N. J. 07733

ABSTRACT

In this lecture the fabrication of thin-film dielectric waveguides with two dimensional confinement is described. A review of etching and implantation techniques that have been used to fabricate glass waveguide bends and couplers is given. Also included is a description of some of the testing procedures that have been used either during the fabrication process or to determine the optical properties of waveguides.

I. INTRODUCTION

It is anticipated that integrated optical circuits will be used in optical communication systems to perform similar functions to those now performed by microwave integrated circuits in microwave communication systems. These include interconnection, summation and filtering. If systems are built employing single mode transmission then compact optical processing of signals will require that single-mode transmission lines be constructed in the terminal circuits. For these applications means must be found to generate structures employing two dimensional guidance since a mode upon exiting from a single-mode guide will rapidly diverge if not confined.

In this lecture dimensional considerations governing the behavior of dielectric waveguides will first be reviewed. Following this the fabrication of glass guiding structures by both sputter-etching and ion-implantation techniques will be described. Finally, a brief review of some of the measurement techniques that are particularly useful in the fabrication and evaluation of guiding circuits will be given.

II. DIMENSIONAL CONSIDERATIONS

A number of parameters are important in the fabrication of integrated optical circuits. The important dimensions are guide width, height, radius of bends, and guide separation for directional couplers. In addition edge roughness must be minimized to hold radiation losses to a tolerable level.

The maximum dimensions for single-mode guidance are a function of the refractive indices of the guide and of its surroundings. The smaller the index difference, the larger the guide can be. Unfortunately, decreasing the index difference not only increases the guide width, it also increases the minimum bending radius for a given loss. Thus, there is a limit to how small the index difference and how large the cross section can be made.

Table I gives the maximum dimensions for single-mode operation for a rectangular guide of index $n_1 = 1.515$ surrounded by a medium of index $n_2 = 1.500$ for a wavelength $\lambda = 0.6328\mu$.* These dimensions scale as $\lambda/\sqrt{n_1^2 - n_2^2} \simeq \lambda/n_2\sqrt{2\Delta}$ where λ is the free space wavelength and $\Delta = \frac{n_1}{n_2} - 1$.

---

* These results were derived by Li[1] from the curves of Goell[2] and Marcatili.[3]

TABLE I

Maximum Dimensions
For Single-Mode Guides

$n_2 = 1.5$, $\Delta = 0.01$, $\lambda = 0.6328\mu$

Aspect Ratio	Maximum Dimensions (μ)
1	2.1 × 2.1
2	2.5 × 1.3
4	3.4 × 0.82 (estimated)
10	6.3 × 0.63 (estimated)

Table II gives some typical values of radius of curvature to give a specified loss in a 90° bend taken from the results of Marcatili.[4] The loss is a very rapid function of radius curvature and index difference.

The effect of edge roughness is a complex function of the statistical variation of the wall roughness, the indices of the guide and its surroundings, the guide dimensions, and wavelength. Marcuse[5] has calculated the attenuation caused by wall roughness for a slab guide, which can be used to give insight into the rectangular guide problem.

Table III gives the loss due to the roughness of one edge of a slab waveguide for the principal mode with an index difference of 0.01, a substrate index of 1.50, an exponential correlation function with the correlation length corresponding to maximum loss, and an rms edge roughness of 500Å, for 0.6328μ wavelength. The loss drops with increasing guide width, decreasing index difference, and the square of the magnitude of the wall roughness. The stringency of the edge roughness requirement is evident from Table III.

TABLE II

Bend Loss

$n_2 = 1.5$, $\lambda = 0.6328\mu$

Height Factor $= 1 - (k_y/k_2)^2 = 0.9$

Bend Loss (90°) dB	Δ	Radius of Curvature (mm)
.1	.005	1.7
.1	.01	.59
.1	.02	.20
1.0	.005	1.3
1.0	.01	.45
1.0	.02	.16

When two dielectric waveguides are brought close together their evanescent fields interact causing their modes to split into one of even symmetry and the other of odd symmetry. The beat effect of these modes can be used to build a directional coupler. Exciting one guide is equivalent to exciting both modes equally, but 180° out of phase in the dark guide. As the modes propagate along the guides, power oscillates from one guide to the other. By terminating one arm of the coupler at the appropriate point, coupling of a desired degree can be achieved.

TABLE III

Wall Loss
for
$\Delta = 0.01$

$n_2 = 1.5$, $\lambda = 0.6328\mu$

500Å rms Edge Roughness

Guide Width μ	Correlation Length μ	Loss dB/cm
1.08	4.2	4.1
2.02	3.8	1.2
3.1	3.3	.5

Since the evanescent modes decay rapidly from the guide edge, for coupling to be achieved (the beat length to be comparable with practical circuit dimensions) the guides must be very close together. Marcatili has derived the coupling length for rectangular dielectric waveguide.[2]

Figure 1 shows the coupling length, L, for half power transfer. For other couplers, the power transferred goes as $\sin^2 \pi \ell/4L$, where ℓ is the coupling length. Table IV gives the coupling coefficient for several values of guide separation, guide width, and index difference. The parameters of Table IV are defined in Figure 1.

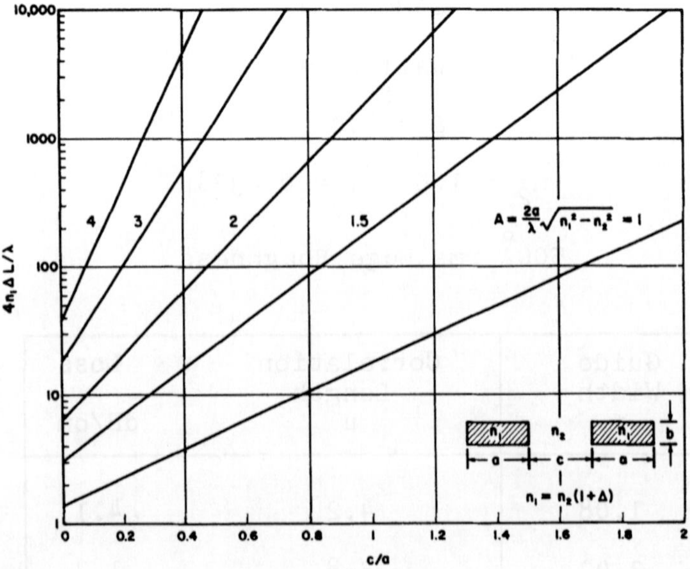

Fig. 1 - 3 dB Coupling Length vs Guide Separation

TABLE IV

3 dB Coupling Length

$n_2 = 1.50$, $\lambda = 0.6328\mu$

a(μ)	c(μ)	Δ	L(mm)
3	1/2	0.01	0.18
3	1	0.01	0.45
3	2	0.01	3.4
6	1/2	0.01	1.2
6	1	0.01	2.9
6	2	0.01	21.
3	1/2	0.02	0.32
3	1	0.02	1.3
3	2	0.02	22.

Due to processing limitations that will be described, it was found that couplers could be produced with guide separations between 1μ and 2μ for guide thicknesses up to about 0.6μ. Thus, for significant coupling to be achieved in a few millimeters the index difference must be less than 2 percent. However, to keep bend losses to acceptable levels (0.1 dB or less) with bend radii of about 0.5 mm the index difference must be greater than about 1 percent. With a 1 percent index difference, the rms edge roughness of the sides of the guide must be held to 500Å for 3μ wide guides to keep the loss below 1 dB/cm.

III. CIRCUIT FABRICATION

In early work on circuit fabrication a guiding structure was achieved by sputter etching a glass film using a glass fiber as a mask.[6] This work demonstrated the feasibility of sputter etching, but the use of fiber masks is not a practical approach for generating complex circuits. Except for the simplest components, a resist technique is required that can serve as a basis for economical batch processing.

The best results with resist have been obtained by Haller and his coworkers at IBM[7] who exposed poly-(methyl-methacrylate), PMM, with an electron beam for transistor and acoustic wave applications. The structures they have been able to generate include a gold grid with 1400Å wide lines and 1μ spacing, and an interdigital line consisting of 0.3μ wide metal stripes separated by 0.7μ. The best optically exposed masks have considerably lower resolution.

Resist techniques using PMM have been used to successfully generate both bends and directional couplers by sputter etching[8] and bends by ion implantation.[9] For the initial attempt at observing the interguide coupling the configuration of Figure 2 was employed. Concentric circle segments were used because they allowed a long interval of tangency and the curvature allows an easy determination of beam guidance. In launching

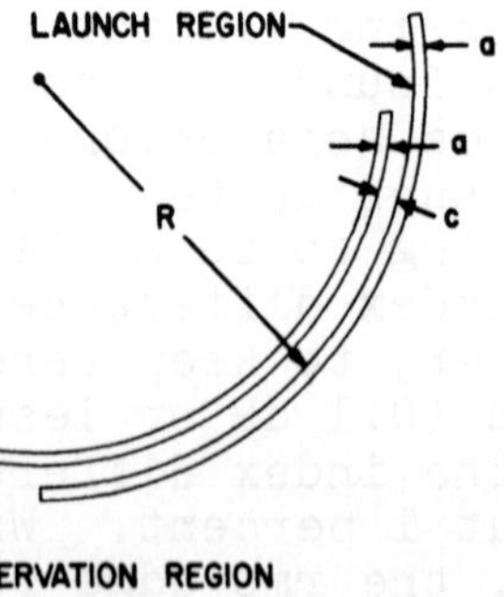

Fig. 2 - Coupler Configuration

a light beam into a coupler it was not possible to hold the input beam to a sufficiently small diameter so that only one guide could be excited and it was difficult to distinguish the beams propagating in adjacent guides. Therefore, launching and observation regions were provided as indicated in Figure 2 by blanking the electron beam. In this way it was possible to be certain light was only being launched into one guide, and to observe the light emanating from the other guide at the end of the coupler.

A Cambridge scanning electron microscope (SEM) was used to expose the resist. The first two condenser lenses and filament current were adjusted to give a beam current of 1.2×10^{-9}A and an undeflected focused spot size of about 1μ.

In order to produce guides 3μ-wide and above using a 1μ beam a signal of the form

$$A(t)\sin(2\pi ft)$$

was applied to the x-deflection coils and of the form

$$A(t)\cos(2\pi ft)$$

to the y-deflection coils. The frequency, f, was 1000 Hz and

$$A(t) = V_0 + \Delta Vg(t)$$

where g(t) had the triangular time dependence shown in Figure 3 and a period of about 0.1 second, V_0 sets the radius of curvature, and ΔV sets the guide width. Thus, the beam traversed a circle with a slowly varying radius of the form of g(t).

It was found that 3.0μ wide guides with bend radii up to 1.5 mm could be produced by this method. However, with larger radii considerable deflection defocusing was observed. The distortion had a dependence on azimuthal position which tended toward the form observed by Amboss and Wolf[10] shown in Figure 4, but was not as symmetric. For bends of greater than 1.5 mm radius, the detrimental effects of deflection defocusing were largely overcome by following the path shown by the dotted line so the major axis of the defocused beam was directed along the guide axis for large deflection.

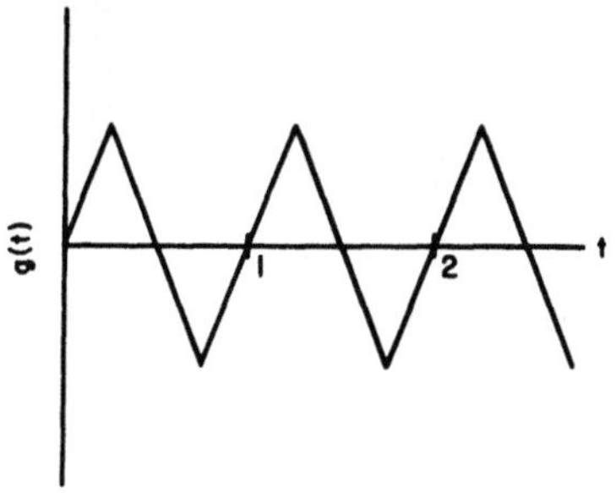

Fig. 3 - Envelope of the Deflection Voltage vs Time

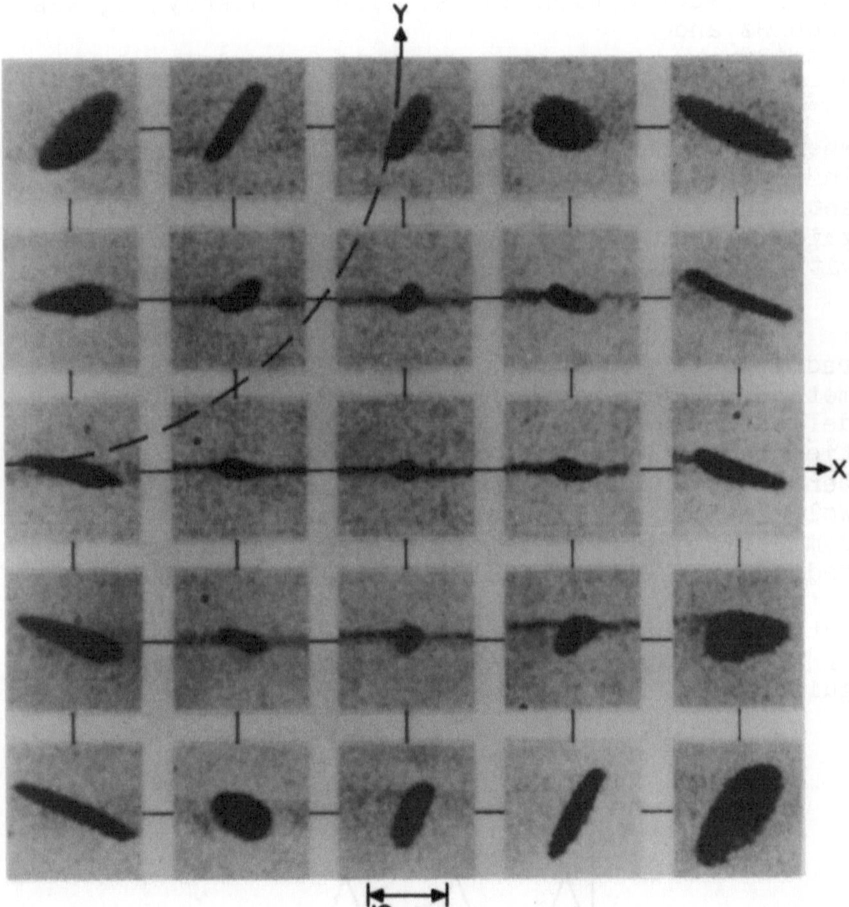

Fig. 4 - Beam aberration vs position as measured by Amboss and Wolf for a 1 cm working distance with a 600μ aperture. (dotted line gives a path a small guide widening)

The problem of writing two guides in close proximity is exacerbated by the undercutting of the resist, as illustrated in Figure 5. This undercutting is caused by scattering of the beam as it travels from the top surface of the resist towards the glass film and by electrons scattered from the glass film back towards the surface. If the guides are too close together, the resist separating them is cut away. The more constant the beam parameters can be held the closer together the guides can be drawn. Thus variations in the guide width and placement must be minimized.

One source of variation in guide width and separation, deflection defocusing, has already been discussed. Another cause is nonlinearity in the scan amplifiers and beam deflection. The deflection circuits of the machine are not designed for a high degree of linearity. Unfortunately, linearity is critical for closely spaced guides. Assuming constant guide width due to scattering, if the incremental gain of the scanning circuits is a factor α at one position on the field relative to another, it can be shown that the guide separations will vary by

$$\alpha(c+a) - a$$

where a is the individual guide width and c the guide separation at the first location, assuming constant guide width. Thus, if $\alpha = 0.9$, $c = 1\mu$

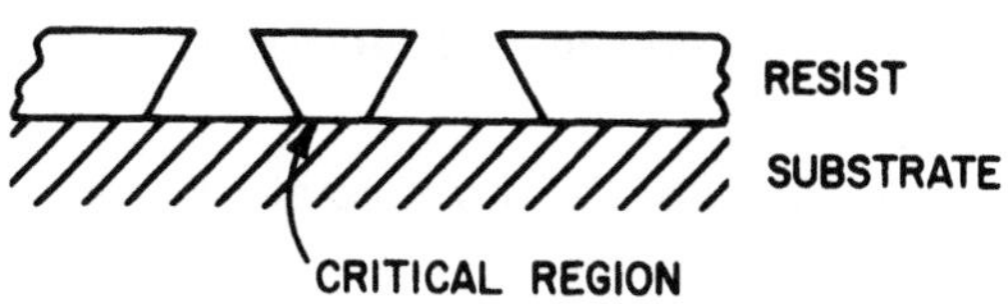

Fig. 5 - Resist Undercutting

and a = 3, the minimum separation will only be 0.6μ.

To correct for deflection nonlinearity and to further reduce the effects of deflection defocusing, when writing couplers a dc shift was added for one of the guides as shown in Figure 6. In this way coupling regions of 3 mm were achieved with a guide separation of 1.25μ in the resist.

A. Etched Waveguide Fabrication

The lift-off technique of Haller and his coworkers was used to form a metal mask. Sputter etching was employed to etch the mask pattern into the glass film. Sputter etching was chosen because, unlike chemical etching, it does not cloud glass surfaces. Furthermore, sputter etching does not undercut the mask.

In early work aluminum was used as the mask. It was found that the aluminum masks could be produced with very smooth edges. However, during sputter etching the edge roughness increased considerably. This increase appeared to be due to an increase of the grain size of the aluminum

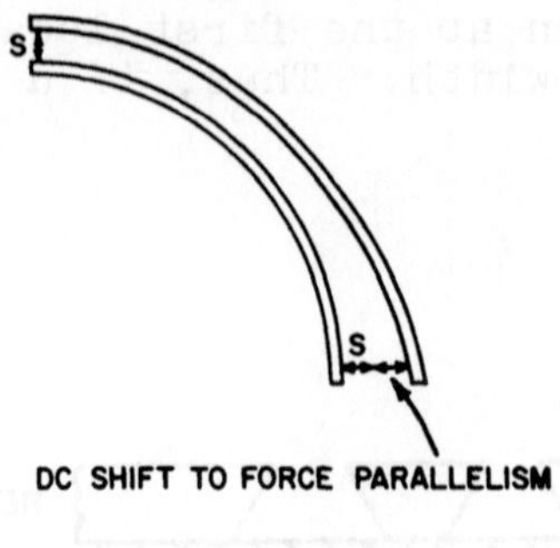

Fig. 6 - DC Shift to Force Parallelism

mask. The increase in grain size of a sputtered source can be readily observed in many materials.[11] It is theorized that a source having a higher recrystallization temperature would be less likely to recrystallize when sputtered. Manganese was selected as the mask material because it had a high recrystallization temperature and could be evaporated fairly easily.[12] It was found that it sputtered at about 1/5 the rate of glass when it was used as a mask.

The process began with a polished glass substrate on which a 0.55μ thick barium silicate glass film[13] had been deposited. The index of the film was 1.54 and the substrate 1.515.

1. Masking and Etching. The substrates were cut into approximately 0.7 cm squares and attached to mounting posts. They were then spin coated with PMM dissolved in methyl isobutyl ketone. Next, the samples were coated with a few hundred angstroms of aluminum which was connected to the mounting post with silver paint to prevent charge buildup on the surface of the sample.

The resist was then exposed at about 5×10^{-5} coulombs/cm^2 with a 10 KV beam. The beam current was 1.2×10^{-9}A and the writing time was about 30 seconds per track (3 mm radius). The working distance was 3.5 cm. Following exposure the aluminum was etched away in sodium hydroxide and the samples developed for 1 minute in a 3:1 mixture of isopropyl alcohol and methyl isobutyl ketone. After development, in order to form an etch resistant mask, the samples were vacuum coated with 2000Å of manganese. Then, they were dipped in acetone to remove the remaining PMM and the overlying metal to leave a mask in the exposed areas. Prior to metallizing, the samples were ultrasonically cleaned with isopropyl alcohol and briefly sputter etched in oxygen to clean the exposed glass to improve the adhesion of the metal mask.

2. Summary of Process. The main steps of the process, which is illustrated in Figure 7, are as follows:

1. A 0.6μ thick film of PMM is spin coated on the substrate (a).
2. The PMM is exposed by an electron beam which is deflected to produce the desired pattern.
3. The PMM is developed (b).
4. A 2000Å-thick coat of magnanese is deposited on the substrate (c).
5. The remaining PMM and unwanted manganese are removed in acetone (d).
6. The sample is sputter etched in argon (e).
7. The remaining manganese is removed with hydrochloric acid.

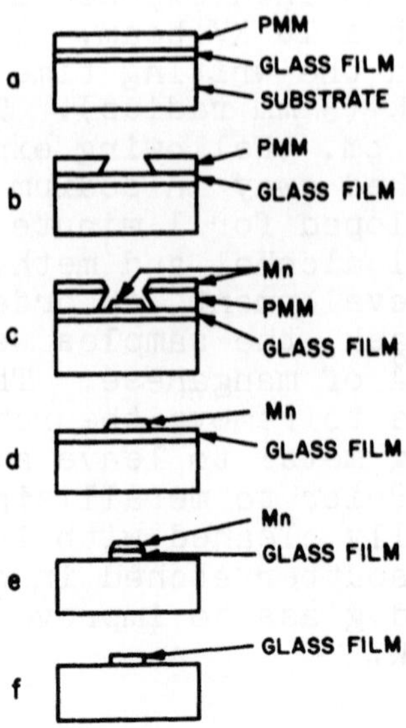

Fig. 7 - Main Steps of the Resist Process

3. <u>Results</u>. Figure 8 shows the intersection of two channels in a PMM film. The viewing angle is 45 degrees. The edge roughness shown is well below 500Å peak-to-peak. The undercutting of the PMM due to electron scattering is visible at the intersection. An end view of a fractured channel in PMM is shown in Figure 9. Before being fractured, the sample was "frozen" in liquid nitrogen. In this figure the undercutting of the resist, which is essential for removing the unwanted manganese in step 5, is clearly visible. The result of step 5 is shown in Figure 10, a micrograph of the surface. The edge roughness again appears to be well below 500Å.

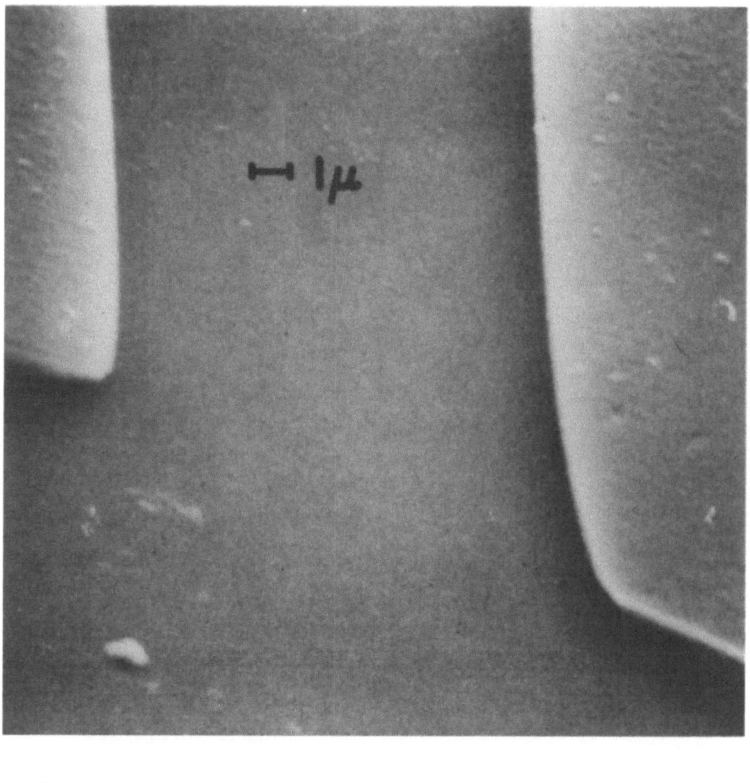

Fig. 8 - Intersection of Two Channels in PMM Film

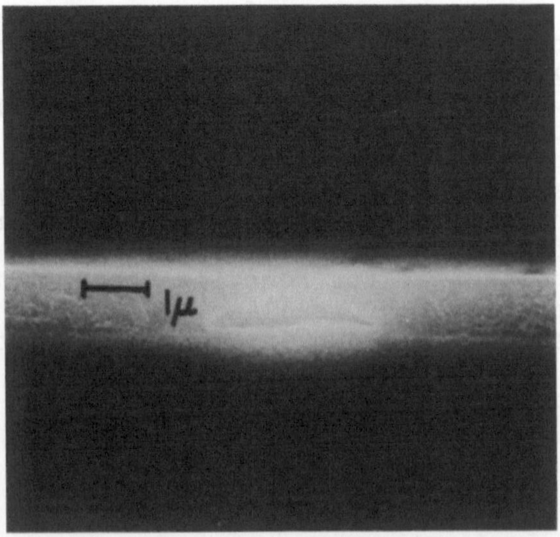

Fig. 9 - End View of Channel in PMM Film

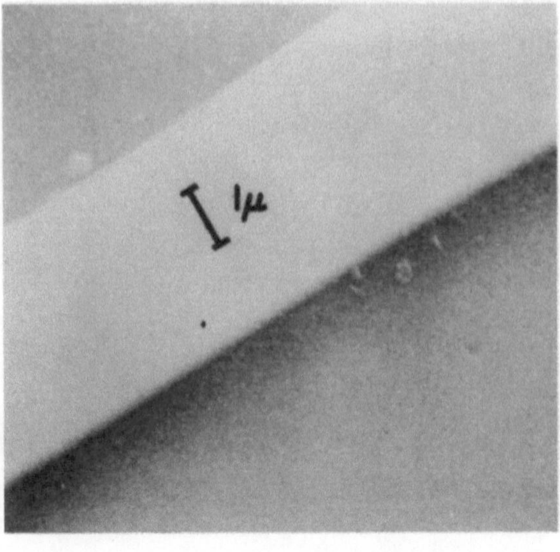

Fig. 10 - Manganese Mask

Figure 11 is an micrograph of a section of a barium silicate glass waveguide. The guide, which was made from a sample that was similar to the one shown in Figure 10, had an rms edge roughness below 500Å. The edge roughness of the sputtered glass film is somewhat greater than the edge roughness of the PMM mask. The increase in edge roughness is caused by graininess and edge tapering of the manganese mask. The mask is tapered for two reasons. First, some of the manganese does not deposit directly, but rather is scattered off the edges of the PMM or bounced off the surface and redeposits, much as sand particles would if dropped onto a surface through a mask. Second, a penumbra similar to that encountered with optical shadows exists. The width of this penumbra is equal to the product of the resist thickness and the ratio of the source diameter to the distance from the source to the film. For example, for 1.0μ-thick mask, 1-cm diameter source, and a source to substrate separation of 10 cm, the penumbra will be 1000Å.

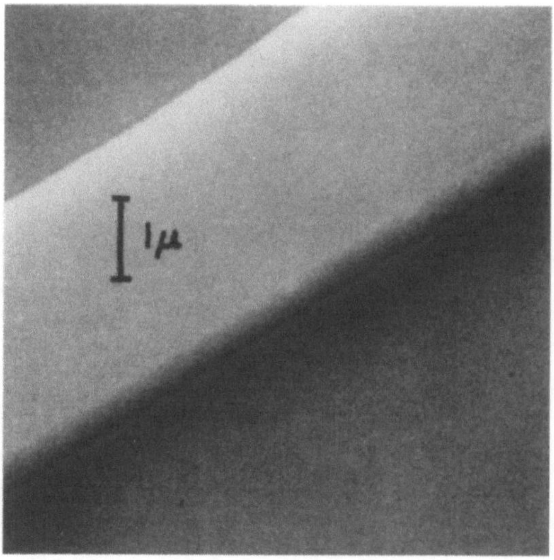

Fig. 11 - Sputtered Glass Guide

The rate of sputtering of the guide edges is much higher than elsewhere because ions strike the surface at grazing incidence and material can be ejected from the edges in the downward direction. Because of the mask edge tapering, thin areas of the mask are etched away first. The guides then rapidly etch down to the level of the surround. As a result, the guide edges tend to be vertical, and are rougher than those of the mask. In addition, the guides are progressively narrowed and their separation increased by sputter etching.

Parallel guides for directional couplers were produced with separation of 1μ for distances up to 2 mm and of 1.25μ up to 3 mm for small etch depths. The guide separations were only repeatable to about 1μ. It is believed that the problem was caused by irregular turn-on transients of the electron writing beam. Figure 12 shows a scanning electron micrograph of two guides that were spaced by about 1μ. These guides were etched to a depth of 0.6μ. The steepness of the edge is clearly visible.

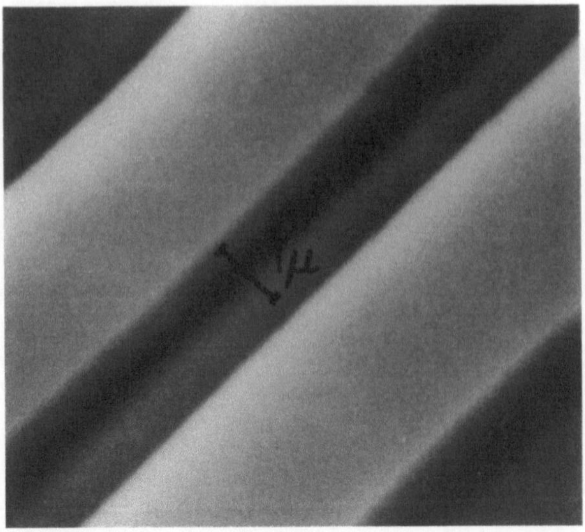

Fig. 12 - Section of a Directional Coupler

In order to observe the optical properties of the waveguide, a beam of 0.6328μ laser light was launched into it from a prism coupler[14] placed directly on the waveguide. Figure 13 shows a sequence of pictures of varying exposure times of a beam propagating in a 0.55μ thick guide immersed in 1.495 matching oil. The light was injected at the top left of the pictures. The bend radius is 3 mm and the guide 3μ-wide. By comparing the exposures an estimate of the loss can be made. Since the exposure times vary by factors of 2 and since the change in intensity from one exposure to another exceeds the change in intensity along the guide, it appears the loss is below 3 dB/cm. For matching oils of index up to 1.53 light could not be observed fanning out from the guide. For the case shown, therefore, it is believed that the bend loss is negligible. This is in agreement with Table II.

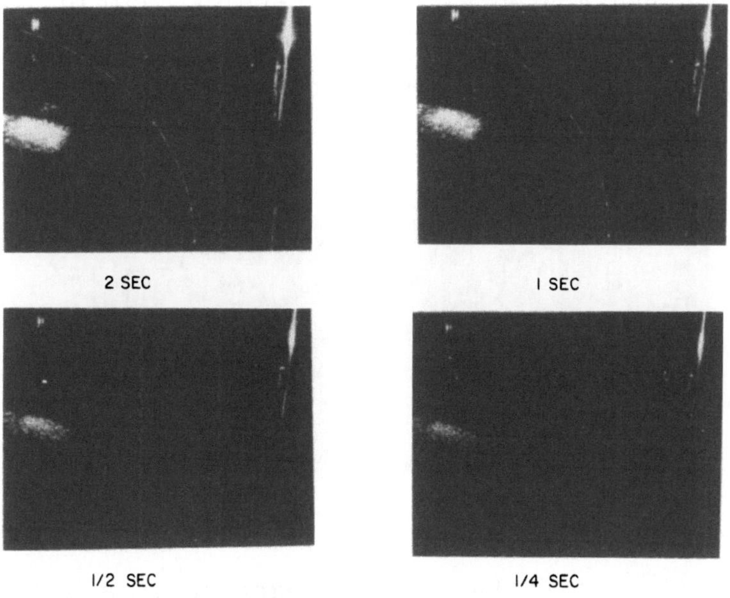

Fig. 13 - Sequence of Exposures of a 0.6328μ Beam Propagating in a Sputtered Glass Guide.

Figures 14a and 14b show light propagation in a coupler of the configuration shown in Figure 2 with matching oils of index 1.515 and 1.522, respectively. The guides were separated by 1.5μ, etched to a depth of 0.55μ (the thickness of the film), and the coupling region was 3 mm long. The point of large attenuation occurs at the termination of the input guide and all of the light visible beyond that point is in the output guide. For the low index matching oil the coupling was quite small while for the higher index oil most of the light was coupled. Radiation loss was not evident for either case, as predicted by Table II.

B. Ion Bombardment Fabrication

In this section guide fabrication by ion implantation through a mask is described. Using this method a curved waveguide has been successfully fabricated. The process starts with a fused quartz substrate which has been carefully cleaned and coated with a 1.5μ-thick film of PMM (see Fig. 15a). The PMM was then exposed by an electron beam (Fig. 15b). The beam was deflected through a half circular arc to expose a strip of

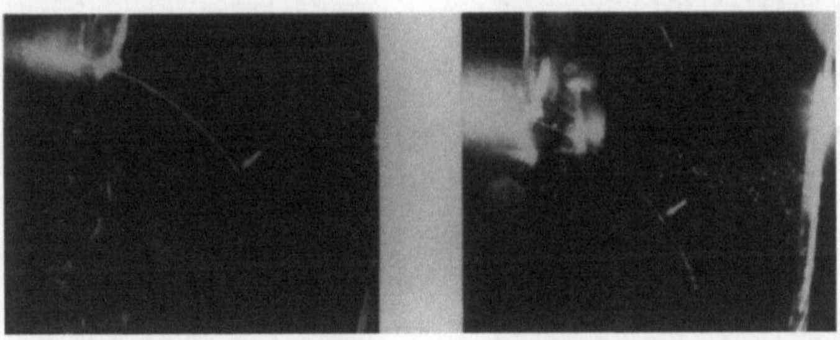

(a) (b)

Fig. 14 - A 0.6328μ Beam Propagating in a Directional Coupler. (a) 1.495 Matching Oil. (b) 1.522 Matching Oil. (Input guide ends at white arrow.)

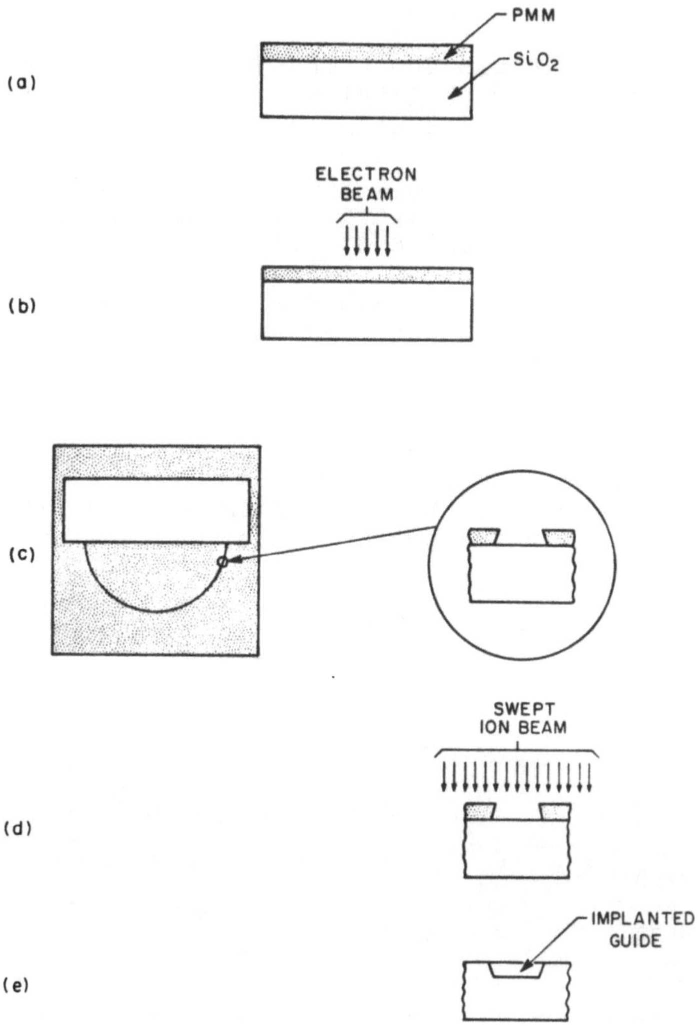

Fig. 15 - Schematic of Ion Bombardment-Electron Resist Guide Fabrication

about 1.7 mm radius and a width of 12μm and was then scanned over an area of a few mm^2 intersecting the ends of the arc to form a beam launching region. Next the exposed pattern was developed in the PMM as shown in Fig. 15c.

The sample was ion bombarded through the mask (Fig. 15d) with lithium ions at four different incident energies with the doses chosen at each energy to give an approximately uniform Li^+ concentration throughout the implanted volume. The object was to create a uniformly bombarded region about 1 μm thick starting from the surface (Fig. 15e). The set of four energies used with the dose implanted at each energy is shown in Table V. It should be noted that the previous study[15] indicates that the observed index change is principally due to disorder produced by the ion beam and not chemical doping. The beam energy and dose schedule used cannot be expected to produce a uniform damage distribution; but since it was the one used in the thin film studies, it was also adopted in the present case.

TABLE V

Energy (keV)	Dose (Particles/cm^2)	
200	10^{15}	(Doping of
108	5.9×10^{14}	2.2×10^{19} Li/cc
58	3.6×10^{14}	over a region
32	2.4×10^{14}	1 μm in depth)

One of the most important contributions of ion implantation in this application, as well as in standard integrated circuit technology, is reduction of the number of processing steps involved compared to sputtering, diffusion or chemical exchange techniques. Another advantage is the fact that only one face of the completed guides is exposed and truly planar geometry can be maintained. At higher ion energies it should be possible to completely bury the guide. This would, of course, require thicker masking to stop the beam which in turn might prove to be impractical for very narrow guides. An attractive but difficult alternative is to write the guide with a very narrow ion beam.

Figure 16 shows 0.6328 μm light scattered from a beam which was prism launched into the guide. The beam enters from the top of the figure, where there is reflected light from the prism, and travels counter-clockwise through the guide until it returns to the launching region. In the launch region the beam continues but now is guided only in the plane. The illusion of an elliptical path is associated with camera angle. This result

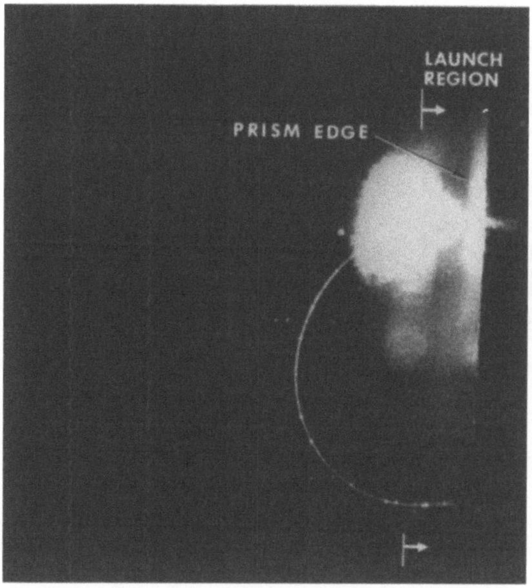

Fig. 16 - Light Scattered From 12 μm-Wide Guide

was obtained with no annealing of the sample. Previous refractive index measurements on films implanted with lithium ions under the same energy and dose conditions have shown a refractive index of 1.505 (3 percent greater than the unimplanted material) and optical loss of 8-10 dB/cm. This is consistent with the observation of guidance in the present sample and an estimated loss of 10 dB/cm. Annealing of the sample reduces the loss but over-annealing will also reduce the refractive index difference between the guide and substrate to the point where radiation losses from the curved guide may become important. Figure 17 shows light scattered from a 5 μm-wide waveguide after annealing to 200°C for 1 hour. The loss in this sample has been reduced by the annealing to approximately 3 dB/cm.

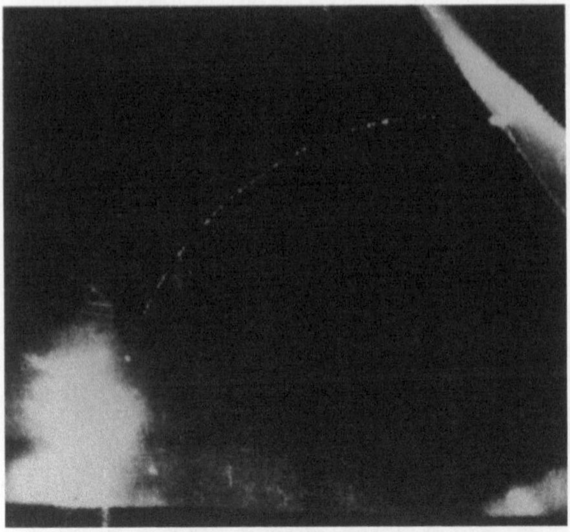

Fig. 17 - Light Scattered from a 5 μm-Wide Guide After 200°C One-Hour Annealing.

IV. MEASUREMENT TECHNIQUES

In this section several techniques that are useful for determining the optical properties of films and substrates, and film thickness will be reviewed.

A. Determination of Refractive Index and Thickness

A number of techniques for measuring thickness and refractive index will now be described. Some of the approaches are suitable for both measurements. In the case of thickness measurement it is often necessary to have a partially coated sample. This can usually be achieved either by masking during deposition or masked etching afterwards.

1. Stylus Measurements. Stylus measurement is one of the useful approaches for thickness determination. Three approaches are in widespread use, the first two with an instrument known as the Talysurf[16] and the third one with an instrument known as the Dektak.[17] With these instruments thickness variations of a few hundred angstroms can be readily measured.

In the first method the configuration of Figure 18 is employed. The position of the stylus

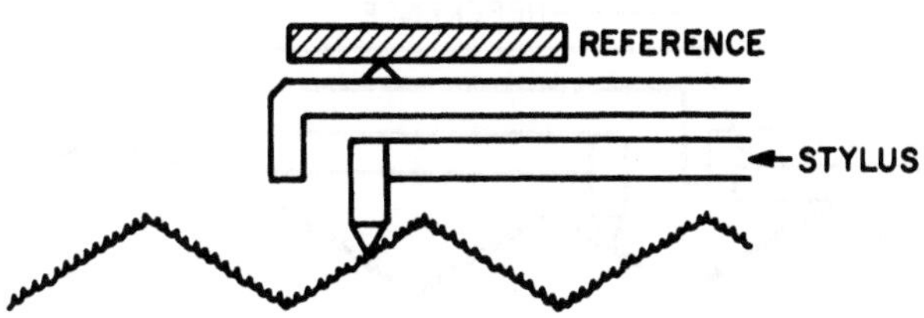

Fig. 18 - Stylus Measurement with Fixed Reference

relative to the reference plane is recorded by a strip recorder which is electrically activated by the displacement of the stylus. Before measurements can be made the reference plane must be adjusted so that it is parallel to the surface of the sample being evaluated. In the second method a reference which moves with the stylus, as shown in Figure 19, is employed. This approach is well suited to measuring abrupt changes in sample thickness such as at the edge of a resist or film. In the third method the sample holder rather than the stylus is moved. This requires leveling of the support, which is readily accomplished by trial and error.

2. <u>Interferometer</u>. Interferometric measurements[18,19] can be made by reflection or transmission. In both cases optical path differences are measured so if either refractive index or thickness is known the other can be determined.

In transmission, for a film of thickness t the fractional fringe displacement across the film-substrate step is given by

$$F = (n_f - n_0)\,\frac{t}{\lambda_0}$$

where λ_0 is the free space wavelength, n_0 the index

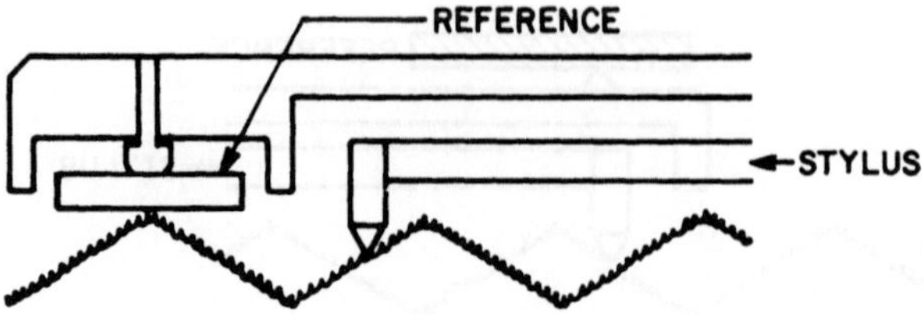

Fig. 19 - Stylus Measurement with Moving Reference

of the surrounding medium and n_f the film index. For reflection measurements the sample is usually coated with a high reflectivity layer to enhance contrast and set the reflection coefficients either as shown in Figure 20a or as shown in 20b. For reflection measurements the fractional fringe displacement is given by

$$F = \frac{n_0 t}{2\lambda}$$

for the configuration of Figure 20a and by

$$F = (n_f - n_0) \frac{t}{2\lambda}$$

for the configuration of Figure 20b.

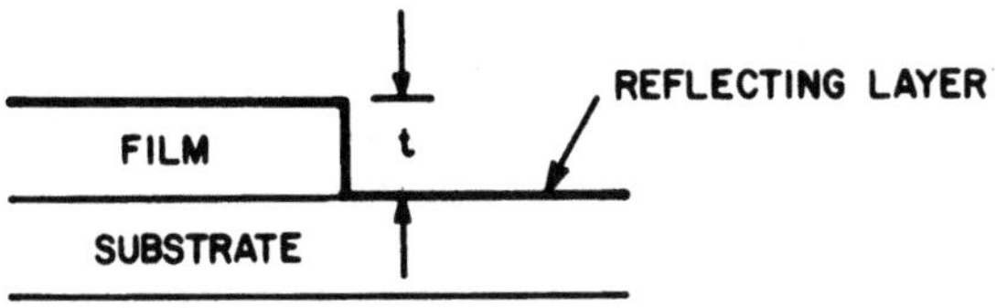

Fig. 20a - Sample for Interferometry With High Reflectivity Layer on Film and Substrate

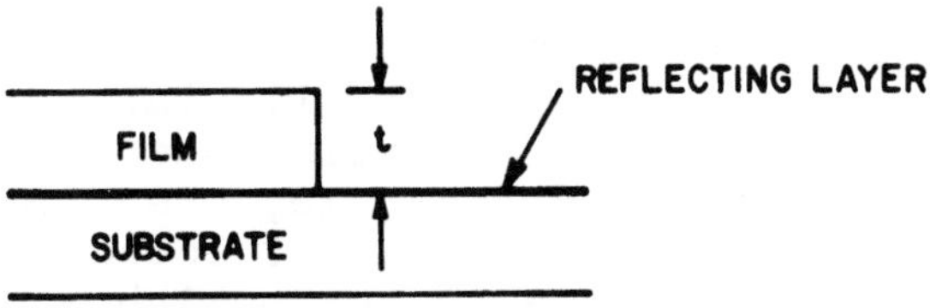

Fig. 20b - Sample for Interferometry With High Reflectivity Layer on Substrate

3. Abbe Refractometer. The Abbe refractometer[20] is well suited to the measurement of refractive index of bulk samples and thick films. Commercial instruments[16] are available which allow refractive index to be read directly from the instrument. Provision is also included for measurement of dispersion.

Figure 21 shows the basis of the instrument. Uncollimated light is injected onto the sample as shown. Tracing backwards through the instrument, light will exit from the prism at all angles α up to an angle

$$\alpha_c = \sin^{-1} \frac{n_1}{n_0} \sin^{-1}\left(\sin^{-1} \frac{n_2}{n_1} - \delta\right)$$

corresponding to the angle of total reflection of light reflecting from the prism substrate interface (β=90). The measureable index is limited by the index of the prism and prism geometry. Commercial instruments are available for the index range 1.25 to 1.70 with an accuracy of 0.001.

4. Abeles Method. Abeles[21] method is an extremely useful one for measuring the refractive index of films. Therefore, it will be described in some detail including the effect on the results of a graded index transition.[22] Similar problems involving refractive index gradients will also be encountered with some other methods. The results described here give insight into the nature and extent of the problem.

When light polarized in the plane in incidence is incident on the air-film interface at the Brewster angle for the film, there is no reflection from the interface, ie, the film behaves as if it did not exist. The reflected signal is then equal in magnitude to that which would be obtained from the substrate alone. More precisely, the reflection coefficient from a film of thickness d and index n_f on an infinitely thick substrate of index n_s is given by

$$R = \frac{r_f + r_s e^{-j2\delta}}{1 + r_f r_s e^{-j2\delta}}$$

where r_f and r_s are Fresnel coefficients, i.e.,

$$r_f = \frac{\cos \Phi_f - n_f \cos \Phi_i}{\cos \Phi_f + n_f \cos \Phi_i}$$

$$r_s = \frac{n_f \cos \Phi_s - n_s \cos \Phi_f}{n_f \cos \Phi_s + n_s \cos \Phi_f}$$

and

$$\delta = \frac{2\pi n_f d}{\lambda_0} \cos \Phi_f$$

where Φ_i is the angle of incidence in air; Φ_f is the angle of refraction in the film; and Φ_s is the angle of refraction in the substrate. At the Brewster angle, Φ_{iB} for the air-film boundary, $r_f=0$, so that

$$\tan \Phi_{iB} = n_f.$$

We then obtain from above

$$|R| = \frac{\cos \Phi_s - n_s \cos \Phi_i}{\cos \Phi_s + n_s \cos \Phi_i} = r_{s0}$$

where r_{s0} is the Fresnel coefficient for the air-substrate boundary.

It can be shown that if $r_s=0$, then $|r_f|=|r_{s0}|$ is also a solution. However, this root exists only in a restricted range of n and n given by

$$\frac{n_f^2 n_s^2}{n_f^2 + n_s^2} \leq 1.$$

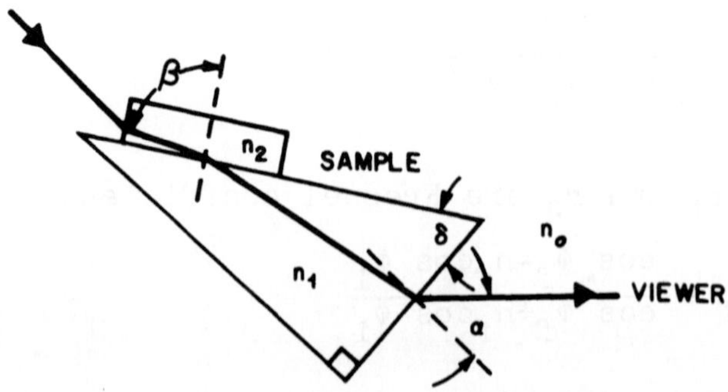

Fig. 21 - Abbe Refractometer

Fig. 22 is a plot of the region for which this latter solution applies. In general, silicate glasses are in the region n_f, $n_s > 1.45$, so that this latter solution is of no concern. However, it could be significant in the evaluation of fluoride glass and other smaller index systems. In practice, the refractive index of a film can be determined to an accuracy better than ±0.002 by this method.

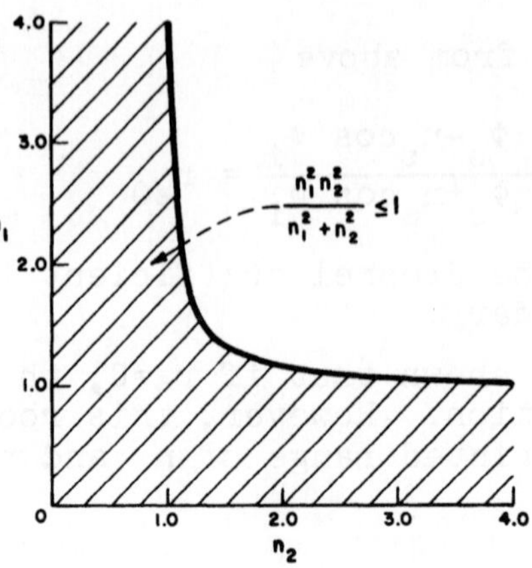

Fig. 22 - Domain of Substrate Brewster Angle Root

The Method described in the preceding paragraphs assumes that the film index is uniform. It is well known that for some materials, such as MgF_2, the refractive index varies with depth. In order to gain insight into the effects of index gradients, consider the index profile of Fig. 23. For a transition region thickness $\Delta t=0$, the analysis of Section III-A applies, so the true film index results from the measurement. However, for Δt finite an error will result, ie, the measured index, $\bar{n}$ will not be n_f. Assuming that the transition from n_f to n_s is continuous, then for Δt sufficiently large it can be shown analytically that for n_f and n_s large or for $n_f \approx n_s$, the measurement described above gives the geometric mean of the film and substrate indices.

In order to obtain further information, a numerical analysis of the problem has been performed. It was assumed that n(x), the index variation in the transition region, took the form

$$n(x) = \frac{n_f+n_s}{2} - \frac{n_f-n_s}{2} \cos\left(\frac{x-d}{\Delta t} \pi\right),$$

$$\text{for } d \leq x \leq d + \Delta t .$$

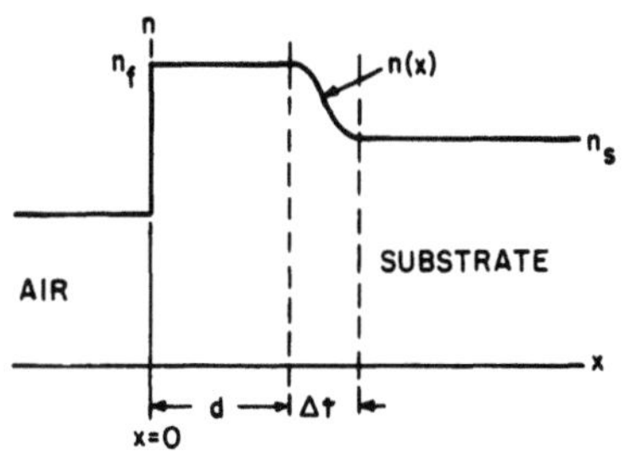

Fig. 23 - Index Gradient Profile

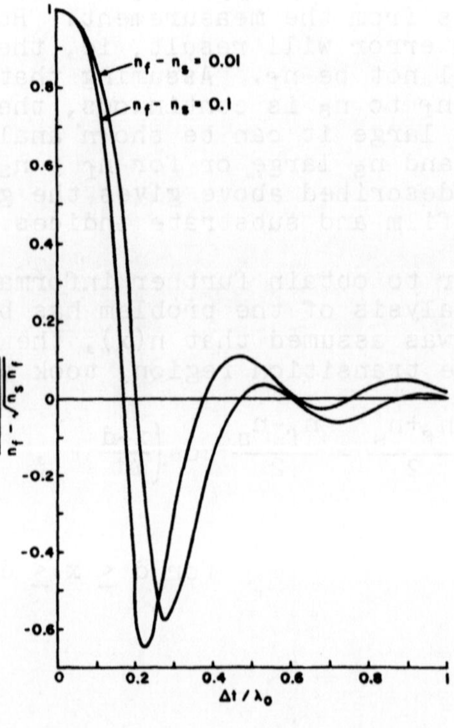

Fig. 24 - Measured Refractive Index vs Graded Region Thickness for n_s=1.5, δ/λ_0=0.01.

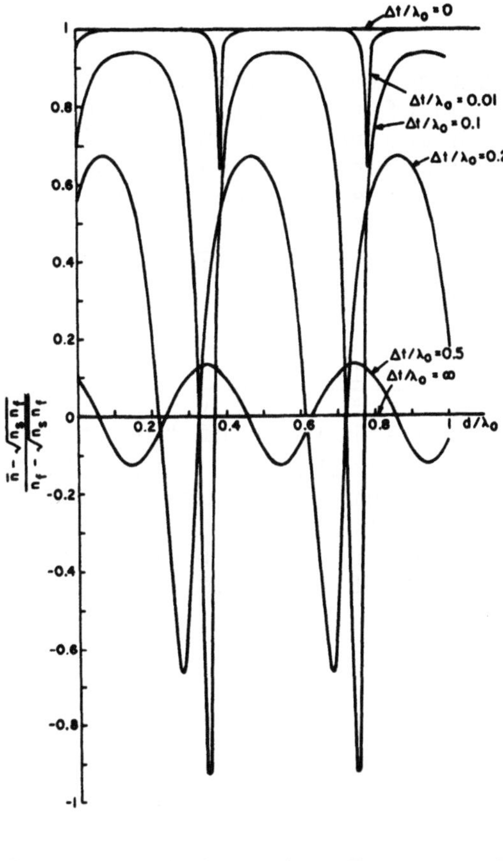

Fig. 25 - Measured Refractive Index vs Film Thickness for n_s=1.5, n_f=1.51.

The problem was solved using a matrix formulation with n(x) approximated by a series of discrete steps. Fig. 24 shows the variation of the measured index $\bar{n}$ as found by Abeles' method as a function of transition region thickness for a substrate index of 1.50 and a film thickness d= d=0.1 λ_0.

For a graded index, the measured index $\bar{n}$ is also a function of film thickness because the angle at which the reflection coefficient of the film on the substrate is equal to the reflection coefficient of the substrate alone is not equal to the Brewster angle for the film alone. This effect is demonstrated in Fig. 25. Clearly, even fairly short graded regions can cause significant error.

5. <u>Prism Coupler Method</u>. Referring to Fig. 26, a high index prism ($n_p > n_f$) is clamped to film. By proper selection of the injection angle in the high index medium, the propagation constant of the incident beam can be made equal to that of the mode in the film. Assuming weak coupling, the incident angle of the beam is related to the propagation constant of the film by

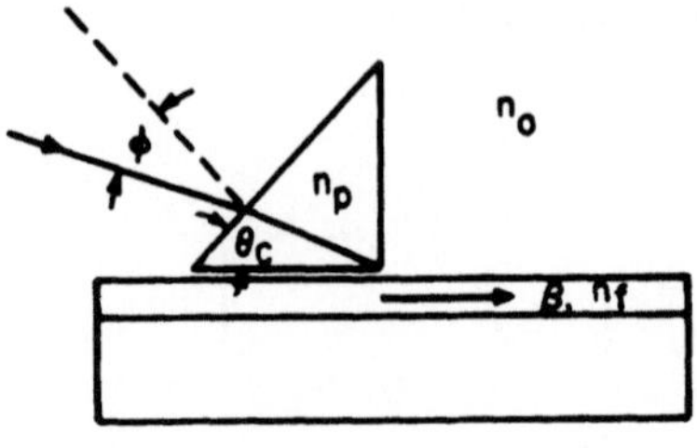

Fig. 26 - Prism Coupler

$$\phi = \sin^{-1}\left\{\frac{n_p}{n_0}\sin\left(\sin^{-1}\left(\frac{\beta}{k_0 n_p}\right) - \theta_c\right)\right\}$$

where n_p is the prism refractive index; β is the mode propagation constant in film; θ_c is the corner angle of the prism; $k_0=2\pi/\lambda_0$; and ϕ is the angle of incidence of light with respect to the normal to the input face of the prism.

This relationship permits calculation of several important properties. For example, knowledge of ϕ, n_p, and θ_c permits calculation of β, which in turn allows determination of n_f or d. If several modes can be launched, the correlation of the various β's with n_f and d allows determination of both parameters. Alternatively, launching efficiency can be optimized by selecting θ_c so that ϕ is the Brewster angle.

B. Attenuation Measurements

The measurement of attenuation of beams propagating in thin films by scattering and transmitted light have received wide attention. In both of the methods the light is injected either by launching into the end of the film or through the surface of the film with a prism, grating or taper coupler. A full description of these methods is given in other chapters.

In the scattered light method a plot of light scattered from the film is made using a probe consisting of an aperture and a photodetector. Assuming the scattering coefficient is constant along the film the attenuation constant can be found. In practice, it is also found that large scattering centers are present. Because of the presence of these centers it is usually necessary to make a least mean square fit to the data. For extreme cases the method cannot be used.

The coupled light method has been found to be a useful approach. In this method the beam is probed by a prism coupler. The coupler extracts all the energy from the film so the light emanating from the output prism is equal to the light in the film less a constant fraction lost because of reflections. Attenuation measurements are performed by measuring the light emanating from the output coupler as a function of position along the propagating beam.

The accuracy to which the attenuation coefficient can be measured is given by the accuracy of the power measurement divided by the path length over which data can be obtained. Power measurements to 0.1 dB accuracy can be achieved. Weber and his coworkers using films 5 cm long have made measurements to 0.02 dB/cm.[23]

Several mechanical configurations have been used. A simple one[13] to implement is shown in Figure 27. In this arrangement the substrate is

Fig. 27 - Two-Prism Attenuation Measurement Apparatus (Input on Right, Output on Left).

supported by a fixed mount and the input and output prisms are clamped to and supported by the substrate. In the figure light enters from the right. The output prism can be adjusted without affecting the input coupler since it can be moved along the film without disturbing the input coupler.

V. CONCLUSION

Thin-film waveguide fabrication and testing techniques have progressed to the point that circuits now appear feasible. Widespread efforts independent of integrated optics are now in progress which will lead to improved electron beam apparatus capable of delineating complex circuits under digital control. At the same time a large selection of both optically passive and active films are being developed for integrated optics. The merger of the pattern delineating techniques described here with advanced election beam writing techniques and new film materials should make feasible the fabrication of numerous complex integrated optical circuits.

REFERENCES

1. T. Li, "Passive Integrated Optical Circuits," Digest of the Topical Meeting on Integrated Optics - Guide Waves, Materials, Devices, Las Vegas, February 7-10, 1972.

2. E. A. J. Marcatili, "Dielectric Retangular Waveguide and Directional Coupler for Integrated Optics," BSTJ, Vol. 48, pp. 2071-2102, September, 1969.

3. J. E. Goell, "A Circular-Harmonic Computer Analysis of Rectangular Dielectric Waveguide," BSTJ, Vol. 48, No. 7, September, 1969, pp 2133-60.

4. E. A. J. Marcatili, "Bends in Optical Dielectric Guides," BSTJ, Vol. 48, pp. 2103-2132, September, 1969.

5. D. Marcuse, "Mode Conversion Caused by Surface Imperfections of a Dielectric Slab Waveguide," BSTJ, Vol. 48, pp. 3187-3217, December, 1969.

6. J. E. Goell, R. D. Standley, "Sputtered Glass Waveguide for Integrated Optical Circuits," BSTJ, Vol. 48, pp. 3445-3448, December, 1969.

7. I. Haller, M. Hatzakis, and R. Srinivasan, "High Resolution Positive Resist for Electron Beam Exposure," IBM J. Res. Develop., Vol. 12, pp. 251-256, May, 1968.

8. J. E. Goell, "Election-Resist Fabrication of Integrated Optical Circuits," Digest of Topical Meeting in Integrated Optics-Guided Waves, Materials, Devices, Las Vegas, February 7-10, 1972, pp. MB8-1 to MB8-4.

9. J. E. Goell, R. D. Standley, W. M. Gibson, J. W. Rodgers, "Ion Bombardment Fabrication of Optical Waveguides Using Electron Resist Masks," Appl. Phys. Lett., Vol. 21, No. 2, July 15, 1972, pp. 72-73.

10. K. Amboss and E. D. Wolf, Double-Deflection Aberrations in a Scanning Election Microscope, 11th Symposium on Election, Ion, and Laser Beam Tech., Boulder, Colorado, 1971.

11. J. L. Vossen, "Control of Film Properties by r-f Sputtering Techniques," J. Vac. Sci. Tech., Vol. 8, pp. 512-530, May, 1971.

12. W. M. Muska, to be published.

13. J. E. Goell, "Barium Silicate Films for Integrated Optical Circuits," Applied Optics, Vol. 12, No. 4, April, 1973, pp. 737-741.

14. P. K. Tien, R. Ulrich and R. J. Martin, "Modes of Propagating Light Waves in Thin Deposited

Semiconductor Films," Appl. Phys. Lett., Vol. 14, pp. 291-294, May 1, 1969.

15. R. D. Standley, W. M. Gibson, and J. W. Rodgers, "Properties of Ion-Bombarded Fused Quartz for Integrated Optics," to be published in Applied Optics.

16. Rank Precision Industries, Inc., Des Plaines, Ill.

17. Sloan Instrument Corp., Santa Barbara, Calif.

18. F. A. Jenkins and H. E. White, Fundamentals of Optics, (McGraw Hill, New York, 1957), pp. 261-285.

19. M. Born and E. Wolf, Principles of Optics, (Macmillan, New York, 1959), pp. 255-368.

20. O. S. Heavens, Optical Properties of Thin Solid Films, (Academic Press, New York, 1955), pp. 127-129.

21. F. A. Abeles, Advanced Optical Techniques (Wiley, New York, 1967), pp. 143-188.

22. J. E. Goell and R. D. Standley, "Effect of Refractive Index Gradients on Index Measurements by the Abeles Method," Applied Optics, Vol. 11, No. 11, November, 1972, pp. 2502-2505.

23. H. P. Weber, F. A. Dunn, and W. N. Leibolt, "Loss Measurements in Thin Film Optical Waveguides," Applied Optics, Vol. 12, No. 4, April, 1973, pp. 755-757.

Chapter 7 - ELECTRON AND ION BEAM MICROFABRICATION OF INTEGRATED OPTIC ELEMENTS

E. D. Wolf

Hughes Research Laboratories
Malibu, California 90265

INTRODUCTION

This lecture dealt with various aspects of electron and ion beam microfabrication of prototype structures suitable for light guiding and coupling. The introductory material on electron and ion beam interactions with solid surfaces was not reproduced here because these subjects have been discussed extensively in the literature. The intent of these notes is to provide the reader with some of the more recent techniques and successes of beam microfabrication of integrated optics elements.

The fundamental attribute of combined beam microfabrication is three-dimensional tailoring at submicrometer linewidths of passive and active elements for light coupling, guiding, modulation, and ultimately for fabrication of distributive feedback surface lasers suitable for planar integration. Two dimensional spatial tailoring is provided by computer controlled electron beam exposure of the desired patterns in an appropriate electron resist, and then subsequent processing of these high resolution patterns by ion beam sputtering and/or ion implantation or even conventional chemical processing, in certain cases, provides the third dimension of tailoring.

Unique demands on microfabrication techniques are imposed by low loss waveguides, couplers, etc. Waveguides less than one micrometer in width must be made with an edge smoothness rms deviation of only a few hundred Angstroms. "Line of sight" beam processing can produce this desired resolution and edge smoothness and, furthermore, offers some control of the cross-sectional shapes of the guides.

ELECTRON BEAM MICROPATTERN DEFINITION AND FABRICATION

High resolution electron beam micropattern definition requires a focused (typically 200 Å to 2500 Å diameter), high energy (10 to 20 keV) electron beam which can be deflected and modulated in a controlled manner. When this electron beam strikes an electron resist coated substrate secondary ionization is produced in the thin resist film and bond scission results. If crosslinking predominates in the recombination of these dangling bonds (free radicals), the resist is negative acting, and the original unexposed resist can be dissolved differentially. If recombination produces predominantly shorter chain lengths (lower molecular weights) than the original polymer, the resist is positive acting and the exposed regions are more soluble.

The selection of a particular area for exposure is accomplished by deflecting the electron beam to a specific point in a two-dimensional scan field. After deflecting the electron beam to the selected location, it is turned on for a given time period. During this time period, a sufficient amount of energy is deposited into the resist to produce the desired linewidth in the resist after development. Thus, while the scanned beam-substrate interactions give a single high resolution line, the ability to place another line precisely relative to the first line depends on the resolution, accuracy and stability of the beam deflection system. In the Hughes electron beam micropattern definition system, the location of the beam within the scan field is determined by 15-bit digital-to-analog converters. The digital information to these converters, as well as beam on-off control, is provided by a 16-bit word computer. During exposure, the computer utilizes software pattern programs to generate data for the digital-to-analog converters as well as performing other control functions associated with pattern exposure. This system is based on a Cambridge Mk IIa scanning electron microscope and a

Varian 620/i minicomputer system (Fig. 1) and has been described previously.[1,2] Several other systems have also been reported.[3]

In order to fabricate high quality R&D prototype test elements suitable for integration into integrated optical systems later, it is important to know the characteristics of electron beam energy dissipation in thin resist films in which the elements are first defined. We have generated experimental data on this effect in Elvacite* 2041, a high molecular weight polymethyl methacrylate (PMM), which is a positive electron resist. The experimental method has been reported previously[4] and, in principle, is quite simple. A series of exposures is made in a thin film of PMM on

*du Pont trade name.

Fig. 1. Computer Controlled Electron Beam Microfabrication System.

silicon wafers which have a few hundred Angstroms of SiO_2. After development the wafer is fractured under liquid nitrogen in the direction orthogonal to the line exposures. The low temperature fracture prevents ductile deformation of the resist edge. The line profiles are then examined end-on from the fracture edge using scanning electron microscopy.

Figure 2 shows a family of curves taken from high magnification scanning electron micrographs (60,000x) of the edge view of the exposed and developed beam track such

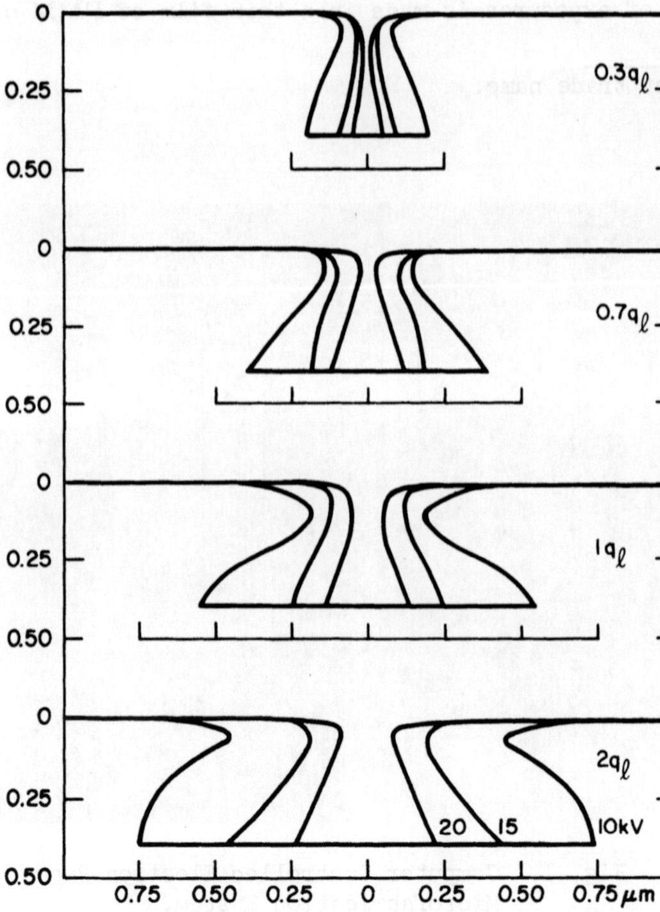

Fig. 2. Family of Profiles for 10, 15, and 20 kV Exposures.

as those shown in Fig. 3 for 10 kV primary beam voltage and several different exposures. The term q_ℓ is defined as 1×10^{-8} C/cm, a line charge density. From this type of

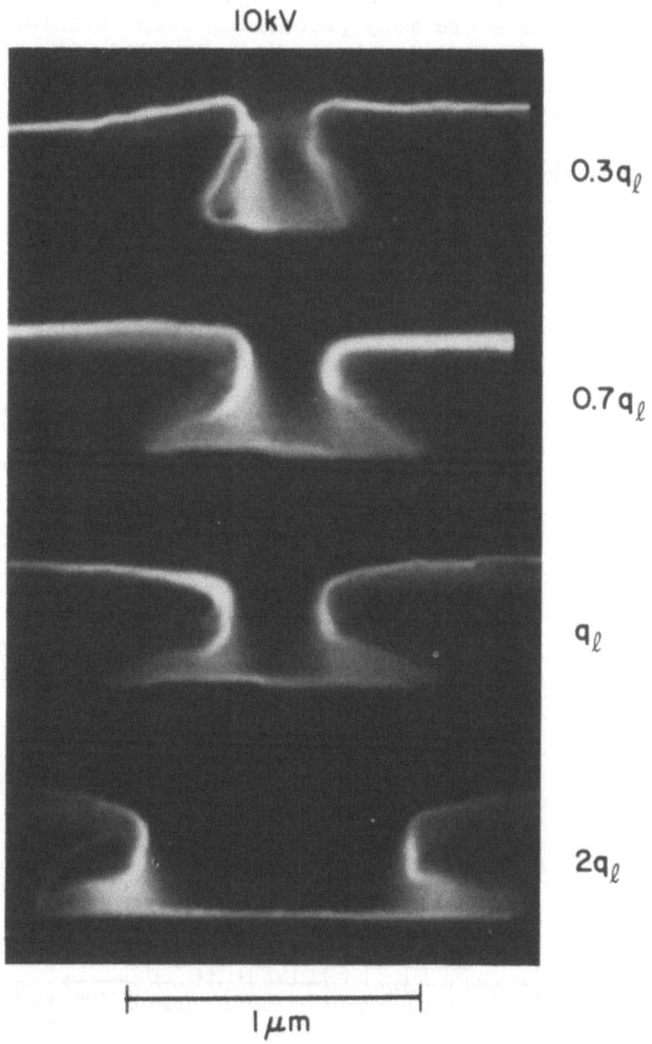

Fig. 3. Energy Dissipation Profiles in Thin (4000 Å) Resist Film. $q_\ell \equiv 1 \times 10^{-8}$ C/cm^{-1}.

information, the desired linewidth can be programmed directly for each resist thickness and substrate composition. The mechanism for lift-off is also immediately evident from the undercut in the resist caused by backscattered electrons.

After the pattern has been exposed properly in the resist film there are many processing steps available to convert the pattern into three-dimensional device structures. Figure 4 illustrates several of these alternatives leading to elements of interest in integrated optics. Not shown are the more conventional planar processes of diffusion, chemical etching, etc., all of which obviously have some merit for integrated optics component fabrication. However, this lecture is emphasizing beam processing which is

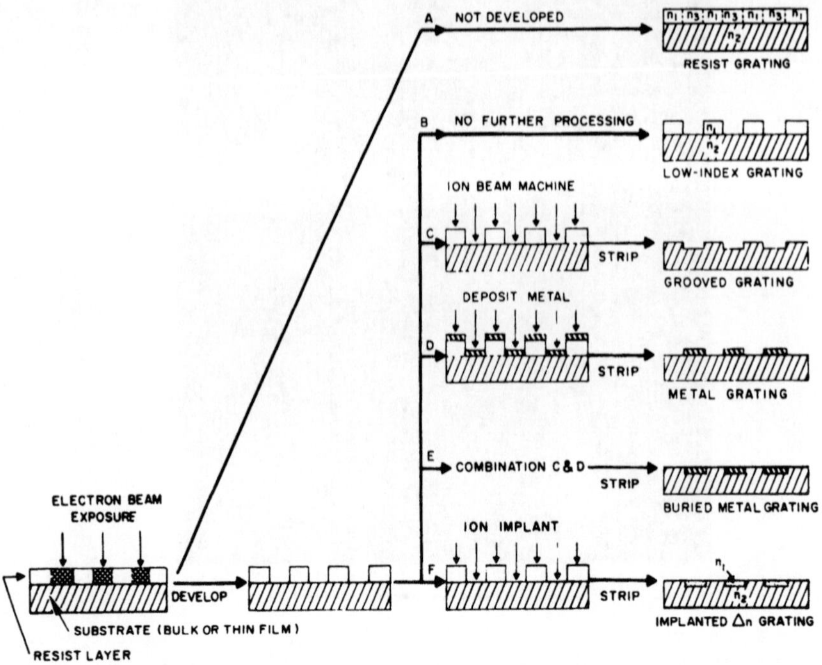

Fig. 4. Several Methods for Microfabrication of Planar Grating Couplers (and/or waveguides in certain cases)

believed to be uniquely well suited for high performance components, particularly in the visible wavelength range.

Process A is the simplest and requires no further processing after exposure. The change in index of refraction is caused solely by the electron beam exposure. The negative version of Process B has been used to produce guides in resist patterns.[6] Processes C and D are most useful and their combination (using chemical etching instead of sputtering) to produce buried grids has been reported previously.[7] Ion implanted waveguides have been made successfully using Process F.[8,9]

ION BEAM SPUTTERING FOR MICROPATTERN PROCESSING

The ion beam sputtering technique discussed in these notes is to be distinguished from plasma sputtering (dc or rf) which has been accepted as a production tool in the microelectronics industry for several years. Ion beam sputtering offers the advantages over plasma sputtering of directionality of the ions, better vacuum environment, and the capability of maintaining the target at any temperature. Generally, the ions extracted from a plasma will impinge a surface normally, whereas an energetic beam of ions may be made to bombard the surface at any desired angle. By means of contact masking or by shadow masking, patterns can be machined in the surface of the target. The directed beam permits etching of patterns whose resultant edge acuity is better than those produced by chemical etching processes. Because most materials have finite sputter yields, high resolution patterns can be produced in a broad variety of thin film materials.

Ion beam sputtering for machining or deposition can be combined with electron beam pattern definition and ion implantation doping toward the eventual realization of a completely automated (computer controlled) beam fabrication process. From experience at Hughes and elsewhere, ion beam machining produces a very accurate copy of the resist mask with negligible distortion and no undercutting. The high vacuum system shown schematically in Fig. 5 has been applied to the investigation of the fabrication of integrated optical components.[10] Argon ions are produced in a duoplasmatron ion source and accelerated either to sputter etch

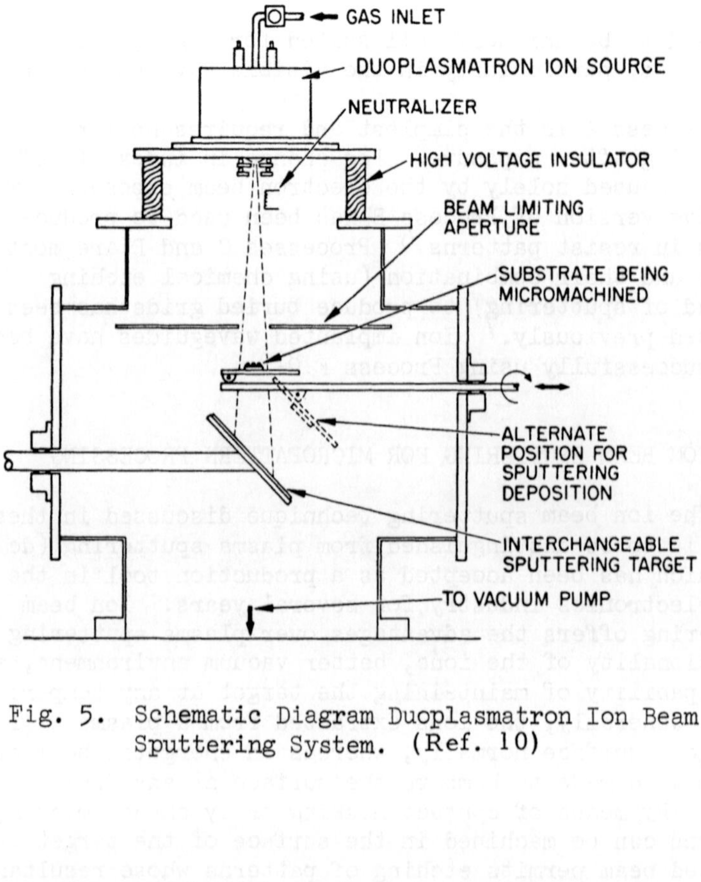

Fig. 5. Schematic Diagram Duoplasmatron Ion Beam Sputtering System. (Ref. 10)

material from the surface of a substrate or to deposit material onto the substrates from an opposing target. In this system,[10] the ion beam current density and energy can be regulated to as low as 10 μA/cm^2 at 500 V. There is little bombardment heating under this condition that can cause distortion or blistering of masking materials such as polymers. Also, the directionality of the ion beam ensures that patterns produced by electron beam exposure and development of polymers are faithfully etched into SiO_2 or other thin film materials employed for guiding or coupling structures. Unfortunately, the removal rate of PMM is considerably higher (see Fig. 6 taken from Ref. 10) than that of the typical glassy or oxide films and a thick PMM masking film is required if etching to a depth of 5000 Å or more is desired. Results are presented later of the etching profile quality produced by this technique.

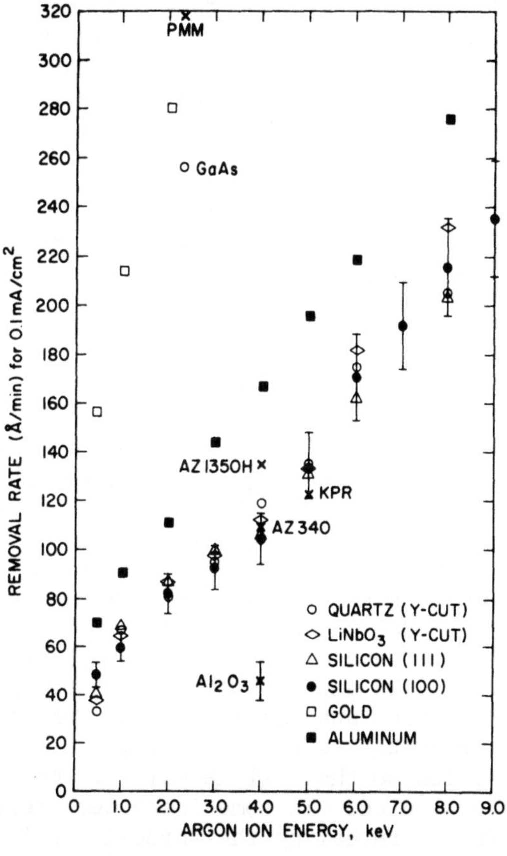

Fig. 6. Sputtering Etch Rates for Monocrystalline and Thin Film Materials. (Ref. 10)

When etching must be carried to depths that are prohibitive with PMM or other resist masking materials, the replacement mask technique is used. In this technique, aluminum is evaporated through a resist (polymer) liftoff mask and the aluminum film serves as the mask against sputtering during subsequent ion etching of the substrates or films. Not only is the sputtering rate of aluminum low enough to be comparable to many of the oxide materials, but by introducing oxygen into the system the rate is further reduced.[11,12] Thus a relatively thin, high resolution aluminum mask can serve for deep etching of guiding film materials.

In the alternative application, ion beam sputtering permits the deposition of a wide variety of thin films into openings produced in electron beam-exposed and developed, liftoff masking patterns.

SCANNING ELECTRON MICROSCOPY

Conventional scanning electron microscopy provides an ideal method for the evaluation of the electron and ion beam fabricated waveguide structures. Most samples were cleaved for the purpose of evaluating the cross section of the waveguide structures at high magnification (e.g., 2000x to 100,000x) using secondary electron emission imaging. Using this mode, the scanning electron micrographs are similar in appearance to photomicrographs but with much greater depth of field and with lighting that appears to emanate in a direction away from the secondary electron collector. It is possible to rotate, tilt, and translate the samples in position so that all details of the surface topology can be studied. In some instances stereographic pairs are useful in revealing the three-dimensional aspects of the waveguide structures.

It should also be noted that considerable care is required in obtaining representative cross sections (such as in Fig. 3) in resist materials which undergo plastic deformation if fractured at room temperature. All of the profiles shown in this report were fractured at liquid nitrogen temperature. Furthermore, scanning a resist profile just once is generally enough to cause slight deformation. Therefore, it is necessary to set up (focus, etc.)

on an adjacent structure and then mechanically shift over and photograph "blind." In this manner we are able to obtain scanning electron micrographs which contain few, if any, artifacts.

RESULTS OF BEAM MICROFABRICATION (Ref. 13)

The results of several combinatory beam techniques for waveguide fabrication are discussed separately and then evaluated collectively at the end of this section. It should be noted that most of the methods discussed are equally well suited for planar coupler fabrication (and two examples are given) and some are usable for high resolution diffusion and ion implantation masks.

Direct Waveguide Exposure Without Development

Exposures can be made directly in PMM without chemical development. The most obvious changes are in the color and thickness of the exposed area (the latter is the predominant cause of the former). The color change associated with a $10^{-4}C/cm^{-2}$ broad area exposure of a 2800 Å thick film is from blue to red (in thicker films little or no color change is observed). Surface roughness and film thickness measurements were made on these 2800 Å films using a Dektak* and revealed a 20% film shrinkage and a significant increase in surface roughness. This surface roughness is thought to be a result of microbubble formation caused by gas evolution during electron beam exposure. A much lighter exposure of $10^{-6}C/cm^{-2}$ at 10 kV primary beam acceleration voltage does produce a very slight perceptible color change to the unaided eye for large area exposures in 1 μm thick PMM but closely spaced waveguide structure (i.e., 5 μm on 7 μm centers) are only barely visible in a Nomarski microscope. No micrographs are reproduced here because of this microphotography difficulty. No measurements of the change in the index of refraction of electron exposed PMM have been reported.

---

*Trade name, Sloan Instrument Company.

Direct Waveguide Exposure with Development

Polymethyl methacrylate, as discussed earlier, is a positive acting electron resist; that is, under electron beam exposure it undergoes bond scission and becomes more soluble than the original film. Using this property and properly programmed electron beam exposure, ridges of resist are left on the surface after development. Figure 7 is a photomicrograph of an array of waveguides formed on a thin SiO_2 film. This photomicrograph demonstrates the lack of surface detail even at high optical magnification (original at 500x), thus, the need for SEM for analyzing micron-size structures. In Figs. 8, 9, and 10 a sequence of scanning electron micrographs shows the same type of waveguide as in Fig. 7. In this sequence the cross-sectional profile is controlled by annealing above the glass transition temperature (~120°C) for PMM. The dynamic balance between surface tension and substrate wetting has actually caused the resist to "ball up" slightly in Fig. 10. Note the extremely smooth edge of the waveguides. At very high magnification (~50,000x) the edge smoothness was observed to be ~200 Å.

Waveguide and/or Masking Structures in Chemically Etched Silicon Dioxide

Even though the structure shown in Fig. 11 is not ideal for guiding because the substrate is silicon (with n ~ 2.3), this method of fabrication offers some possibilities. In this series of experiments, PMM resist structures such as shown in Figs. 8 to 10 were used as chemical etch-resistant masks to a buffered HF solution that removed the chemical vapor deposited SiO_2 from the electron beam exposed areas. The most significant feature is the smoothness (~300 Å roughness) of the waveguide edges. The change in etch angle was not studied, but may be caused by variations in the CVD deposited SiO_2.

Waveguide and/or Masking Structures in Ion Beam Sputter-Deposited SiO_2

The resist liftoff technique is readily applied to sputter-deposited films provided that the thermal radiation

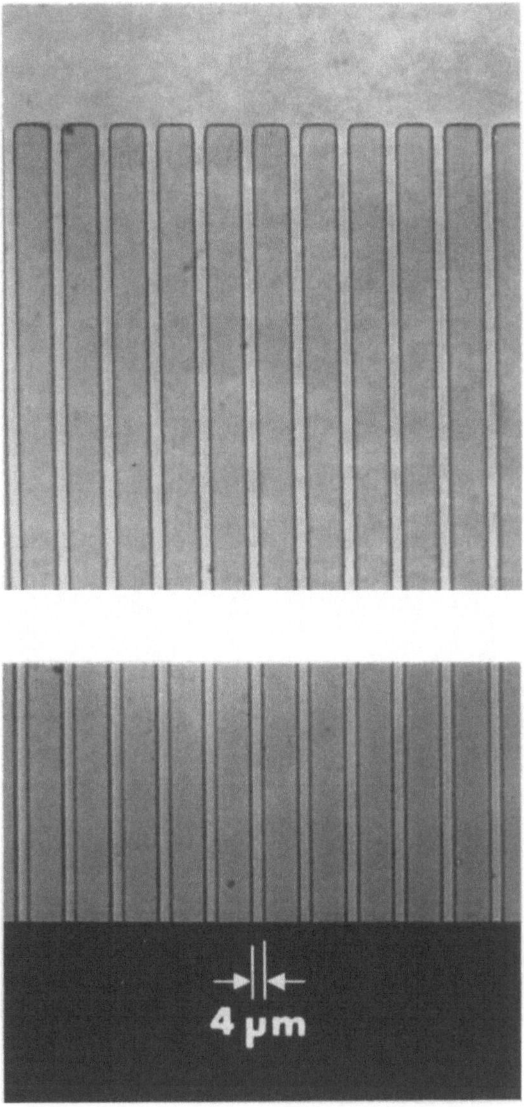

Fig. 7. Photomicrograph of Waveguide Structures Formed in Polymethyl Methacrylate After Development.

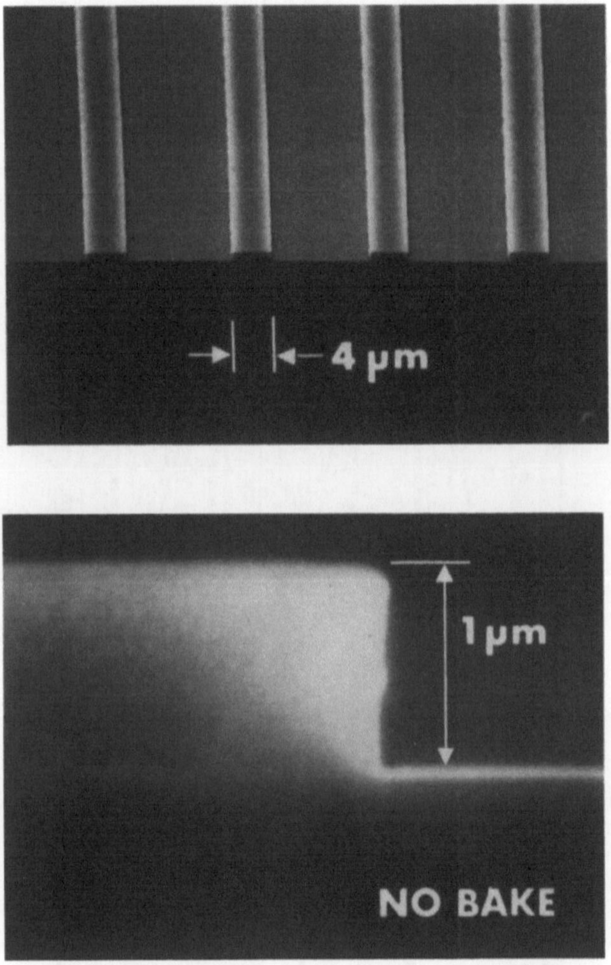

Fig. 8. Scanning Electron Micrographs of Waveguide Structures in Polymethyl Methacrylate with No Bake.

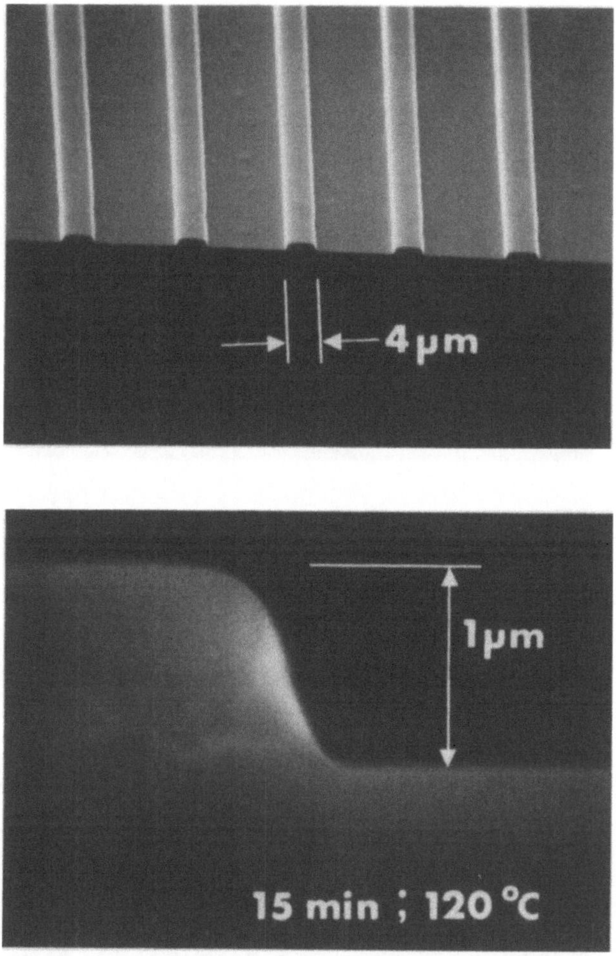

Fig. 9. Scanning Electron Micrographs of Waveguide Structures in Polymethyl Methacrylate After a 15 Minute Anneal at 120°C.

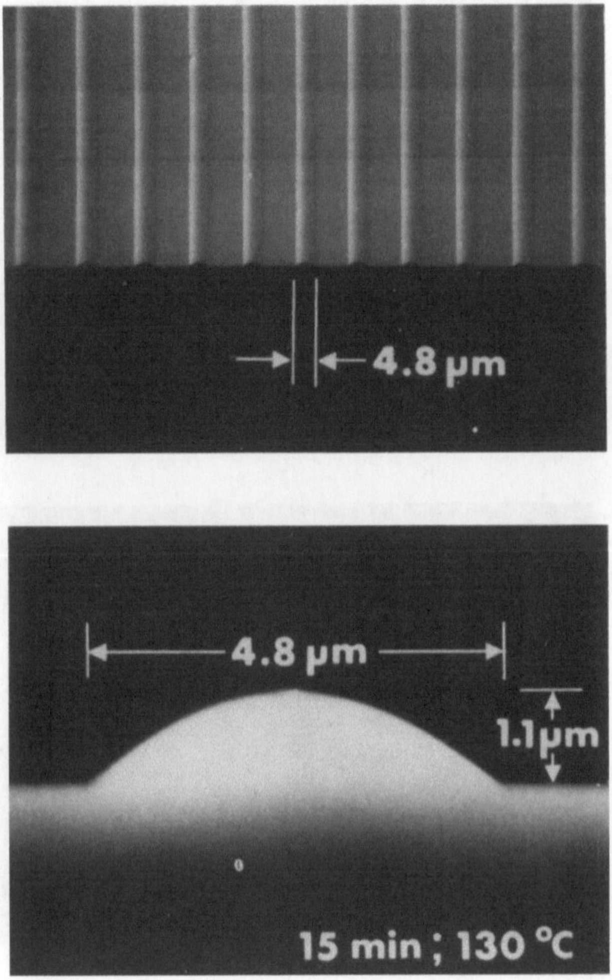

Fig. 10. Scanning Electron Micrographs of Waveguide Structures in Polymethyl Methacrylate After a 15 Minute Anneal at 130°C.

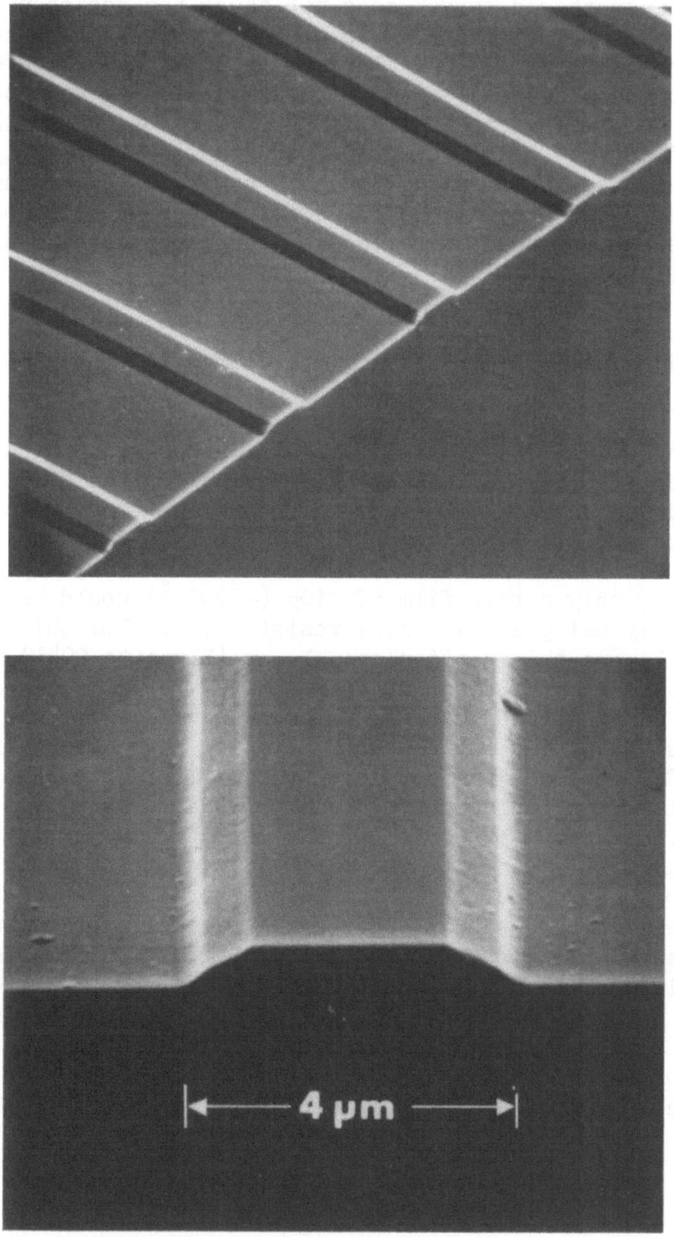

Fig. 11. Scanning Electron Micrographs of Chemically Etched Waveguide and/or Masking Structures in Chemical-Vapor-Deposited SiO_2.

from the sputtered target does not cause the resist profile to change. By using a water-cooled specimen holder and a low-deposition rate, 7059 glass was sputter deposited on electron beam exposed and developed masks. The unwanted resist and glass were lifted off using hot trichloroethylene to produce 7059 glass waveguide structures on thermally grown SiO_2 as shown in Fig. 12. In general, this method appears to be adaptable to guide formation. Edge smoothness is somewhat degraded compared with the other methods and is not fully understood.

Waveguide and/or Masking Structures in Ion Beam Sputter-Etched

In Fig. 13 a series of waveguide structures is shown that were fabricated in chemical-vapor-deposited SiO_2 using exposed and developed PMM as an ion beam sputtering mask. The edge profile is quite sharp and the edge smoothness is very good. However, because PMM has a rather high sputtering rate, only a thin film of SiO_2 (~5000 Å) could be sputtered away using a 1 μm thick resist layer. For thicker resist layers this particular polymer (Elvacite 2041) suffers from microcracking after development. A resist replacement method was mentioned earlier which, we believe, offers tremendous advantages for electron beam/ion beam microfabrication of waveguides and couplers and for microprocessing in general. This replacement scheme provides a mask type change. That is, a "positive" PMM mask becomes a "negative" aluminum mask after liftoff. Second, much thicker layers of SiO_2 (or ZnSe, CdSe, etc.) can be sputtered away because of the dramatic reduction in sputtering rate of the aluminum compared with the sputtering rate of PMM. Figure 14 shows a preliminary result using an aluminum replacement mask to shield a 3000 Å thick gold film from ion beam sputtering to form a simulated waveguide structure. In this example the aluminum was not dissolved off after sputtering. A slight angle change at the aluminum/gold interface produces the barely visible demarcation line between the two films.

Planar Couplers

Planar grating couplers for wavelengths in the visible demand very high resolution linewidths and minimum center-to-center spacing and, indeed, begin to press to the

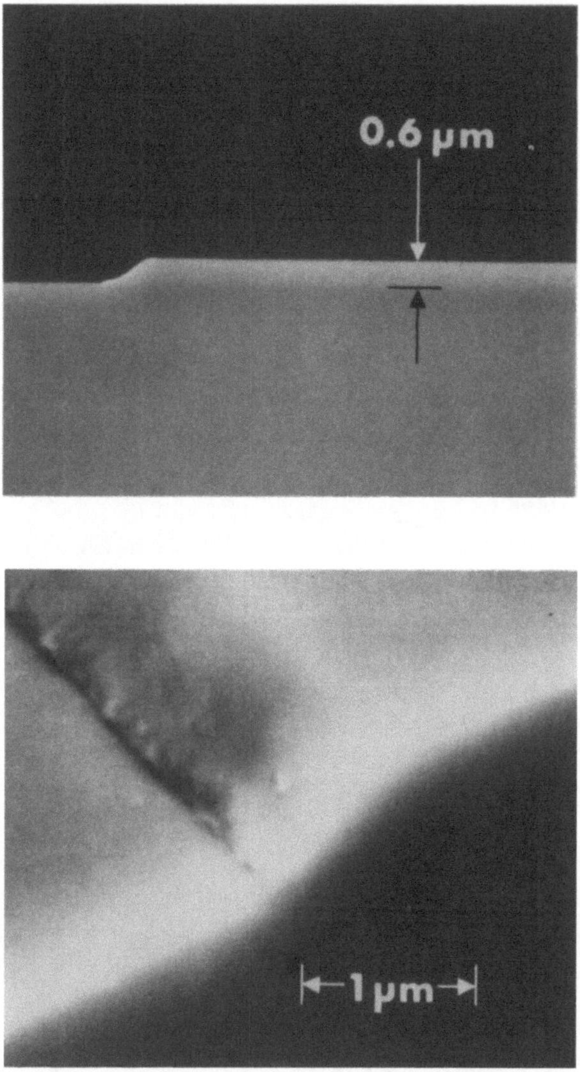

Fig. 12. Scanning Electron Micrographs of Sputter-Deposited 7059 Glass Waveguide Edge and/or Masking Structures.

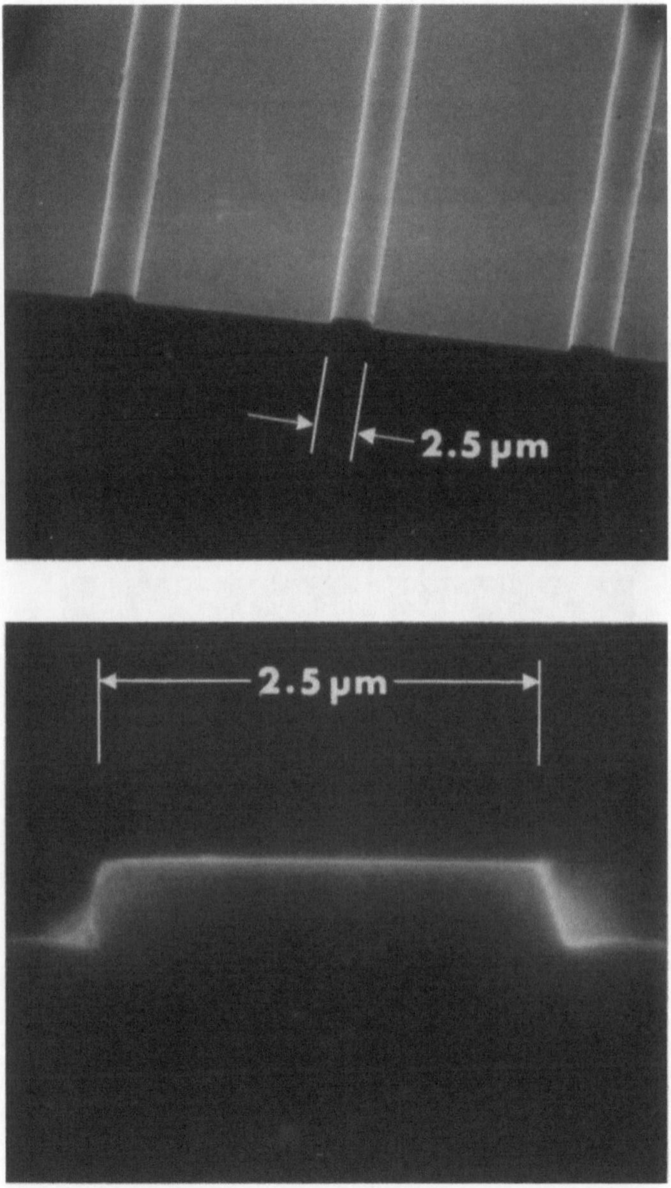

Fig. 13. Scanning Electron Micrographs of Ion Beam Sputter Waveguide and/or Masking Structures Using Polymethyl Methacrylate Resist Masks.

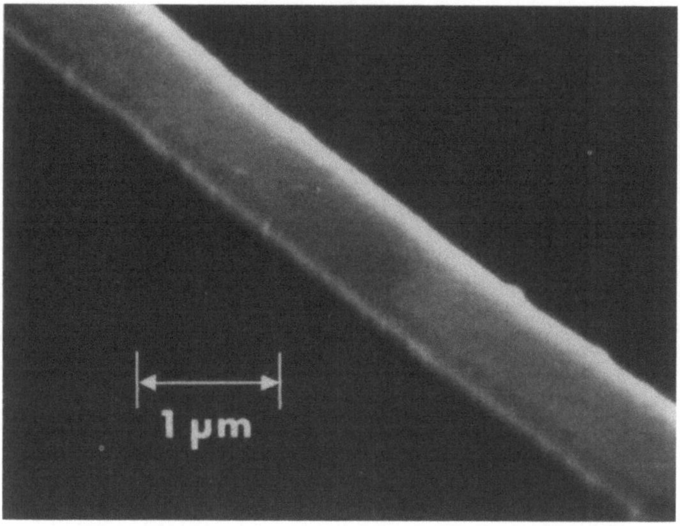

Fig. 14. Gold Line Structure Using the Aluminum Resist Replacement Mask Technique.

available state of art for electron beam lithography. This is particularly true for the case of a distributed feedback GaAs surface laser where a periodicity of 1100 Å is required (e.g., 550 Å wide liner or 1100 Å centers). The principal problem is the contribution of scattered electrons from adjacent line exposures. However, this problem can be attacked by optimization of the resist characteristics.

Two examples of gratings with periodicities about four times that required for a GaAs distributive feedback laser are shown in Fig. 15. The first type of grating, shown in Fig. 15 (a), was an amplitude planar grating with 1700 Å aluminum lines and a periodicity of 4200 Å. This grating was fabricated using method D in Fig. 4. The second type of grating, shown in Fig. 15 (b), was a planar phase grating with ion beam micromachined grooves which were ~1000 Å

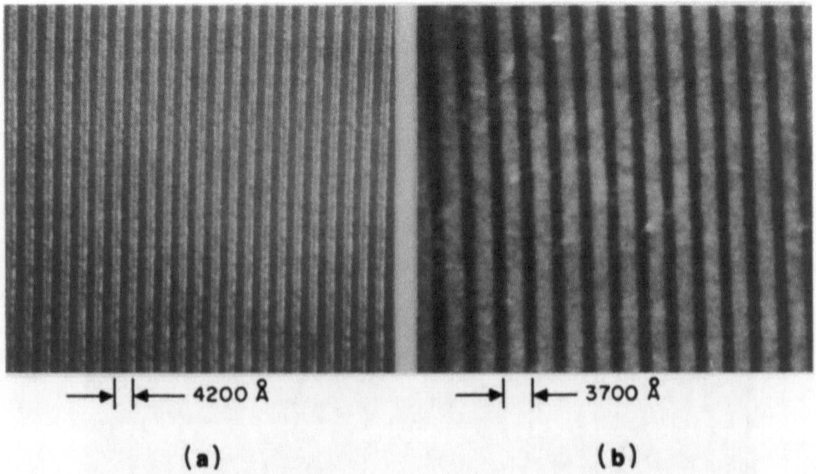

Fig. 15. Electron and Ion Beam Fabricated Planar Gratings. (a) Aluminum Amplitude Grating (b) Grooved Phase Grating on ZnS.

deep by ~1000 Å wide and with a periodicity of 3700 Å.[14,15] This grating coupler was fabricated using method C in Fig. 4.

CONCLUSIONS AND SUMMARY

Several other structures not based on initial pattern definition by electron beam exposure were presented during this lecture and are briefly summarized here. Optical guiding and coupling in proton bombarded gallium arsenside has been demonstrated.[9] A 1.5-μm thick gold mask which had a periodicity of 6 μm was used to mask against 300 keV protons. Other applications of fine gold lines are for wire-grid polarizers for infrared radiation.[16] Gold wire grid polarizers with a 1.1 μm-interval on CdTe have been fabricated using photolithography and ion beam sputtering.[17]

The results shown in this lecture have in part demonstrated the value of beam fabrication techniques in a long-range effort to develop processes for the fabrication of integrated optical elements ultimately into monolithic integrated optical circuitry using dominant mode guides for light of λ from ~0.4 to 1 μm; that is, guides of width of 1 μm and below, as well as wider (overmoded) guides. Scanning electron microscopy has revealed that several combinations of electron beam and ion beam microfabrication techniques can produce smooth and small waveguide structures in a variety of materials.

These submicrometer dimensions (~0.2 to 1 μm) and edge smoothness (~200 to 500 Å) are well beyond the capability of photolithography techniques. Beam fabrication processes being developed at Hughes and elsewhere will allow the fabrication of dominant mode integrated optics guides and elements that should have low transmission losses resulting from scattering and mode conversion. Waveguiding and coupling have been demonstrated in electron and ion beam fabricated waveguides. The task now lies in refining these thin film pattern definitions and fabrication techniques as well as in improving the control of interfacial smoothness of deposited and grown thin films.

ACKNOWLEDGMENTS

The author expresses appreciation to his colleagues F. S. Ozdemir, H. L. Garvin, and R. G. Wilson for use of technical information presented in the lecture. Also P. J. Coane, D. D. Loper, and R. Yim provided invaluable technical assistance.

REFERENCES

1. F. S. Ozdemir, E. D. Wolf, and C. R. Bucky, "Computer-Controlled Scanning Electron Microscope System for High-Resolution Microelectronic Pattern Fabrication," IEEE Transactions on Electron Devices ED-19, 624-628 (May 1972).

2. E. D. Wolf, F. S. Ozdemir, and R. D. Weglein, "Precision Electron Beam Microfabrication of Acoustic Surface Wave Devices," Invited Paper (J1), Presented at the 1973 IEEE Ultrasonics Symposium Monterey, California, Nov. 5-7, 1973.

3. More than 25 papers on electron beam systems and microfabrication applications are contained in the Proceedings of the 12th Symposium on Electron, Ion, and Laser Beam Technology, E. D. Wolf, Special ed., J. Vac. Sci. Technol. Vol 10, No. 6, Nov./Dec. 1973.

4. E. D. Wolf, F. S. Ozdemir, W. E. Perkins, and P. J. Coane, "Response of the Positive Electron Resist Elvacite 2041 to Kilovolt Electron Beam Exposure," in Rec. 11th Symp. Electron, Ion, and Laser Beam Technol., R. F. M. Thornley, ed., San Francisco, Calif., San Francisco Press, pp. 331-336, 1971.

5. M. Hatzakis, "Electron Resists for Microcircuit and Mask Production," J. Electrochem. Soc. 110, 1033-1037 (1969).

6. D. B. Ostrowsky and J. C. DuBois, "Electron Beam Masking for Optical Waveguide Fabrication," Digest of the Optical Society of America Topical Meeting on Integrated Circuits (Optical Society of America, Washington, D.C., 1972), p. MB7.

7. M. Hatzakis, S. Magdo and C. H. Ting, "Electron Beam Fabrication of Micron Transistors," Proceedings of 4th International Conference Electron and Ion Beam Science and Technology, Los Angeles, California, May 1970, p. 524-530.

8. J. E. Goell, Digest of the Optical Society of America Topical Meeting on Integrated Circuits (Optical Society of America, Washington, D.C., 1972, p. MB8, and J. E. Goell, R. D. Standley, "Ion Bombardment Fabrication of Optical Waveguides Using Electron Resist Masks," Appl. Phys. Lett. 21, 72-73 (1972).

9. S. Somekh, E. Garmire, A. Yariv, H. L. Garvin, and R. G. Hunsperger, Appl. Phys. Lett. 22, 46 (1973).

10. H. L. Garvin, "High Resolution Fabrication by Means of Ion Beam Sputtering," Paper Presented at Kodak Microelectronics Seminar, San Diego, Calif., December 11-12, 1972.

11. E. H. Hasseltine, F. C. Hurlbut, N. T. Olson, and H. P. Smith, "Cesium Ion Bombardment of Al_2O_3 in a Controlled Oxygen Environment," J. Appl. Phys. 38, 4313 (1967).

12. M. Marcal, Thomson CSF, France, Private Communication.

13. The results reported here were obtained from work supported in part by the Naval Electronics Laboratory Center, Contract N00123-73-C-0231.

14. E. D. Wolf, "Computer-Controlled Electron Beam Microfabrication," Invited Paper in Proceedings Seventh National Conference on Electron Probe Analysis, San Francisco, Calif., July 17-21, 1972, p. 24A-24G.

15. H. L. Garvin, E. Garmire, S. Somekh, H. Stoll, and A. Yariv, "Ion Beam Micromachining of Integrated Optics Components," Appl. Optics 12, 455-459 (1973).

16. J. P. Auton, Appl. Opt. 6, 1023 (1967).

17. H. L. Garvin and J. E. Kiefer, "Wire-Grid Polarizers for 10.6 μm Radiation," Digest of Technical Papers 1973 IEEE/IOA Conference on Laser Engineering and Applications, Washington, D.C., May 30-June 1, 1973, Paper 20.9 p. 100.

Chapter 8 - INTRODUCTION TO OPTICAL WAVEGUIDE FIBERS

Robert D. Maurer

Corning Glass Works
Corning, New York 14830

The greatest potential for fiber waveguides is long distance transmission of optical signals (10 to 10,000 meters). Therefore, this chapter will primarily treat their properties as transmission devices. The three practical factors for a signal link are its information carrying capacity (bits per second), its attenuation (how far the signal carries) and its ability to withstand the system environment. The information capacity will be emphasized with only a few remarks on attenuation since it has been covered in reviews.[1] Environmental factors are still in very early stages of investigation. However, it is not possible to cover thoroughly all the background theory that has been published on information carrying capacity of dielectric waveguides. Instead the main topics will be introduced in a way that hopefully provides a lasting foundation for more rigorous and technical development. New aspects are still being uncovered by research and it is not possible to foresee how important all these will become and which should be emphasized now. The theory is similar to that for other waveguides discussed in this book which should provide some helpful overlap for the reader. An important difference is the geometry (cylindrical) and the necessity for emphasizing wave propagation over long distance.

I. PROPAGATION

1.1 Ray Theory

The use of ray theory for optical waveguides seems incongruous as, in many ways, it is. An ultimate goal is a mode theory for these devices. However, ray theory provides a good introduction to the physical phenomena occurring in a multimode waveguide and thus provides many experimentalists with a more easily understood description. In addition, a ray description of the light source is the most straightforward for excitation at the entrance face of the waveguide. If one attempts to use modes for this problem, it is necessary to expand the incident light source intensity in the mode field distributions--a generally difficult approach for multimode waveguides. Some remarks will be made from time to time which try to tie the concepts of modes and rays together.

Two types of waveguides will be considered: the step-refractive index and gradient-refractive index. The former is the older and is in many ways the easier to understand. It contains a core material of one refractive index and a cladding of another. The latter contains a high refractive index along the axis which decreases to the edge of the core. The core may or may not be covered by a cladding. Ray trajectories are curves in the gradient waveguide. The description of these is provided by a differential equation which is derived in many standard texts.[(2),(3)] Let ds be a unit of length along the ray trajectory and $\underline{r}$ a vector from an arbitrary origin to ds (Figure 1). Then,

$$\frac{d}{ds}\left(n\,\frac{d\underline{r}}{ds}\right) = \text{grad } n \qquad (1)$$

For axially symmetric problems, Eq. (1) is sometimes written with ds replaced by dz, where z is the direction of the axis. This is the paraxial approximation which often greatly simplifies calculation, although it will not be necessary here. Eq. (1) is evaluated through unit vectors $\underline{h}$ (r-direction), $\underline{k}$, (z direction) and $\underline{m}$ (θ direction). The three relations

$$\underline{r} = r\underline{h} + z\underline{k} \qquad \frac{d\underline{h}}{d\theta} = \underline{m} \quad \frac{d\underline{m}}{d\theta} = -\underline{h}$$

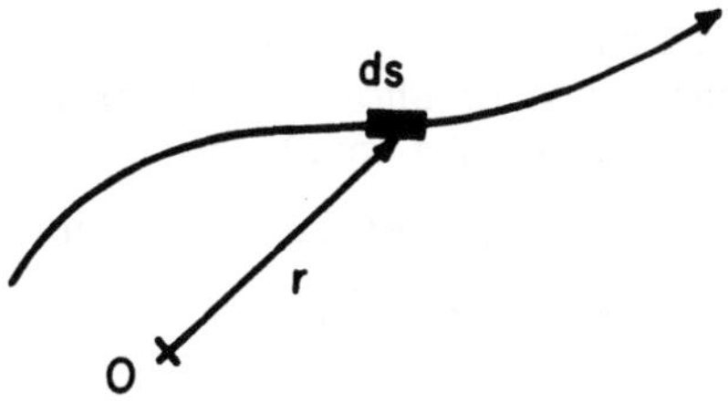

Fig. 1 Path of a ray in a medium of inhomogeneous refractive index.

are used in reducing Eq. (1) to

$$\frac{d}{ds}\left(n\,\frac{dr}{ds}\right) - nr\left(\frac{d\theta}{ds}\right)^2 = \frac{dn}{dr} \qquad r \text{ component} \qquad (2)$$

$$n\,\frac{dr}{ds}\,\frac{d\theta}{ds} + \frac{d}{ds}\left(nr\,\frac{d\theta}{ds}\right) = 0 \qquad \theta \text{ component} \qquad (3)$$

$$\frac{d}{ds}\left(n\,\frac{dz}{ds}\right) = 0 \qquad z \text{ component} \qquad (4)$$

which must be solved to obtain the coordinates of the ray path.

1.2 Rays in Step Refractive Index Waveguides

Rays are often classified into two categories--meridional rays which pass through the axis of the waveguide (at some point in their path) and skew rays which do not intersect the axis. The optical power carried by meridional rays is minimal but they are continually used for explaining the properties of waveguides, perhaps because they are simplest to analyze. An illustration for a typical meridional ray is shown in Figure 2. The velocity of the ray along the guide is $c \cos \varphi / n_1$. Since the critical angle gives a maximum value $\cos \varphi = (1 - \frac{\Delta n}{n_1})$ with $\Delta n \equiv (n_1 - n_2)$, the slowest ray travels with a phase velocity

$$v_m = \frac{c}{n_1}\left(1 - \frac{\Delta n}{n_1}\right) \tag{5}$$

where c is the velocity of light in vacuum.

An illustration for a typical skew ray is shown in Figure 3. The reflection of the ray is equivalent to reflection from a plane tangent to the fiber core at the point of intersection. This plane intersects the core along a line parallel to the fiber axis. A little consideration shows that a skew ray maintains its angle with respect to this line of intersection, and therefore to the axis of the fiber, on successive reflections. Those having an angle with respect to the axis within the permissible range of angles for meridional rays (Figure 2) will travel at the same phase velocity as the meridional ray at that angle. Thus, the consideration of meridional rays alone often gives approximately the correct answer even though they carry little optical power. On the other hand, skew rays propagating near the boundary of the core require more

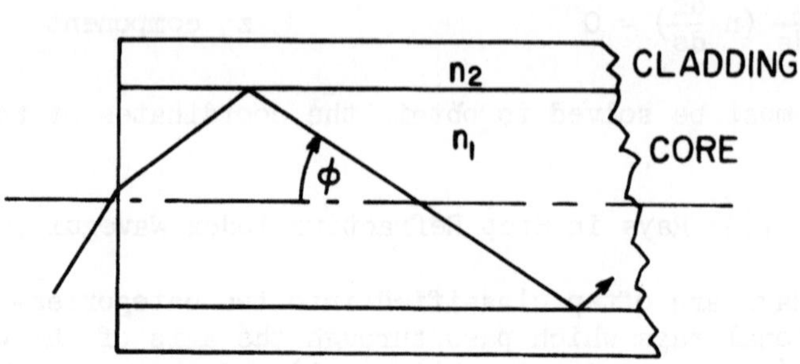

Fig. 2 Typical path of a meridional ray in a step refractive index fiber.

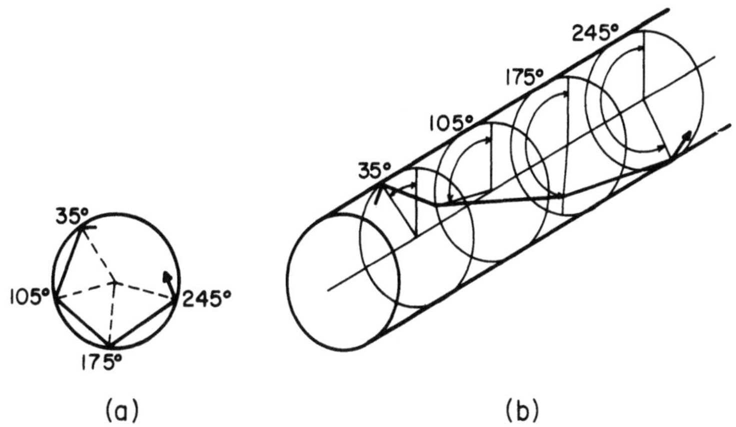

Fig. 3 Typical path of a skew ray in a step refractive index fiber a) end view b) perspective view.

consideration. The farther a skew ray is from the fiber axis, the higher the angle with respect to the axis at which it may still be below the critical angle and reflect. In the thin shell adjacent to the fiber cladding, rays can propagate at right angles to the fiber axis--their phase velocity down the guide is zero. Therefore the phase velocity for skew rays varies from zero to c/n_1. Figure 4 shows the fractional power of a plane wave at angle φ to the axis which is transmitted down the fiber. The decreasing segments at high angles are skew rays.

In practice, high angle rays can often be neglected because they are so highly attenuated. The ray theory has been used to calculate pulse shapes in step refractive index guides without attenuation for various input conditions.[4] High angle skew rays were omitted but, as shown by Figure 4, in most experiments they carry little power.

1.3 Ray Theory for Gradient Refractive Index Waveguides

The ray theory for gradient refractive index is somewhat more difficult to understand intuitively. Basically, what follows is an examination of focussing conditions in

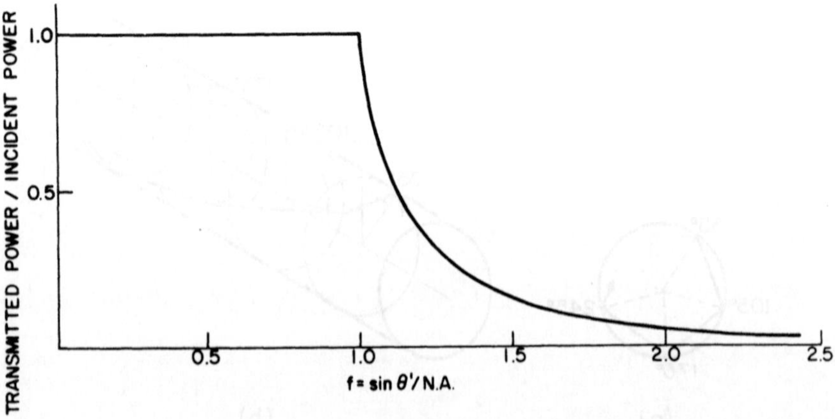

Fig. 4 The fractional power transmitted for plane waves incident on the core of a step refractive index fiber. This is plotted against the universal factor f where N.A. is the numerical aperture and θ' is the angle between the fiber axis and the plane wave propagation direction. Plane waves excite both meridional and skew rays but only skew rays are excited beyond the numerical aperture (f=1) of the fiber.

these fibers. It is well known that focussing is equivalent to equal phase[2] so a gradient focussing all incident rays at the same point would be a dispersionless structure. This condition can only be approximated although it can be reached exactly for certain classes of rays.

Equations (3) and (4) can be directly integrated, where the integral of (4) permits the elimination of ds with

$$ds = \frac{n(r)}{n(r_o)} \left.\frac{ds}{dz}\right|_o dz$$

and the integral of (3) substituted in (2) gives

$$\frac{d^2r}{dz^2} = \frac{d}{dr}\left[b^2n^2(r) - \frac{a^2}{2r^2}\right]$$

with $a \equiv r_o^2 \frac{d\theta}{dz}\Big|_o$ and $b \equiv \frac{\sqrt{2}}{n(r_o)} \frac{ds}{dz}\Big|_o$

Noting that the left hand side is $\frac{1}{2} d(\frac{dr}{dz})^2$ gives

$$\frac{dz}{dr} = \pm \{\frac{n^2(r)}{n^2(r_o)} [r_o^2 (\frac{d\theta}{dz}\Big|_o)^2 + (\frac{dr}{dz}\Big|_o)^2 + 1)]$$

$$- \frac{1}{r^2} [r_o^2 \frac{d\theta}{dz}\Big|_o]^2 - 1\}^{-1/2} \tag{6}$$

The double sign shows that the ray path is always periodic in r. For the same value of r the slope has both positive and negative values. Integrating Eq. (6)

$$z(r) = \int_{r_o}^{r} \{\frac{n^2(r)}{n^2(r_o)} [r_o^2 (\frac{d\theta}{dz}\Big|_o)^2 + (\frac{dr}{dz}\Big|_o)^2 + 1]$$

$$- \frac{1}{r^2} [r_o^2 \frac{d\theta}{dz}\Big|_o]^2 - 1\}^{-1/2} dr \tag{7}$$

Equations (6) and (7) can be used to illustrate all the special cases that have been solved. The solutions will not be derived but they will be shown consistent with these equations.

a) Skew Rays(5)(6) A special class of skew rays which enter with $\frac{dr}{dz}\Big|_o = 0$ are to be focussed. To solve for the index distribution which keeps these in a helical path, set $dr/dz = 0$ in Eq. (6). Then

$$\frac{n^2(r)}{n^2(r_o)} = \frac{1 + \frac{r_o^4}{r^2} (\frac{d\theta}{dz}\Big|_o)^2}{1 + r_o^2 (\frac{d\theta}{dz}\Big|_o)^2} \tag{8}$$

In order to have a bound ray with the refractive index higher in the center than at large r_o, the expression consistent with Eq. 8 must be

$$n^2(r_o) = \frac{\text{constant}}{1 + \frac{r_o^4}{r^2}\left(\frac{d\theta}{dz}\Big|_o\right)^2}$$

This is independent of r only when $r = r_o$ as postulated for the helical path. Then

$$n^2(r) = \frac{\text{constant}}{1 + r^2\left(\frac{d\theta}{dz}\Big|_o\right)^2}$$

but $r^2\left(\frac{d\theta}{dz}\Big|_o\right)^2$ is the skewness of the ray so that any particular waveguide can focus only one class of skew rays although at all values of r.

b) Meridional rays(7)(8) There is a refractive index distribution which focusses all meridional rays, $\left(\frac{d\theta}{dz}\Big|_o\right) = 0$, given by $n(r) = n(0)\ \text{sech}\ \frac{\pi r}{2F}$. To show this, consider rays at $z = 0$ parallel to the axis so $\frac{dr}{dz}\Big|_o = 0$. These intersect the axis at a focus F. We substitute $n(r) = A\ \text{sech}\ Br$ in Eq. (7) and see whether the focus F is independent of r_o.

$$F = \text{sech}\ Br_o \int_o^{r_o} \frac{dr}{[\text{sech}^2 Br - \text{sech}^2 Br_o]^{1/2}}$$

which gives $F = \frac{\pi}{2B}$ as required.

A second interesting case for meridional rays is the index distribution that gives sinusoidal paths, $R = A\ \sin Bz$. the integral in Eq. (7) can be analyzed to give a solution of this type when $n^2(r) = \alpha^2(1 - \frac{Br^2}{\cos\gamma_o})$ where γ_o is the initial angle of the ray with respect to the axis. Thus, this type trajectory gives the same focussing distance for all rays only in the paraxial approximation when $\cos\gamma_o \simeq 1$.

There is an important point to be deduced from all these special cases. When the refractive index distribution changes only a little across the waveguide so that the distributions can be expanded to first order, then

$$\text{Skew rays} \qquad n^2(r) \cong C\,[1 - (\tfrac{2\pi}{F'})^2 r^2] \qquad (F \tfrac{d\theta}{dz}\Big|_o = 2\pi)$$

Meridional rays

$$\text{Perfect focus} \qquad n^2(r) \cong C\,[1 - (\tfrac{2\pi}{F'})^2 r^2] \qquad (F' = 4F) \qquad (9)$$

$$\text{Sinusoidal path} \qquad n^2(r) \cong C\,[1 - (\tfrac{2\pi}{F'})^2 r^2] \qquad (2\pi F' = B)$$

where F' is the distance for the ray to return to its original point in the guide, or make a full cycle. This series termination is similar to using the paraxial approximation. It is not surprising that a parabolic refractive index contour is almost perfectly focussing and some waveguides are fabricated with this as a goal.

1.4 Mode Theory

A knowledge of modes in optical waveguides is a more complete approach to understanding these structures. This knowledge is derived from solutions of Maxwell's equations, explained in Chapter I. However, a complete treatise on this topic is not possible here and it has been covered widely in other publications. Instead, this section will outline the approach to the problem, provide salient results along the way and sketch the final results obtained. References will indicate where more details can be found. This is intended to give an overview and enough grasp of the theory to work with the results.

Initial simplification of Maxwell's equations is carried out in rectangular coordinates. For dielectrics with $\mu = \mu_o$

$$\text{curl } \underline{E} = -\frac{\partial \underline{B}}{\partial t} \qquad \text{div } \underline{B} = 0$$

$$\text{curl } \underline{H} = -\frac{\partial \underline{D}}{\partial t} \qquad \text{div } \underline{D} = 0$$

$$\underline{D} = \epsilon \underline{E} \qquad \underline{B} = \mu_o \underline{H}$$

These are reduced to the wave equation by taking curl curl $\underline{E}$ and using a vector identity to give[3]

$$\nabla^2 \underline{E} + \text{grad}\ (\underline{E}\cdot \frac{\text{grade}\ \epsilon}{\epsilon}) = \epsilon\, \mu_o \frac{\partial^2 \underline{E}}{\partial t^2} \qquad (11)$$

Only when grad $\epsilon = 0$ do we have the standard wave equation

$$\nabla^2 \underline{E} = \epsilon\, \mu_o \frac{\partial^2 \underline{E}}{\partial r^2} \qquad (12)$$

Thus, the standard equation is valid in both core and cladding individually for a step-refractive index waveguide. It will also be used for the gradient-refractive index waveguide as an approximation. Consider the z component of Eq. (11). The gradient of the refractive index has only an r component, so that the wave equation for E_z has a term like $\text{grad}_z\ (E_r \frac{1}{\epsilon}\frac{\partial\epsilon}{\partial r})$ which is dropped. Since E_r varies in the z direction as the guide wavelength, it is the ratio of the E_r coefficient to the wavelength which determines the total magnitude of the term. Therefore, this approximation is similar to that giving Eq. (9) in that $\frac{1}{\epsilon}\frac{\partial\epsilon}{\partial r}$ is small in both cases. Eq. (11) can, if desired, be solved by iteration in which E_r from the solution of Eq. (12) is used.

The transformation to cylindrical coordinates is straightforward for E_z and H_z since they remain the same as E_z and H_z in rectangular coordinates. These components are also particularly suitable for expressing the others. Assuming wave solutions of the type

$$\underline{E} = \underline{E}_o\ (r,\theta)\ e^{i(\omega t - kz)}$$

for $\underline{E}$ and $\underline{H}$, substituting these in the curl equations (10) changed to cylindrical coordinates, and simplifying gives

$$E_r = -\frac{i}{K^2}\ (\frac{\mu\omega}{r}\frac{\partial H_z}{\partial\theta} + \beta\,\frac{\partial E_z}{\partial r}$$

Similarly:

$$E_\theta = -\frac{i}{K^2}\left(\beta\frac{1}{r}\frac{\partial E_z}{\partial\theta} - \mu\omega\frac{\partial H_z}{\partial r}\right)$$

$$H_r = -\frac{i}{K^2}\left(\beta\frac{\partial H_z}{\partial r} - \epsilon\omega\frac{1}{r}\frac{\partial E_z}{\partial\theta}\right)$$

$$H_\theta = -\frac{i}{K^2}\left(\frac{\beta}{r}\frac{\partial H_z}{\partial\theta} + \omega\epsilon\frac{\partial E_z}{\partial r}\right)$$

which can be used to determine E_r, E_θ, H_r and H_θ. Here, $K^2 = \omega^2 \epsilon\mu - k^2$ and $\omega^2 \epsilon\mu = \kappa^2$ where $\kappa(r)$ is a wave vector in a bulk medium of dielectric permittivity ϵ.

The wave Equation (12) is expressed in cylindrical coordinates (9) and a solution $E_z = F(r)\, e^{i\nu\theta}$ substituted. The result becomes

$$\frac{d^2F}{dr^2} = \frac{1}{r}\frac{dF}{dr} + \left(\omega^2\epsilon\,\mu_o - k^2 - \frac{\nu}{r^2}\right) F = 0 \qquad (13)$$

This is to be solved to obtain k, the waveguide propagation constant, and F in order to obtain the electric and magnetic fields. It is apparent that a similar equation will follow for H_z.

1.5 Mode Theory of Step Refractive Index

The step refractive index waveguide is one of the few dielectric waveguide problems which has been solved completely.(10)(13) All the mode fields and propagation constants, k, can be found. The total results are extensive and their salient features can be used for understanding the character of solutions for other, more complicated, refractive index profiles.

With K a constant in the core and in the cladding, the solutions to Eq. (13) are Bessel functions. The particular type of Bessel function is chosen to assure a finite solution at r = 0 and a solution approaching zero as $r \to \infty$.

$$E_z = AJ_\upsilon(\frac{ur}{a})e^{i\upsilon\theta}$$

core $$K^2 \equiv (\frac{u}{a})^2 = \varkappa_1{}^2 - k^2$$

$$H_z = BJ_\upsilon(\frac{ur}{a})e^{i\upsilon\theta}$$

(14)

$$E_z = CH_\upsilon{}^{(1)}(\frac{iwr}{a})e^{i\upsilon\theta}$$

cladding $$-K^2 \equiv (\frac{w}{a})^2 = k^2 - \varkappa_2{}^2$$

$$H_z = DH_\upsilon{}^{(1)}(\frac{iwr}{a})e^{i\upsilon\theta}$$

Note that $u^2 + w^2 \equiv v^2 = a^2(\varkappa_1{}^2 - \varkappa_2{}^2) = (\frac{2\pi a}{\lambda_o})^2(n_1{}^2 - n_2{}^2)$ is a constant for all modes and is characteristic of the waveguide. This characteristic constant is often used as a parameter to graphically portray properties of the waveguide.

Several important points can be deduced from this solution as it stands. First, the solution in the cladding goes as exp $-r(k^2-\varkappa_2{}^2)^{1/2}$ for large r. In order for the wave to disappear, the phase velocity k must be greater than $\varkappa_2$. When $k \leqq \varkappa_2$ the wave oscillates at large distance, i.e., it is no longer bound to the core and the energy is lost. Thus, $k = \varkappa_2$ is the cut-off condition for any given mode. Inside the core, u/a is real in order to have oscillatory solutions ($k \leq \varkappa_1$). Therefore the phase velocity of the wave varies from $\varkappa_2$ to $\varkappa_1$ depending on how far the mode is from cut-off. The "phase refractive index," defined from the ratio of k to the k in free space at the same frequency, varies in a corresponding way. Far from cut-off, $k > \varkappa_2$, the wave dies rapidly in the cladding and most of the power is carried in the core.

A classification of the various modes follows from the nature of the electromagnetic fields. Substituting E_z and H_z in the equations obtained from the curl gives the other component of $\underline{E}$ and $\underline{H}$. Equating E_r, E_θ, H_r, H_θ at the boundary r = a gives four equations in A, B, C and D (plus k). Thus, all can be determined in terms of a single given (initial) field amplitude. Since the four equations are homogeneous, they only have a solution if the determinant of the coefficients vanishes--this gives the eigenvalue equation that determines k. After much algebra this characteristic equation emerges:

$$[(\frac{\epsilon_1}{\epsilon_2}\frac{w^2}{u}\frac{J_\upsilon'(u)}{J_\upsilon(u)} + (iw\frac{H_\upsilon^{(1)'}(iw)}{H_\upsilon^{(1)}(iw)})]$$

$$[\frac{w^2}{u}\frac{J_\upsilon'(u)}{J_\upsilon(u)} + (iw\frac{H_\upsilon^{(1)'}(iw)}{H_\upsilon^{(1)}(iw)})] \tag{15}$$

$$= [\upsilon(\frac{\epsilon_1}{\epsilon_2} -1)\frac{k\kappa_2 a}{u^2}]^2$$

Consider first those modes with $\upsilon = 0$. The characteristic Eq. (15) is zero on the right hand side and setting each expression in brackets equal to zero results in two sets of modes. Analysis of the amplitudes shows that these are transverse electric (TE with $E_z = 0$) and transverse magnetic (TM with $H_z = 0$). The Bessel function J_o is an oscillating function with r. The modes have an increasing number of radial maxima as "higher order" is included. At some greatest number of maxima, this set of modes will terminate. Since $\upsilon = 0$, the modes are radially symmetric. They are the analog of a uniform radial density of meridional rays. At higher angle the uniform density of rays (waves) interferes through phase differences to form radial rings of maxima corresponding to higher order modes.

For $\upsilon \neq 0$ the modes are hybrid with both H_z and E_z components. This introduces a θ dependence (Eq. 13) which must be generated by skew ray contributions. Higher order Bessel functions have small amplitude over a large range surrounding the fiber axis. Thus, these modes have more power near the cladding but with many maxima and minima in the θ direction. Such modes are a good illustration of the high angle skew ray characteristics discussed under Ray Theory.

Figure 5 shows the result of detailed calculations. There is one mode, the HE_{11} which has no cut-off. When a waveguide is constructed so that $V < 2.4$ at the operating light wavelength, only this one mode propagates and the waveguide is referred to as single mode or monomode.

1.6 Mode Theory of Gradient Refractive Index Fibers

The step refractive index waveguide is the only type which has been solved in all its details. The gradient refractive index is more difficult. Equation (13) is still

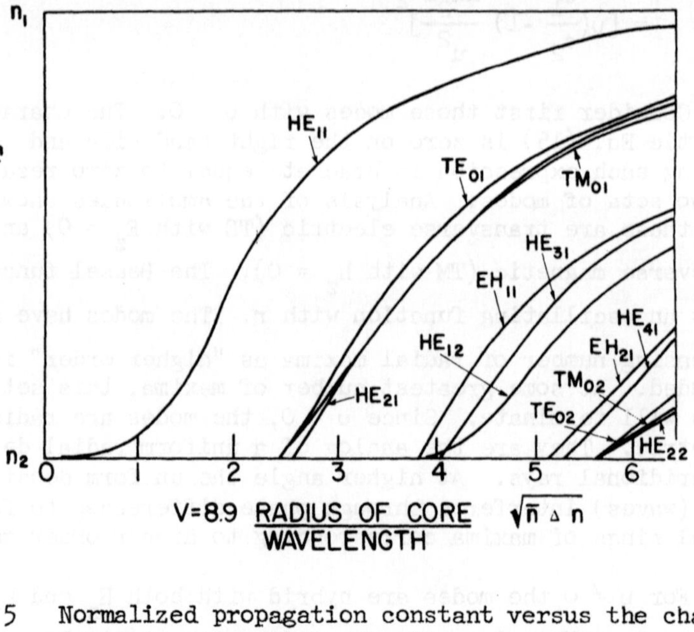

Fig. 5 Normalized propagation constant versus the characteristic parameter v for low order modes in a step refractive index waveguide.

used as a starting point but ϵ is now a function of r. Approximate solutions of this equation have been obtained in the form of Hermite polynomials when the refractive index varies in a parabolic fashion.[3]

An important purpose for solving Eq. (13) is to obtain the phase velocities of the various modes in order to predict pulse broadening (see below). This can be done without finding the electric and magnetic fields. Instead, Eq. (13) is analyzed by the WKB method.[11] For gradients which vary as some power of the radius, an approximate relation has been found for an arbitrary mode. These modes, m, are ordered from 1 to M. For the parabolic variation given by

$$n(r) = n_1 \; [1-2\Delta \; (\frac{r}{a})^2]^{1/2},$$

it is found that

$$k^2 = \varkappa_1^{\;2} \; [1-2\Delta \; (\frac{m}{M})^{1/2}]$$

Assuming the second term in the bracket is small compared to unity (small refractive index change), gives

$$n(m) \cong n_1 \; -(\frac{m}{M})^{1/2} \; \Delta n$$

for the phase refractive index for any particular mode.

II. INFORMATION CAPACITY

Information capacity will be considered from the standpoint of digital transmission. The important point to analyze is broadening of the digital pulses as they travel down the transmission line. This causes overlap between successive pulses at the receiver and encumbers discrimination between them. The degree of tolerable overlap will depend upon system performance in desired signal-to-noise ratio and upon the pulse shape. A representative overlap will be chosen and the shape assumed as gaussian.

The first section discusses single mode waveguides. It will be shown that the pulse broadening depends on the spectral width of the source. Both origins of spectral width will be included, the spectral width due to modulation and the continuous wave width. The second section deals with multimode waveguides in which broadening is due to group velocity differences between the various modes. The broadening in each individual mode follows the principles of the first section but is, in general, neglected along with any coupling, or transfer of energy between modes.

2.1 Pulse Broadening in Single Mode Fibers

A constant frequency source travels down the guide as a wave given by the solutions of Eq. (12). The variation of the phase velocity, k, for a signal mode of many frequencies must be properly considered. As definitions for Fourier transforms used in this analysis

$$F(\omega) = \frac{1}{\sqrt{2\pi}} \int_{-\infty}^{\infty} f(t)e^{i\omega t}\, dt$$

$$f(t) = \frac{1}{\sqrt{2\pi}} \int_{-\infty}^{\infty} F(\omega)e^{-i\omega t}\, d\omega$$

For a gaussian pulse in time used to modulate a carrier of frequency ω_o:

$$f(t) = e^{-i\omega_o t}\, e^{-\frac{1}{2}(\sigma t)^2} \quad \text{and } F(\omega) = \frac{1}{\sigma}\, e^{-\frac{1}{2}\left(\frac{\omega-\omega_o}{\sigma}\right)^2} \tag{16}$$

Consider a light source which has a broad frequency distribution when run continuously. Let the amplitude sum in any small frequency range $d\omega$ be

$$\frac{1}{s\sqrt{2\pi}}\, e^{-\frac{1}{2}\left(\frac{\omega-\omega_o}{s}\right)^2}$$

When this is modulated by

$$f(t) = e^{-i\omega t} e^{-\frac{1}{2}(\mu t)^2}$$

the resultant is

$$f(t) = \frac{e^{-\frac{1}{2}(\mu t)^2}}{\sqrt{2\pi}} \int_{-\infty}^{\infty} e^{-i\omega t} \frac{1}{s} e^{-\frac{1}{2}\left(\frac{\omega-\omega_o}{s}\right)^2} d\omega$$

$$= e^{-\frac{1}{2}(\phi t)^2} \quad \text{where } \phi^2 = \mu^2 + s^2 \qquad (17)$$

with the transform of this representing the frequency spectrum of the modulated source.

A digression is necessary here. Equation (17) gives a time dependence to the source even in the absence of modulation ($\mu = o$) which is contrary to our assumption. This is artificial and will have to be ignored in the final answer even though it has physical significance. If a constant frequency source shifts phase at some particular time, this shift is equivalent to introducing other frequencies which add to change the wave at that time. For an incoherent light source, Eq. (17) thus represents the random phases within the spread of frequencies of the source. This can be demonstrated just by the transform of waves with phases $\varphi(\omega)$

$$\int e^{i\omega t} e^{i\varphi(\omega)} d\omega = e^{-\frac{1}{2}(st)^2}$$

where $e^{i\varphi(\omega)}$, comparing Eq. (17) is the same as the source gaussian spectral form.

The time dependent gaussian pulse of Eq. 17 is to be transmitted down an optical waveguide of length L. The output, q(t), is

$$q(t) = \frac{1}{\sqrt{2\pi}} \int_{-\infty}^{\infty} F(\omega) e^{ik(\omega)L} e^{-i\omega t} d\omega \qquad (18)$$

where attenuation is neglected.

The approximation used for $k(\omega)$ is the expression

$$k(\omega) = k_o + \frac{dk}{d\omega}\Big/_{\omega_o} (\omega-\omega_o) + \frac{1}{2}\frac{d^2k}{d\omega^2}\Big/_{\omega_o} (\omega-\omega_o)^2 + \ldots .$$

with only these terms retained. Substituting the transform of (17) for $F(\omega)$ and evaluating (18) gives

$$q(t) = e^{i(\omega_o t + k_o L)} \exp\left[\frac{t-L/\frac{d\omega}{dk}}{2(iL\frac{d^2k}{d\omega^2} - \frac{1}{\phi^2}}\right]^2$$

The gaussian pulse is now centered at $t = L/\frac{d\omega}{dk}$. Thus, $d\omega/dk$ is the group velocity of the pulse. The pulse width has increased and can be calculated from the power $|q(t)|^2$. If we consider the initial pulse width as $\tau_1 \equiv \frac{1}{\phi}$ from Eq. (17), then the output pulse width is

$$\tau_2 = [\frac{1}{\phi^2} + (L\phi \frac{d^2k}{d\omega^2})^2]^{1/2} \tag{19}$$

In the absence of dispersion in the waveguide ($\frac{d^2k}{d\omega^2} = 0$), the output pulse is the same as the input. Thus, we drop the s^2 part of ϕ^2 in the first term, for reasons noted above.

$$(\tau_2)^2 = \frac{1}{\mu^2} + (\mu^2+s^2)\,(L\frac{d^2k}{d\omega^2})^2 \tag{20}$$

The factor $d^2k/d\omega^2$ depends upon the combined contribution of guide characteristics and dispersion in the material. The curves of Fig. 5 which show the guide characteristics were calculated neglecting the dispersion of the material. However, the material effect dominates in the carrier frequency region of interest, as will be shown below.

It is appropriate to estimate the effect of the material alone since it represents a limit to the waveguide performance. A typical low refractive index glass, fused silica, has $\frac{d^2k}{d\omega^2} \cong 3 \times 10^{-23} sec^2 km^{-1}$ at 800 nm wavelength. Consider first a light source with no spectral width when continuously operated ($s = 0$). The broadening then comes entirely from the frequency spectrum imposed by the pulse modulation, μ in Eq. (20). Figure 6 shows the result for a 1 km path and the minimum output pulse width of about 8×10^{-12} sec with an input pulse width of 5.5×10^{-12} sec. Figure 6 also shows the result for an LED source. A typical width of 20 nm at a center frequency of 800 nm was assumed for the spectral width when continuously operated. Then, $s \cong 3 \times 10^{13}$ sec^{-1}. The minimum output pulse width appears at the same input pulse, 5.5×10^{-12} sec. However,

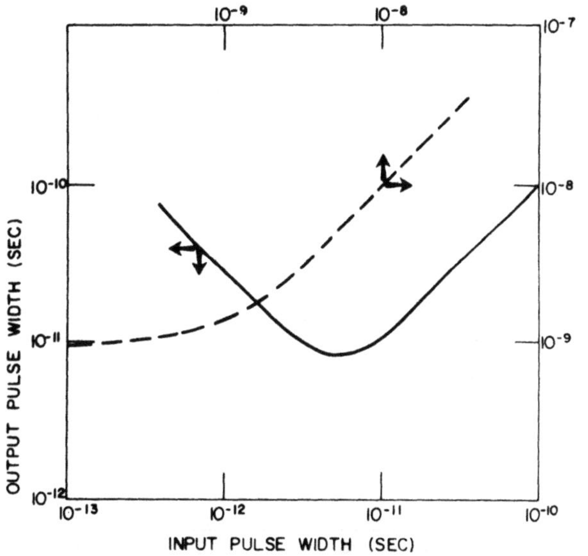

Fig. 6 Pulse output breadth versus pulse input breadth for material dispersion over one kilometer of path. Pure silica glass and gaussian pulses are assumed. The solid line is for a light source which is purely monochromatic in unmodulated operation. The dashed line is for a light emitting diode with spectral breadth of 20 nm in unmodulated operation.

the source broadens the pulse so much that at an input pulse width of 10^{-9} sec, the output pulse is almost at its minimum. Thus, little would be gained by modulating an LED faster than this when using transmission lines longer than about 1 km.

Figure 7 shows the result for more exact computations of the light source with modulation spectral width.(12) Equation (15) was used to calculate k with numerical

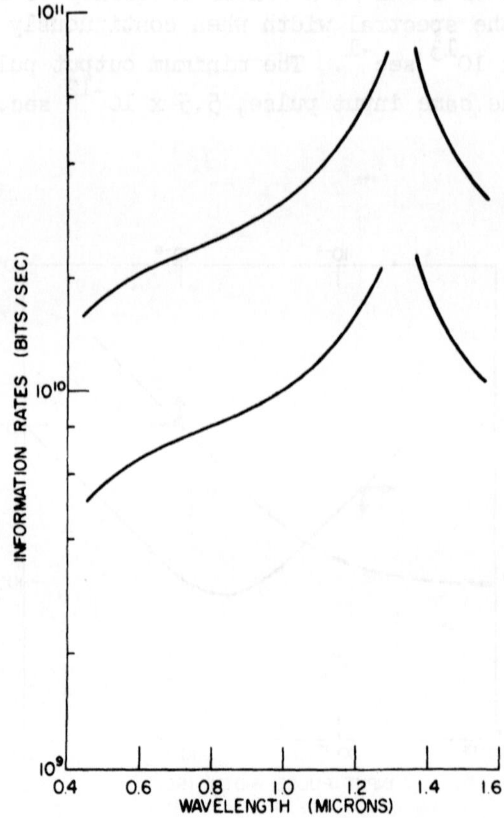

Fig. 7 Calculated maximum information rate for a wave-guide of high silica glass. Successive gaussian pulses are assumed to overlap at a point 20 dB down from their peak. The input pulse is a breadth to give minimum broadening at the output. The upper curve is for a fiber 1 km long and the lower curve for a fiber 10 km long (after Kapron and Keck).

values of the measured refractive index at various free space wavelengths. The dispersion of ω vs k is thus determined under the influence of both the material dispersion and the waveguide characteristic dispersion (Fig. 5). These two dispersions have opposite sign and tend to cancel. However, as previously stated, the material dispersion dominates up to about 900 nm as the figure shows. Then, for longer wavelengths the material dispersion has dropped and the two tend to cancel, becoming equal at about 1300 nm to the accuracy of this calculation. The two curves are for distances of one and ten kilometers with output pulses overlapping at a point 20 dB down from their peak. Maximum bit rates are calculated from the optimum input pulse width at the distance considered (Eq. 20). At 800 nm, 1 km, and width 20 dB down, the band pass is 3×10^{10} sec^{-1}, in rough agreement with our estimate of $\sim 10^{11}$ sec^{-1} above. The difference is due to inclusion of the waveguide characteristic dispersion and different assumptions regarding point of overlap.

2.2 Pulse Broadening in Multimode Fibers

In a multimode fiber, pulse broadening is caused primarily by the different group velocities of the excited modes. A pulse transmitted in any one of the modes will broaden in the manner discussed in the previous section. However, this is usually smaller than the multimode group velocity differences, at least for conventionally narrow band light sources. In what follows, the pulse broadening in a single mode will be ignored although qualitative indication of its influence will be given.

One major complication with multimode fibers is the degree to which the various modes carry the optical power. Each type of light source will have its own characteristic excitation effectiveness. Since the attenuation of the different modes varies, the measured attenuation of the waveguide will also depend on excitation. In the analysis below, the equal excitation of all modes will be assumed. It is supposed that all the individual modes, m, can be ordered with respect to their phase velocity and that the result is some smooth function that is continuously varying. The slope of this curve and the variation in the number of modes with frequency will give the group velocity. Let the signal power, P_s, be considered a function of time, τ.

Then

$$P_s = \frac{E}{d\tau} = \frac{E}{dm}\ \frac{1}{\frac{d\tau}{dm}} \quad (21)$$

where E/dm is the energy per mode, which is assumed constant. The number of modes, dm, arising in any interval $d\tau$ will depend upon their group velocity. Thus, we desire m as a function of the group velocity so that we can obtain $dm/d\tau$ as a function of τ.

It is convenient to introduce two parameters, the phase refractive index of the mode, n, and the group refractive index of the mode, N. These are related respectively to the phase velocity

$$\frac{\omega}{k} = \frac{\omega}{nk_o} = \frac{c}{n} \quad (22)$$

with k_o the free space propagation constant, ω/c, and to the group velocity

$$\frac{d\omega}{dk} = \frac{c}{N} = \frac{c}{\frac{dk}{dk_o}} = \frac{c}{\frac{d(nk_o)}{dk_o}} \quad (23)$$

The group refractive index is thus derived from the phase refractive index for the mode. The procedure is to get n(m), derive N(m) and then $\tau(m)$ from $\tau = LN/c$. The reason for starting with n(m) is its accessibility through solution of the mode problem.

A previous section showed how the phase velocity of the modes varied from that of the core material to that of the cladding material (cut-off). Thus, we assume a form

$$n = n_1 - \left(\frac{m}{M}\right)^p \Delta n \quad (24)$$

where p is an arbitrary number. For the step refractive index waveguide, a good approximation is p = 1 and[13][14]

$$M \cong \frac{a^2 k_o^{\,2}}{2}\ (n_1^{\,2} - n_2^{\,2}). \quad (25)$$

Equation (25) is physically plausible (except for the constant 1/2) for any dielectric waveguide. In many practical cases the number of modes, M, varies from 10^3 to 10^4.

$$N = \frac{d}{dk_o} [k_o n_1 - (\frac{m}{M})^p k_o \Delta n] \quad (26)$$

using Eqs. (23) and (24). Substituting in Eq. (25),

$$N(m) = n_1 - (\frac{m}{M})^p (1-2p)\Delta n \quad (27)$$

where the variation of n_1 and n_2 with k_o are neglected, although this variation can be included in a straightforward way and shown to be negligible. From Eq. (27)

$$\frac{dm}{d\tau} = \frac{M}{p} \frac{c}{L(2p-1)\Delta n} [\frac{Ln_1 - c\tau}{L(1-2p)\Delta n}]^{\frac{1}{p} - 1} \quad (28)$$

For the step refractive index waveguide (p=1), the received power is a constant from Eq. (21). Using Eq. (27) with m=o and m=M gives the arrival times of the fastest (lowest order) mode and the slowest (highest order) mode. Figure 8 shows the result in the upper part. If the single mode pulse width is not negligible, but is comparable to the variation in arrival times due to group velocity dispersion, a pulse shape similar to the bottom of Figure 8 will result.

It is interesting to note, within the approximations of this analysis, that a waveguide with a mode density of $p = \frac{1}{2}$ in Eq. (24) has exact group velocity cancellation from Eq. (27). The signal power is then infinite, Eq. (28). This is in accord with the mode theory for gradient index waveguides discussed earlier. From Eq. (27) and Fig. 8, the step refractive index waveguide with $\Delta n = 0.02$ gives a pulse broadening of about 70×10^{-9} sec over a kilometer. Waveguides with this same refractive index difference on the axis but approximately parabolic variation ($p = \frac{1}{2}$) have been made with less than 5 percent of this broadening.

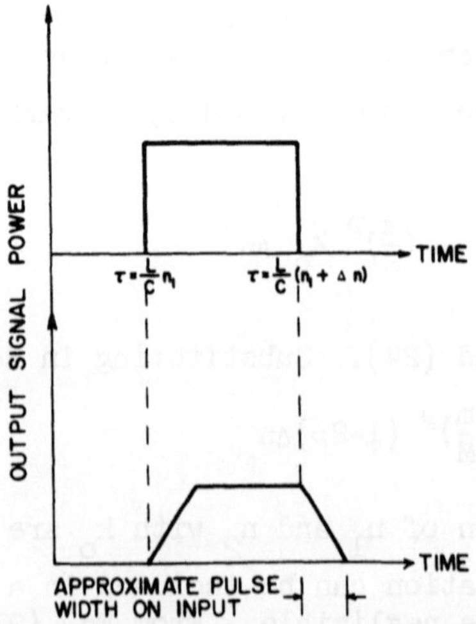

Fig. 8 Top: Output pulse shape in a multimode step refractive index waveguide. All modes are equally excited and attenuated with an input pulse much shorter than the output pulse.

Bottom: Approximate output pulse when input pulse width is comparable to output pulse width.

If the optical waveguide is not perfect in geometrical distribution of the refractive index, internal light scattering shifts the spatial power distribution as the signal propagates down the waveguide. This may be viewed as a coupling between modes since, for any two modes with characteristic spatial power distribution, the power transfer coefficient from one mode to the other is reciprocal. This has been shown to reduce the pulse broadening due to group velocity differences.[15]

III. ATTENUATION

Considerable progress has been made in understanding the causes of attenuation in optical fibers. Two types of mechanism are important, scattering and absorption. Absorption is due primarily to impurities and atomic structural defects. Much effort on careful purification and glass composition choices have reduced absorption loss to a small contribution over a large part of the wavelength region. Scattering, thus, is often a major source of attenuation, especially in the region of 600 nm wavelength. In fact, optical waveguides have been made in which the major contribution to the loss in some wavelength regions is due primarily to the intrinsic scattering of the material itself. The intrinsic material properties represent an ultimate lower limit to the attenuation. It is therefore helpful to understand their origin.

All transparent materials scatter light. This intrinsic property is due to thermal agitation of the constituent atoms and can be calculated from a straightforward statistical fluctuation theory. For most glasses the scattering is not complicated by the intricate structural details that affect the scattering by other materials, such as polymers, composites, etc. This permits the straightforward calculation below which can yield some representative scattering values appropriate for waveguides.

The theory of scattering by bulk materials begins by subdividing the sample into small volumes much less than a wavelength in size but otherwise of arbitrary magnitude. Each of these acts as a small scattering dipole of magnitude $\bar{p}_i = \bar{E}\,(\epsilon_i - \epsilon_o)\,\delta v_i$. This is used with the field from a radiating dipole[9]

$$E_\theta = \frac{|\bar{p}|k^3}{4\pi\epsilon}\, e^{i(\omega t - kr')} \sin\theta' \left[-\frac{1}{kr'}\right]$$

$$H_\varphi = \frac{i\omega|\bar{p}|k^2}{4\pi}\, e^{i(\omega t - kr')} \sin\theta' \left[\frac{i}{kr'}\right] \qquad (29)$$

The square of the magnitude of either E or H is proportional to the intensity of the scattered wave. All the scattered waves from the small volumes are added with consideration of their phases $e^{ik\underline{s}\cdot\underline{r}_i}$. Here, $\underline{r}_i$ is a vector from the origin to the volume δv_i and $\underline{s} = \underline{S} - \underline{S}_o$ with $\underline{S}$ a unit vector in the direction of the observation and $\underline{S}_o$ in the direction of the incident wave. Choose the dipole axis along the polar axis so that $\sin\theta' = 1$ for light polarized in the direction of this axis and $\sin\theta' = \cos\theta$ at right angles. The scattered intensity, I_s, is then

$$\frac{I_s}{I_o} = \frac{\pi^2(1+\cos^2\theta)}{2\epsilon^2\lambda^4R^2} \sum_{ij}\sum (\epsilon_i-\epsilon_o)(\epsilon_j-\epsilon_o)e^{ik\underline{s}\cdot(\underline{r}_i-\underline{r}_j)}\delta v_i\,\delta v_j \quad (30)$$

Let $\bar{\epsilon}$ be the average dielectric permittivity so that $\epsilon_i = \bar{\epsilon} + \Delta\epsilon_i$. Terms linear in $\Delta\epsilon_i$ vanish and Eq. (30) reduces to a quadratic term and a term representing the scattering by the shape of the bulk sample. The latter leads to the waveguide boundary value problem. The former can be treated with a correlation function[16] and utilized for some types of inhomogeneity scattering within the waveguide. However, for fluctuation scattering there is no correlation between $\Delta\epsilon_i$ and $\Delta\epsilon_j$, so that

$$\frac{I_s}{I_o} = \frac{\pi^2(1+\cos^2\theta)}{2\epsilon^2\lambda^4R^2}\; V\; \overline{(\Delta\epsilon)^2}\;\delta v \quad (31)$$

where V is the total volume of bulk material. The quantity $\overline{(\Delta\epsilon)^2}$ is a fluctuation quantity inversely proportional to δv, which assures a scattering expression independent of this arbitrary volume size.

The quantity $\overline{(\Delta\epsilon)^2}$ can be evaluated in terms of measureable parameters. A variation of the dielectric constant with volume (or density) must be chosen. Fortunately, the final expression is not very sensitive to this choice and the following proportionality will be used,

$$\epsilon-\epsilon_o = \eta\alpha = \frac{\rho\alpha}{m}$$

where η is the number of atoms per unit volume, α their polarizability and m their mass. Then,

$$\overline{(\Delta\epsilon)^2} = \frac{(\epsilon-\epsilon_o)^2}{(\delta v)^2} \overline{(\Delta V)^2} \tag{32}$$

with ΔV the fluctuation of the small volume δv. Using the relation from statistical thermodynamics,[17]

$$\overline{(\Delta V)^2} = \frac{kT}{\left(\frac{\partial^2 A}{\partial V^2}\right)_{V = V_o}} \tag{33}$$

and $\frac{\partial^2 A}{\partial V^2} = \frac{1}{\beta \delta v}$ gives for Eq. (32) a tractable dependence that can be substituted in Eq. (31). Thus,

$$\frac{I_s}{I_o} = \frac{\pi^2(1+\cos^2\theta)}{2\epsilon^2\lambda^4R^2} V(\epsilon-\epsilon_o)^2 kT\beta$$

which is integrated over all angles to give the attenuation due to scattering.

$$\tau = \frac{8\pi^3(n^2-1)}{3\lambda^4} kT\beta \tag{34}$$

This important expression is appropriate for various types of optical waveguide and bears closer inspection. Consider liquid core first. The isothermal compressibility for carbon tetrachloride, as an example, is $\beta \cong 105 \times 10^{-12}$ cm^2/dyne. Substituting in Eq. (34), $\tau \cong 11 \times 10^{-6}$ cm^{-1} ≡ 4.7 dB/km at 800 nm. The very rapid variation of Eq. (34) with wavelength gives liquids high loss at shorter wavelengths but, conversely, eliminates intrinsic scattering as a matter of importance at long wavelengths. This general wavelength consideration is of importance in other materials as well. Next, consider silica glass which is composed of a single oxide. The room temperature compressibility is the same as a typical solid, 2.7×10^{-12} cm^2/dyne, which means that Eq. 34 gives a negligible attenuation for room temperature fluctuations. Since glass is a solidified liquid, the fluctuations frozen in the annealing range constitute most of the scattering. Appropriate temperatures and compressibilities must be used in (34). The latter is usually unknown for high temperature and must be derived from observation. Using a τ = 1500°K, Eq. (34) leads to a compressibility matching the observed scattering which is about three times the room temperature compressibility. Alternatively, the small angle x-ray scattering leads to the same liquid compressibility.[18] This is not to say that the derived compressibility is correct, only that the two types of scattering measurement are consistent. The fused silica bulk scattering loss is measured to be about 1.7 dB/km at 800 nm, with all its features consistent with the theory above.

In glasses made with more than one oxide Eq. (34) still applies to compressive fluctuation scattering if T and β are those of the relevant glass. However, another form of scattering arises from fluctuations in concentration of the added oxide. From Eq. (31), this leads to

$$\tau \cong \frac{16\,\pi^3 n}{3\lambda^4} \left(\frac{\partial n}{\partial c}\right)^2 \overline{(\Delta c)^2}\ \delta v \qquad (35)$$

which usually cannot be further simplified because these complicated solutions do not lead to straightforward calculation of $\overline{(\Delta c)^2}$. However, as a matter of interest, the observed scattering can be used to calculate the magnitude of the refractive index fluctuation. For a soda lime glass, at $\delta v = (14\text{Å})^3$ the magnitude of $\sqrt{(\Delta n)^2}$ = 1.2×10^{-2}, at $(\delta v) = (55\text{Å})^3$ 1.5×10^{-3}, etc.

To show how these factors contribute in a typical case, consider the total attenuation at 633 nm for the waveguide loss illustrated in Fig. 9. The total loss of 9 dB/km is made up approximately as follows.

Material fluctuation scattering, Eq. (34) ...	4 dB/km
" " " , Eq. (35) ...	1
Scattering from unidentified cause	3
Absorption by impurities	1
	9 dB/km

This illustrates the importance of scattering loss at this particular carrier wavelength.

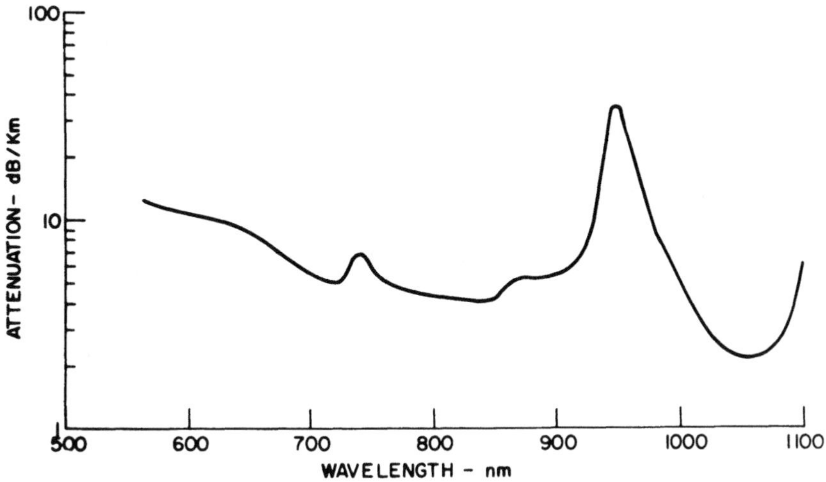

Fig. 9 Spectral attenuation of a low loss optical waveguide. Very high order modes can have larger attenuation. (Courtesy Mr. J. F. Frazier, Corning Glass Works).

CONCLUSION

This chapter has intended to introduce many, but not all, of the important principles for future optical transmission lines. Present waveguides seem adequate for new, advanced systems with respect to information capacity and attenuation. Early results from environmental tests are encouraging. Thus, they seem certain to succeed as useful devices. Present knowledge will at least provide the base on which future technology is built.

Representative References

1. R. D. Maurer, "Glass fibers for optical communication," Proc. IEEE, Vol. 61, p. 452, April 1973.

2. M. Born and E. Wolf, Principles of Optics, Pergamon Press, New York, 1959.

3. D. Marcuse, Light Transmission Optics, Van Nostrand Reinhold, New York, 1972.

4. J. P. Dakin, W. A. Gambling, H. Matsumura, D. N. Payne, and R. D. Sunak, "Theory of dispersion in lossless multimode optical fibers," Optics Communications, Vol. 7, p. 1, Jan. 1973.

5. S. Kawakami and J. Nishizawa, "An optical waveguide with optimum distribution of the refractive index with reference to waveform distortion," IEEE Trans. Microwave Theory Tech., Vol. 16, p. 814, 1968.

6. E. C. Rawson, D. R. Herriott and J. McKenna, "Analysis of Refractive Index Distributions in Cylindrical, Graded-Index Glass Rods Used as Image Relays," Appl. Optics, Vol. 9, p. 753, March 1970.

7. A. Fletcher, T. Murphy and A. Young, "Solutions of two optical problems," Proc. Roy. Soc., Vol. 223, p. 216, April 1954.

8. E. W. Marchand, "Raytracing in Cylindrical Gradient-Index Media," Appl. Optics, Vol. 4, p. 1104, May 1972.

9. J. C. Slater and N. H. Frank, Electromagnetism, McGraw Hill, New York 1947.

10. E. Snitzer, "Cylindrical Dielectric Waveguide Modes," J. Opt. Soc. Am., Vol. 51, 491, May 1961.

11. W. Streifer and C. N. Kurtz, "Scalar analysis of radially inhomogeneous guiding media," J. Opt. Soc. Am., Vol. 57, p. 779, June 1967.
D. Gloge and E. A. J. Marcatili, "Multimode theory of graded core fibers," B.S.T.J.

12. F. P. Kapron and D. B. Keck, "Pulse Transmission through a Dielectric Optical Waveguide," Appl. Optics, Vol. 10, p. 1519, July 1971.

13. D. Gloge, "Dispersion in Weakly Guiding Fibers," Appl. Optics, Vol. 10, p. 2442, Nov. 1971.

14. Allan W. Snyder, "Asymptotic Expression for Eigenfunctions and Eigenvalues of a Dielectric or Optical Waveguide," IEEE Trans. Microwave Theory Techniques, Vol. MTT-17, p. 1130, Dec. 1969.

15. S. D. Personick, "Time dispersion in dielectric waveguides," B.S.T.J., Vol. 50, p. 843, 1971.

16. P. Debye and A. M. Bueche, "Scattering by an inhomogeneous solid," J. Appl. Phys., Vol. 20, p. 518, 1949.

17. E. Schroedinger, Statistical Thermodynamics, Cambridge University Press, Cambridge 1948.

18. D. L. Weinberg, "X-ray scattering measurements of long range thermal density fluctuations in liquids," Phys. Lett., Vol. 7, p. 324, Dec. 15, 1963.

Chapter 9 — FIBER OPTICS APPLICATIONS

M. K. Barnoski

Hughes Research Laboratories
3011 Malibu Canyon Road
Malibu, California 90265

1. Systems Applications

Applications of optical integrated circuits can be categorized into near term and long range applications. As an aid in assessing the long range engineering research and development goals it is often of value to be aware of the more current requirements and developments. An understanding of these requirements can provide a solid basis on which to define the long range device development goals.

Interest in optical data transmission, gathering and processing systems, both in the telecommunications industry and the government is rapidly increasing. While the advantages of optical communications have been recognized for some time it is the recent breakthroughs in the fabrication of low loss optical fiber waveguides (4 dB/km loss) which have made the possibility of utilizing these advantages practical. This in turn has spurred engineering research and development on the other components such as sources, modulators, detectors, and couplers which are required to complete an optical communications system.

Thin film optical circuits coupled to single mode fiber waveguides in principle can provide communication systems with length-bandwidth (bit rate) products in the gigabit-kilometer range. Low loss single mode glass fiber waveguides are currently available at the research level. Much effort is being expended on the key components which need to be fabricated and integrated on a "single chip" of appropriate material(s) to comprise a micro-optical circuit. The obvious long range goal is to interconnect these integrated optical circuits with low loss, single mode fiber waveguides in such a way as to assemble a functional subsystem.

Thin film optical devices are currently being fabricated in a variety of materials which include both amorphous and crystalline materials such as glasses, polymers, semiconductors, garnets, and single crystal dielectrics. The feasibility of fabricating devices in these materials has been demonstrated. Individual components such as waveguides, couplers, distributed feedback lasers, modulators, detectors, and mode converters have all been fabricated. The next step in the process is to integrate these elements into an operational circuit. Although there is a large research effort directed toward achieving an operational integrated optical system a great deal (5 to 10 years) of time will elapse before these are truly practical. In the interim there is increased concern,[1] particularly in military systems, about the effects of electromagnetic (EMI) and environmental interference, electrical shorts and ground loops, security, redundancy, size, weight, and cost considerations which have increased interest in utilizing fiber optic communication links for use in airborne and shipboard telecommunications systems, information transfer between on-board computers, and between remote information gathering terminals and a data processor. These near term applications have moderate length-bandwidth products. Typical lengths and bandwidth requirements are 30 to 300 ft with bit rates of 1 to 200 Mbits/sec. Compared with the projected capability of the thin film optical circuitry these requirements are modest and can be met utilizing multimode, multistrand, fiber waveguides and conventional light-emitting diodes (LED's) or room temperature injection lasers and conventional silicon photodiode detectors. As a consequence, a rapid development of this technology is anticipated. Brief consideration will now be given to some of the characteristics of fiber optic data links.

2. Near Term Fiber Optic Data Links

A schematic diagram of a short point-to-point fiber optic data link is shown in Fig. 1. In the case shown, information is imposed on the optical energy by direct modulation of the GaAs light source drive current. The source may either be an LED or room temperature injection laser. For the purpose of calculation it will be assumed that the source is an LED. The total insertion loss of this system can be computed once the LED drive current to photodiode output current ratio is known. The ratio of the LED to photodiode output current is given by the expression

$$I_{Pd}/I_{LED} = \eta_{LED}\eta_{Pd}C_{in}C_{out}e^{-\alpha z} \qquad (1)$$

where η_{LED} is the external quantum efficiency of the LED, η_{Pd} is the external quantum efficiency of the photodetector, C_{in} and C_{out} are the input and output coupling losses between the LED and fiber and the fiber to photodiode, respectively. The term $e^{-\alpha z}$ results from the optical loss in the fiber bundle of length z. Typical values for η_{LED} and η_{Pd} are 3% and 70%, respectively. Plots of the current transfer ratio as a function of link length for total fiber bundle losses of 20 dB/km and 300 dB/km are shown in Fig. 2. The ratio was computed for input-output ($C_{in}C_{out}$) coupling losses of 100%, 50%, 10%, and 1%. If the circuitry is such that the electrical input impedance of the light emitting diode circuit is matched to the output load impedance attached to the photodetector then the electrical power transfer ratio of the link can be computed by multiplying the current ratio in decibels (listed in Fig. 2) by two.

An estimate of the signal-to-noise ratio of the link under consideration can be obtained by assuming that the dominant noise source is thermal (Johnson) noise in the silicon photodiode load resistor (R_L). The mean squared Johnson noise current is given[2] by

$$\overline{i_N^2} = \frac{4kTeq}{R_L}\Delta\nu \qquad (2)$$

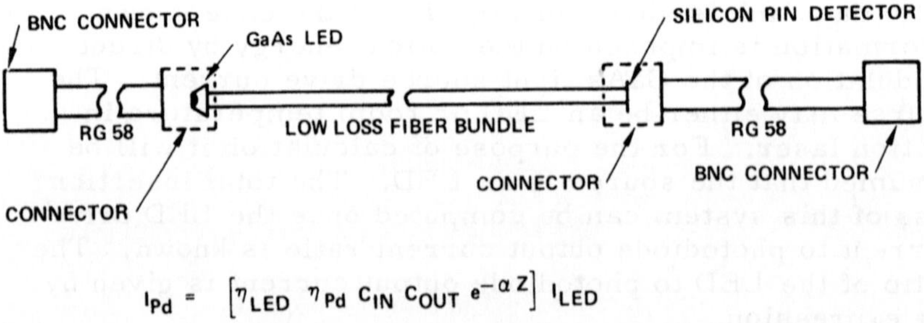

Fig. 1. Short fiber optic point-to-point data link.

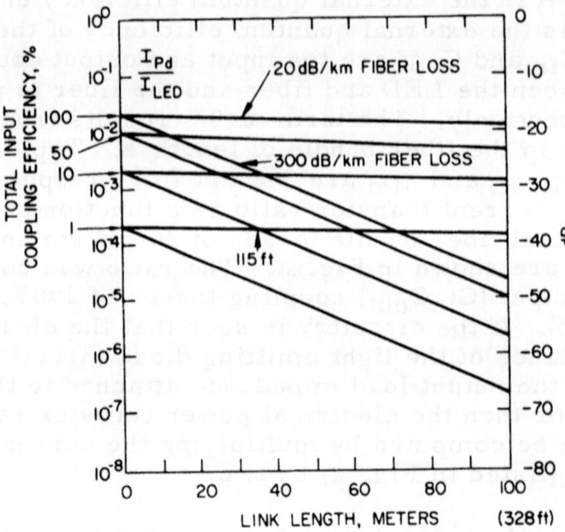

Fig. 2. Current transfer ratio as a function of link length for fiber losses of 20 dB/km and 300 dB/km with various total input coupling efficiency as a parameter.

where $\Delta\nu$ is the bandwidth, k is the Boltzmann constant and T_{eq} is the equivalent temperature chosen to include the equivalent input noise power of an amplifier which normally would follow the photodetector. For an amplifier with a noise figure of ~ 7 dB the equivalent temperature is[2] ~ 1200°K. Under these conditions the signal-to-noise power ratio is

$$(S/N) = (\eta_{LED}\eta_{Pd}C_{in}C_{out}e^{-\alpha z})^2 \frac{R_L I_{LED}^2}{4kT_{eq}\Delta\nu} . \quad (3)$$

Plots of this equation using 20 dB/km and 300 dB/km cable with $R_L = 50\Omega$, $T_e = 1200°K$, $I_{LED} = 100$ mA, and $\Delta\nu = 100$ MHz are shown in Fig. 3.

Figures 2 and 3 indicate that not only are the fiber attenuation losses important but also the total input-output interface losses at the interfaces of the fiber-LED and fiber photodetectors. The input coupling loss is dominant since essentially all the optical energy emerging from the fiber will be incident on the active area of the detector provided the detector area matches that of the fiber and that the reflection losses are minimized by antireflection coatings.

3. Input Coupling Losses

An estimate of the input coupling efficiency of the source to the fiber can be obtained by considering the power transfer efficiency between a radiating surface of area A_S and a receiving surface of area A_R. The power transfer can be computed if the goniometric characteristics of the two surfaces are known.[3] Consider the two arbitrarily oriented surfaces separated by distance vector $\overline{r}_{sc}$ as shown in Fig. 4. The power transfer from an element of area dA_S to an element of area $\cos\theta_R \, dA_R$ normal to $\overline{r}_{sc}$ is

$$dP_{SR} = R_\Omega(\overline{X}_s, \theta_s) \, dA_S \, d\Omega_R \text{ where } R_\Omega(\overline{X}_s, \theta_s) \quad (4)$$

is the radiance of the source in units of watts per square centimeter steradian. For sources with nonuniform radiation distributions, R_Ω is a function of $\overline{X}_s$ which is the

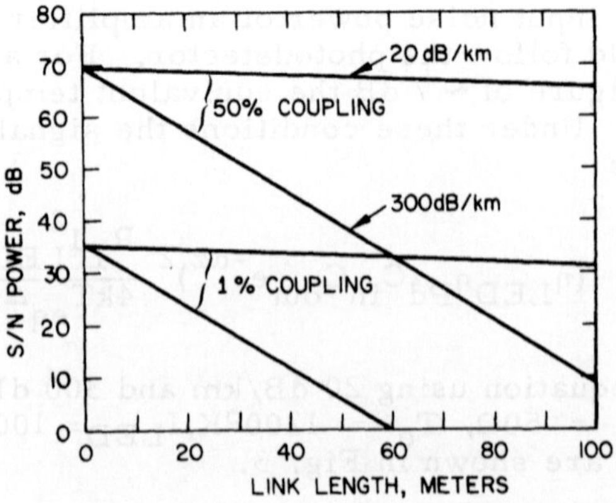

Fig. 3. Signal-to-noise power ratio as a function of link length.

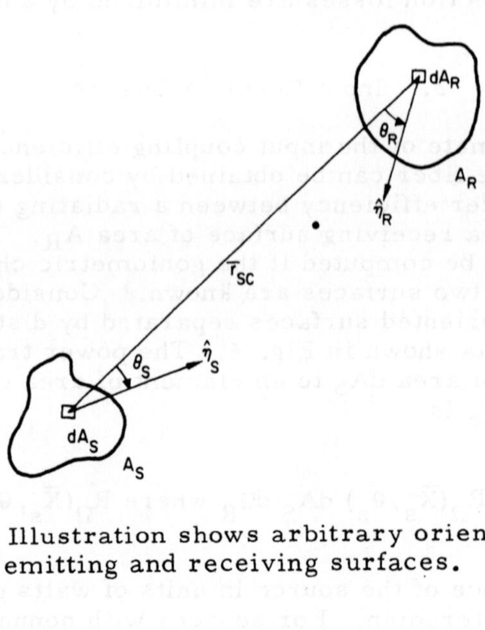

Fig. 4. Illustration shows arbitrary orientation of emitting and receiving surfaces.

position vector on the surface A_S. The quantity $d\Omega_R$ is the solid angle subtended by ($\cos\theta_R\, dA_R$) at the source point dA_S that is,

$$d\Omega_R = \frac{\cos\theta_R dA_R}{r_{sc}^2} \quad . \tag{5}$$

The total power transfer between A_S and A_R is therefore

$$P_{SR} = \int_{A_S}\int_{A_R} R_\Omega(\overline{X}_s, \theta_s) \cos\theta_R r_{sc}^{-2}\, dA_R dA_S. \tag{6}$$

For simplicity it will now be assumed that the source has a uniform radiation distribution across its area and that the extent of the source and collector areas are small compared with the separation vector r_{sc}. In addition, it will be assumed that the collector collects only those rays that are contained within a cone whose half angle is denoted as θ_{NA}. The ratio of the power radiated into this cone to the power radiated into the full hemisphere is

$$\eta_{NA} = \frac{\int_0^{\theta_{NA}} R(\theta)\sin\theta\, d\theta}{\int_0^{\pi/2} R(\theta)\sin\theta\, d\theta} \quad . \tag{7}$$

For a source whose radiance varies as $(\cos\theta)^n$ this becomes

$$\eta_{NA} = 1 - (\cos\theta_{NA})\, n+1 \quad . \tag{8}$$

The amount of energy coupled into a fiber is greatly dependent on its numerical aperture (NA) which can be defined with the aid of Fig. 5. For total internal reflection to occur, the internal reflection angle must be greater

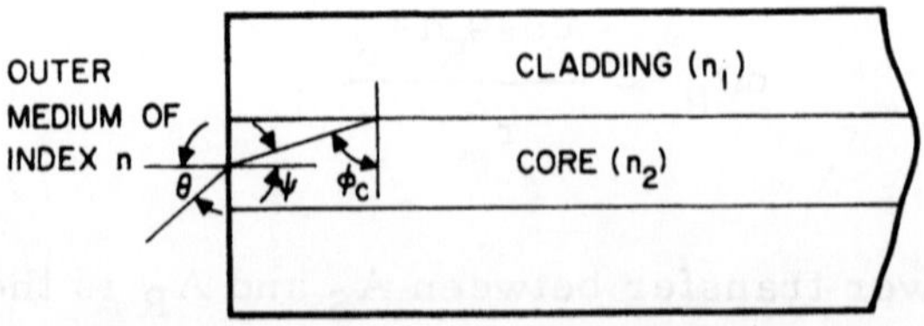

Fig. 5. Illustration shows entrance angle of a cladded fiber for total internal reflection at the core-cladding interface.

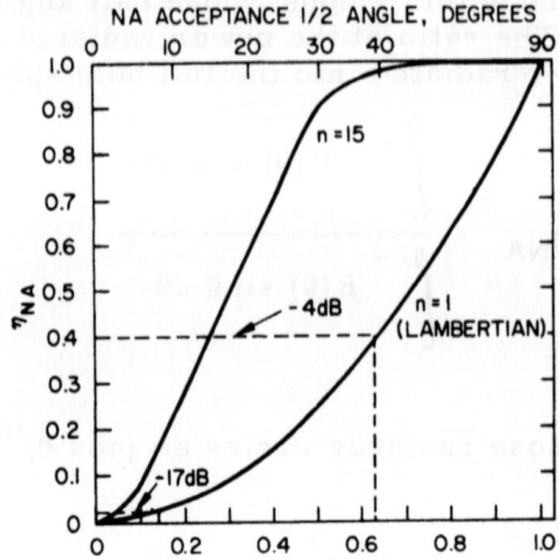

Fig. 6. Numerical aperture coupling efficiency as a function of numerical aperture.

than the critical angle $\sin \phi_c = n_1/n_2$. This can be related to the incident angle θ via Snell's law

$$+ n \sin\theta = n_2 \sin\psi = n_2 \sin(90-\phi_c) = (n_2^2 - n_1^2)^{1/2} \quad . \tag{9}$$

The quantity $n \sin\theta$ is termed the numerical aperture (NA) of the fiber.

Since $\theta_{NA} = \arcsin(NA/n)$ the expression for the numerical aperture coupling efficiency can be written as

$$\eta_{NA} = 1 - \left[1 - \left(\frac{NA}{n}\right)\right]^{\frac{n+1}{2}} \quad . \tag{10}$$

Plots of the numerical aperture coupling efficiency as a function of NA are shown in Fig. 6 for $n = 1.0$ (Lambertian radiator) and $n = 15$. Typical multimode low-loss glass fibers developed by Corning Glass have NA $\cong 0.14$ which results in η_{NA} of -17 dB. Typical values for the numerical aperture of high loss fibers is NA $\cong 0.63$ which results in a η_{NA} of -4 dB.

The angular distribution of the radiation emitted by a flat geometry LED is Lambertian; that is, it is independent of the azimuth angle and varies with the angle of elevation as $R(\theta) = \cos\theta$. This distribution can be altered by external intervening optics and/or different cross-sectional design of the LED itself.[4]

In addition to the numerical aperture, the input coupling efficiency is dependent on the fiber packing fraction. The packing fraction is defined as the ratio of active core area to the total area of the fiber bundle. For a fiber with a core diameter of 0.003 in. and a total diameter (core plus cladding) of 0.005 in. the packing fraction of each individual strand in the bundle is 0.36. For a bundle consisting of 61 individual fibers contained within a 0.045 in. diameter circle the packing fraction of the bundle is 0.27; that is, only 27% of the total bundle area is active. The input coupling packing fraction loss is therefore -5.7 dB.

The input coupling efficiency is also dependent on the area mismatch of the fiber bundle and the cross-sectional area of the radiation pattern incident on the face of the fiber bundle. If coupling is achieved by directly contacting the fiber bundle to a LED whose diameter is smaller than that of the bundle then area mismatch loss does not exist. If, however, the active area of the LED is larger than that of the fiber or if LED is separated from the fiber so that the cross-sectional area of the radiation pattern incident on the bundle is larger than that of the fiber bundle itself, then the area mismatch loss can be significant. If the radiation were uniformly distributed across the cross-sectional area of the pattern then the area mismatch loss is equal to the ratio of this area to the total area of the bundle. For nonuniform radiation distributions the area mismatch and numerical aperture losses are not separable but must be determined by evaluating the integral in Eq. 6.

For the case when the LED area is less than that of the fiber (no area mismatch loss) and the angular distribution of the radiation is Lambertian, the input coupling efficiency is -23 dB for NA = 0.14 and -10 dB for NA = 0.63 assuming that the packing fraction losses are the same for both the low loss fiber (NA = 0.14) and the high loss fiber (NA = 0.63). If this is the case then inspection of Fig. 2 reveals that the transfer coefficient for the 300 dB/km fiber with 10% input coupling efficiency drops below the coefficient for the 20 dB/km fiber with 1% input coupling efficiency at link length of 35 m (115 ft). This indicates that for links greater than 35 m it is more advantageous to use low loss cable with low input coupling efficiency. However, for link lengths less than 35 m the high loss, high numerical aperture cable is the better choice. It should be kept in mind that these results are only valid for Lambertian radiation, no area mismatch loss, and identical packing fraction losses.

REFERENCES

1. H.F. Taylor, "Transfer of Information on Naval Vessels via Fiber Optics Transmission Lines," NELC Technical Report TR 1763 (1971).

2. A. Yariv, "Introduction to Optical Electronics," Holt, Rinehart, and Winston, Inc., New York (1971).

3. P.W. Kruse, L.D. McGlauchlin, and R.B. McQuistan, "Elements of Infrared Technology," John Wiley and Sons, New York (1963).

4. S.M. Sze, "Physics of Semiconductor Devices," Wiley-Interscience, New York (1969).

Chapter 10 – COUPLED MODE FORMALISM FOR GUIDED WAVE INTERACTIONS

Amnon Yariv

California Institute of Technology
Pasadena, California

1. Coupled Mode Formalism

Consider two electromagnetic modes with, in general, different frequencies whose complex amplitudes are A and B. These are taken as the eigenmodes of the unperturbed medium so that they represent propagating disturbances

$$a(z,t) = A\,e^{i(\omega_a t - \beta_a z)}$$

$$b(z,t) = B\,e^{i(\omega_b t \pm \beta_b z)} \qquad (1)$$

with A and B constant.

In the presence of a perturbation power is exchanged between modes a and b. The complex amplitudes A and B in this case are no longer constant but will be found to depend on z. They will be shown below to obey relations of the type

$$\frac{dA}{dz} = \kappa_{ab}\, B\, e^{-i\Delta z}$$

$$\frac{dB}{dz} = \kappa_{ba}\, A\, e^{+i\Delta z} \tag{2}$$

where the phase mismatch constant Δ depends on the propagation constants β_a and β_b as well as on the spatial variation of the coupling perturbation. The coupling coefficients κ_{ab} and κ_{ba} are determined by the physical situation under consideration. Consider first the case where modes a and b carry (Poynting) power in the same direction (codirectional). It is convenient to define A and B in such a way that $|A(z)|^2$ and $|B(z)|^2$ correspond to the power carried by mode a and b, respectively. The conservation of total power is thus expressed as

$$\frac{d}{dz}\left(|A|^2 + |B|^2\right) = 0 \tag{3}$$

which using Eq. (2) is satisfied when (1)

$$\kappa_{ab} = -\kappa_{ba}^{*} \quad . \tag{4}$$

If boundary conditions are such that a single mode, say b, is incident at z = 0 on the perturbed region z > 0, then

$$b(0) \equiv B_o \ , \quad a(0) = 0 \ . \tag{5}$$

Subject to these conditions the solutions of Eq. (2) become

$$A(z) = B_o \frac{2\kappa_{ab}}{(4\kappa^2 + \Delta^2)^{1/2}}\, e^{-\frac{i\Delta z}{2}} \sin\left[\frac{1}{2}(4\kappa^2 + \Delta^2)^{1/2} z\right]$$

$$B(z) = B_o\, e^{\frac{i\Delta z}{2}} \left\{\cos\left[\frac{1}{2}(4\kappa^2 + \Delta^2)^{1/2} z\right]\right\}$$

$$- i \frac{\Delta}{(4\kappa^2 + \Delta^2)^{1/2}} \sin\left[\frac{1}{2}(4\kappa^2 + \Delta^2)^{1/2} z\right] \tag{6}$$

where $\kappa^2 \equiv |\kappa_{ab}|^2$. Under phase-matched condition $\Delta = 0$, a complete spatially periodic power transfer between modes a and b takes place with a period $\pi/2\kappa$

$$a(z,t) = B_o \frac{\kappa_{ab}}{\kappa} e^{i(\omega_a t - \beta_a)} \sin(\kappa z)$$

$$b(z,t) = B_o e^{i[\omega_b t - \beta_b z]} \cos\kappa z \ . \tag{7}$$

A plot of the mode intensities $|a|^2$ and $|b|^2$ is shown in Fig. 1. This figure demonstrates the fact that for phase mismatch $\Delta \gg |\kappa_{ab}|$ the power exchange between the modes is negligible.

For contradirectional coupling the propagation in the unperturbed medium is described by

$$a = A\, e^{i(\omega_a t + \beta_a z)}$$

$$b = B\, e^{i(\omega_b t - \beta_b z)} \tag{8}$$

where A and B are constant. Mode a corresponds to a left (-z) traveling wave while b travels to the right. Conservation of total power can be expressed as

$$\frac{d}{dz}(|A|^2 - |B|^2) = 0 \tag{9}$$

which is satisfied by Eq. (2) if

$$\kappa_{ab} = \kappa_{ba}^* \tag{10}$$

so that

$$\frac{dA}{dz} = \kappa_{ab} Be^{-i\Delta z} \ , \quad \frac{dB}{dz} = \kappa_{ab}^* Ae^{i\Delta z} \ . \tag{11}$$

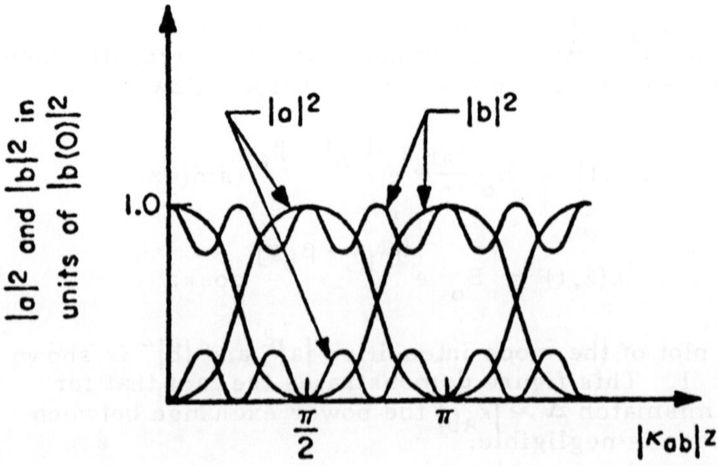

Fig. 1. Variation of the mode power in the case of codirectional coupling for phase matched and unmatched operation.

Assume that mode b with an amplitude B(0) is incident at z = 0 on the perturbation region which occupies the space between z = 0 and z = L. Since mode a is generated by the perturbation it must be true that a(L) = 0. With these boundary conditions the solution of Eq. (11) is given by

$$A(z) = B(0) \frac{2i\kappa_{ab} e^{-i\frac{\Delta z}{2}}}{-\Delta \sinh \frac{SL}{2} + iS \cosh \frac{SL}{2}} \sinh\left[\frac{S}{2}(z - L)\right]$$

$$B(z) = B(0) \frac{e^{i\frac{\Delta z}{2}}}{-\Delta \sinh \frac{SL}{2} + iS \cosh \frac{SL}{2}} \left\{\Delta \sinh\left[\frac{S}{2}(z - L)\right] + iS \cosh\left[\frac{S}{2}(z - L)\right]\right\} \quad (12)$$

$$S \equiv \sqrt{4\kappa^2 - \Delta^2}, \qquad \kappa \equiv |\kappa_{ab}| \ . \quad (13)$$

Under phase matching conditions $\Delta = 0$ so that

$$A(z) = B(0)\left(\frac{\kappa_{ab}}{\kappa}\right)\frac{\sinh[\kappa(z - L)]}{\cosh(\kappa L)}$$

$$B(z) = B(0) \ \frac{\cosh[\kappa(z - L)]}{\cosh(\kappa L)} \ . \quad (14)$$

A plot of the mode powers $|B(z)|^2$ and $|A(z)|^2$ for this case is shown in Fig. 2. For sufficiently large arguments of the cosh and sinh functions in Eq. (14), the incident mode power decays exponentially along the perturbation region. This decay, however, is not caused by absorption but is a result of reflection of power into the backward traveling mode a.

2. The Coupling Equation

The wave equation obeyed by the unperturbed modes is

$$\nabla^2\vec{E}(\vec{r}, t) = \mu\epsilon \frac{\partial^2\vec{E}}{\partial_t^2} \ . \quad (15)$$

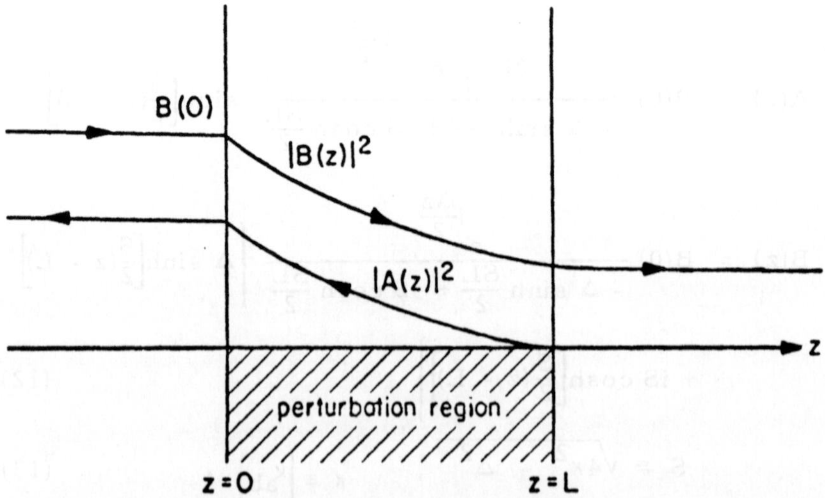

Fig. 2. Transfer of power from an incident forward wave B(z) to a reflected wave A(z) in the case of contradirectional coupling.

In most of the experiments of interest the perturbation can be represented as a distributed polarization source $\vec{P}_{pert}(\vec{r}, t)$ which accounts for the deviation of the medium polarization from that which accompanies the unperturbed mode. The wave equation for the perturbed case follows from Maxwell's equations and is

$$\nabla^2 E_y(\vec{r}, t) = \mu\epsilon \frac{\partial^2 E_y}{\partial t^2} + \mu \frac{\partial^2}{\partial t^2} [\vec{P}_{pert}(\vec{r}, t)]_y \quad . \qquad (16)$$

with similar equations for the remaining cartesian components of $\vec{E}$.

Since the eigenmodes of Eq. (15) form an orthonormal set, $^2 E_y$ can be expressed as

$$E_y = \sum_{\ell} \frac{A_\ell(z)}{2} E_y^{(\ell)}(x)\, e^{i(\omega t - \beta_\ell z)} + \text{c.c.}$$

$$+ \int_{-n_3 k}^{n_3 k} \frac{A(\beta)}{2}\, e^{i(\omega t - \beta z)}\, E_y^{(\beta)}(x)\, d\beta$$

$$n_3 < n_2 \ . \tag{17}$$

where ℓ extends over the discrete set of confined modes ($E_y(x)$ for TE modes) and includes both positive and negative traveling waves. The integration over β takes in the continuum[2] of radiation modes. In Eq. (17) c.c. denotes complex conjugation. Chief interest lies in perturbations that couple only discrete modes so that, in what follows, the second term on the right side of Eq. (17) will be neglected. Problems of coupling to the radiation modes arise in connection with waveguide losses[3] and grating couplers[4].

Substituting Eq. (17) into Eq. (16), assuming "slow" variation so that $d^2A_m/dz^2 \ll \beta_m\, dA_m/dz$ and recalling that $E_y^{(m)}(x)\, e^{i(\omega t - \beta_m z)}$ obeys the unperturbed wave Eq. (15), gives

$$\sum_{\ell}\left(- i\beta_\ell \frac{dA_\ell}{dz} E_y^{(\ell)}(x)\, e^{i(\omega t - \beta_\ell z)}\right) + \text{c.c.} = \mu \frac{\partial^2}{\partial t^2} (P_{pert})_y \ . \tag{18}$$

Multiplying Eq. (18) by $E_y^{(m)}(x)$, integrating and making use of the orthogonality of the confined modes yields

$$\frac{dA_m^{(-)}}{dz} e^{i(\omega t + \beta_m z)} - \frac{dA_m^{(+)}}{dz} e^{i(\omega t - \beta_m z)} + \text{c.c.}$$

$$= \frac{-i}{2\omega} \frac{\partial^2}{\partial t^2} \int_{-\infty}^{\infty} \left[P_{pert}(\vec{r}, t)\right]_y E_y^{(m)}(x)\, dx \tag{19}$$

where $A_m^{(-)}$ is the complex normal mode amplitude of the negative traveling TE mode while $A_m^{(+)}$ is that of the positive one. Equation (19) is the main starting point for the following discussion in which a number of special cases will be considered.

3. Nonlinear Interactions

In this section consideration will be given to the exchange of power between three modes of different frequencies brought about through the nonlinear optical properties of the guiding or bounding layers. The relevant experimental situations involve second harmonic generation, frequency up-conversion and optical parametric oscillation. To be specific, consider the case of second harmonic generation from a TE input mode at $\omega/2$ to a TE output mode at ω. The perturbation polarization is taken as

$$P_i^{(\omega)}(\vec{r},t) = \frac{1}{2}\left[P_i^{(\omega)} e^{i(\omega t - \beta z)} + c.c.\right] . \qquad (20)$$

The complex amplitude of the polarization is

$$P_i^{(\omega)} = d_{ijk}^{(\omega)}(\vec{r})\, E_j^{\omega/2}\, E_k^{\omega/2} \qquad (21)$$

where $d_{ijk}^{(\omega)}$ is an element of the nonlinear optical tensor (which may depend on position vector $\vec{r}$) and summation over repeated indices is understood. The coordinate system for the slab waveguide considered here is identical with that of Fig. 1 in Chapter III.

Neglecting, for simplicity, considerations involving crystalline orientation, and assuming that an optical field parallel to the waveguide y direction will generate a second harmonic polarization along the same direction results in the perturbation polarization

$$P_y^{(\omega)} = dE_y^{(\omega/2)}\, E_y^{\omega/2} \qquad (22)$$

where P and E represent complex amplitudes, and d

corresponds to a linear combination of d_{ijk} which depends on the crystal orientation. In this special case an input TE mode at $\omega/2$ will generate an output TE mode at . Using Eq. (17) in Eq. (22) gives

$$P_y(\vec{r}, t) = \frac{1}{2} d(\vec{r}) \sum_n \sum_p A_n^{(\omega/2)} A_p^{(\omega/2)} E_y^{(n,\omega/2)} E_y^{(p,\omega/2)}$$

$$\times e^{i\left[\omega t - \left(\beta_n^{\omega/2} + \beta_p^{\omega/2}\right)z\right]} + c.c \tag{23}$$

For the case of a single input, say, mode n, the double summation of Eq. (23) collapses to a single term n = p. Using $P_y(\vec{r}, t)$ as $[P_{pert}(\vec{r}, t)]_y$ in Eq. (19) results in

$$\frac{dA_m^{(\omega)}}{dz} = -\frac{i\omega d(z)}{4}\left[A_n^{(\omega/2)}\right]^2 e^{-i\left(2\beta_n^{\omega/2} - \beta_m^{\omega}\right)z} S^{(n,n,m)} \tag{24}$$

with

$$S^{(n,n,m)} \equiv \int_{-\infty}^{\infty} E_y^{(n,\omega/2)} E_y^{(n,\omega/2)} E_y^{(m,\omega)} f(x)\, dx \tag{25}$$

where $d(\vec{r}) \equiv d(z) f(x)$.

In the interest of conciseness consider the case where the inner layer 2 is nonlinear ($f(x) = 1$ for $0 \le x \le t$ and zero elsewhere and $d(z) = d$), and where both the input and output modes are well confined. For this case the coefficients p, q, and h appearing in the expressions (Eqs. (19) through (23) of Chapter III) for the TE mode field components are such that q_m, $p_m \gg h_m$ and $h_m \simeq m\pi/t$. Here m is the mode number. The well confined TE electric field component is therefore

$$E_y^{(m,\omega)} = A_m\left(\frac{q_m}{h_m}\right) \sin\frac{m\pi}{t}x \qquad 0 \le x \le t \ . \tag{26}$$

The arbitrary constant A_m can be chosen so that the field $E_y(x)$ corresponds to a power flow of one watt (per unit width in the y direction) in the mode. A mode with field $E_y = A\,E_y(x)$ corresponds to a power flow of $|A|^2$ W/m. The normalizing condition is therefore

$$-\frac{1}{2}\int_{-\infty}^{\infty} E_y H_x^* \, dx = \frac{\beta_m}{2\omega\mu}\int_{-\infty}^{\infty}\left[E_y^{(m)}(x)\right]^2 dx = 1$$

which results in

$$A_m = 2h_m\left[\frac{\omega\mu}{|\beta_m|\left(t+\frac{1}{q_m}+\frac{1}{p_m}\right)\left(h_m^2+q_m^2\right)}\right]^{1/2}$$

and since the modes $E_y^{(m)}$ are orthogonal,

$$\int_{-\infty}^{\infty} E_y^{(\ell)} E_y^{(m)} \, dx = \frac{2\omega\mu}{\beta_m}\,\delta_{\ell,m} \, .$$

Using this expression for A_m and the fact that the field is well confined results in

$$E_y^{(m,\omega)} = 2\sqrt{\frac{\omega\mu}{\beta_m t}}\,\sin\frac{m\pi x}{t} \, , \quad 0 \le x \le t \, .$$

The overlap integral $S^{(n,n,m)}$ is maximum for n = m = 1, i.e., fundamental mode operation both at ω and ω/2. For this case the overlap integral becomes

$$S^{(1,1,1)} = \int_{-\infty}^{\infty} E_y^{(1,\omega/2)}\, E_y^{(1,\omega/2)}\, E_y^{(1,\omega)}\, f(x)\, dx$$

$$= \int_0^t E_y^{(1,\omega/2)}\, E_y^{(1,\omega/2)}\, E_y^{(1,\omega)}\, dx$$

$$= \frac{1.\,2\sqrt{2}}{\sqrt{t}}\,\frac{(\omega\mu)^{3/2}}{(\beta^{\omega})^{1/2}\,\beta^{\omega/2}} \qquad (27)$$

and Eq. (24) can be written as

$$\frac{dA^{(\omega)}}{dz} = -i\,\frac{2^{1/2} \times 1.2}{4}\,\frac{d}{\sqrt{t}}\,\frac{\mu^{3/2}\,\omega^{5/2}}{(\beta^{\omega})^{1/2}\beta^{\omega/2}}\,(A^{(\omega/2)})^2\,e^{-i\Delta z} \quad (28)$$

with

$$\Delta \equiv \beta^{\omega} - 2\beta^{\omega/2} \quad (29)$$

and where the now superfluous mode number subscripts have been dropped. Integrating Eq. (28) over the interaction distance ℓ gives

$$|A^{(\omega)}(\ell)|^2 = \frac{\omega^5\mu^3(1.2)^2 d^2}{8\beta^{\omega}(\beta^{\omega/2})^2 t}\,|A^{\omega/2}|^4\,\ell^2\,\frac{\sin^2(\Delta\ell/2)}{(\Delta\ell/2)^2} . \quad (30)$$

The normalization condition was chosen so that $|A^2|$ is the power per unit width in the mode. For a waveguide of width w, Eq. (30) can be rewritten as

$$\frac{P^{\omega}}{P^{\omega/2}} = 0.72\left(\frac{\mu}{\epsilon_o}\right)^{3/2}\frac{\omega^2 d^2 \ell^2}{n^3}\left(\frac{P^{\omega/2}}{wt}\right)\frac{\sin^2(\Delta\ell/2)}{(\Delta\ell/2)^2} \quad (31)$$

where we used $\beta^{\omega} \simeq \omega\sqrt{\mu\epsilon}$, $\epsilon/\epsilon_o = n^2$. Note that $(P^{\omega/2}/wt)$ is the intensity (W/m^2) of the input mode. Except for a numerical factor of 1.44, this expression is similar to that derived for the bulk crystal case[5]. Efficient conversion results when the phase matching condition

$$\Delta \equiv \beta^{\omega} - 2\beta^{\omega/2} = 0 \quad (32)$$

is satisfied. In this case the factor $\sin^2(\Delta\ell/2)/(\Delta\ell/2)^2$ is unity.

It follows from Eq. (31) that a necessary condition for second harmonic generation is $\Delta\ell/2 \ll \pi$ so that the factor $\sin^2(\Delta\ell/2)/(\Delta\ell/2)^2$ is near unity. In this case the conversion efficiency is proportional to ℓ^2. This phase matching condition can be satisfied by using the dependence of the propagation constants β of the various modes on the waveguide dimensions.[5] An alternate approach is to introduce a space periodic perturbation into the waveguide with a period Λ satisfying

$$\Lambda = \frac{2\pi}{\Delta} . \tag{33}$$

Schemes based on waveguide corrugation and on modulating the nonlinear coefficient d have been proposed.[7] As an example, consider the case where the nonlinear coefficient is made periodic in z with period Λ. If d(z) varies as a square wave between d(z) = d then by Fourier expansion

$$d(z) = \frac{d}{2} + \sum_{\substack{q \text{ odd} \\ \text{integer}}} \frac{2d}{q\pi} \sin \frac{2\pi q}{\Lambda} z . \tag{34}$$

Substituting this expression for d(z) into Eq. (24) results in

$$\frac{dA_m^{(\omega)}}{dz} = -\frac{i\omega}{4}\left[\frac{d}{2} - \sum_{q \text{ odd}} \frac{id}{q\pi}\left(e^{\frac{i2\pi qz}{\Lambda}} - e^{-\frac{i2\pi qz}{\Lambda}}\right)\right] \tag{35}$$

$$\times \left[A_n^{(\omega/2)}\right]^2 e^{-i\left(2\beta_n^{\omega/2} - \beta_m^{\omega}\right)z} S^{(n,n,m)} .$$

The period Λ can be chosen so that for some value of the integer q

$$-\frac{2\pi q}{\Lambda} + \beta_m^{\omega} - 2\beta_n^{\omega/2} = 0 . \tag{36}$$

This results in a synchronous term (i. e., one with a zero exponent) on the right side of Eq. (35) so that

$$\frac{dA_m^{(\omega)}}{dz} = \frac{\omega d}{4q\pi}\left[A_n^{(\omega/2)}\right]^2 S^{(n, n, m)} \tag{37}$$

when the nonsynchronous terms are neglected. A comparison with Eq. (24) shows that the effective nonlinear coefficient is now reduced to

$$d_{eff} = \frac{d}{q\pi}$$

and that instead of Eq. (31)

$$\frac{P^{\omega}}{P^{\omega/2}} = 0.72 \left(\frac{\mu}{\epsilon_0}\right)^{3/2} \frac{\omega^2 d_{eff}^2 \ell^2}{n^3} \left(\frac{P^{\omega/2}}{wt}\right). \qquad (38)$$

Thus operation based on $q = 1$ is most efficient, leading to a reduction by a factor of π^2 in the conversion efficiency to that obtained for the case where $d(z) = d$. Note, however, that the factor $\sin^2(\Delta\ell/2)/(\Delta\ell/2)^2$ is now unity making it possible to take advantage of the ℓ^2 dependence of the conversion efficiency.

4. Photoelastic Coupling

The possibility of coupling dielectric waveguide optical modes through the intermediary of sound waves has been demonstrated.[8]

The photoelastic effect is defined by relating the effect of strain $S_{k\ell}$ on the constants of the index ellipsoid through[9]

$$\Delta\left(\frac{1}{n^2}\right)_{ij} = P_{ijk\ell}\, S_{k\ell} \qquad (39)$$

where $P_{ijk\ell}$ is the photoelastic tensor. Taking the strain field in the form of

$$S^{(\Omega)}_{k\ell}(\vec{r}, t) = \frac{1}{2} S^{(\Omega)}_{k\ell}\, e^{i(\Omega t - Kz)} + \text{c. c.} \qquad (40)$$

results in

$$[P_{pert}(\vec{r}, t)]_i = \frac{\epsilon_i \epsilon_j P_{ijk\ell}}{4\epsilon_0} \Big[S^{(\Omega)}_{k\ell} E^{(\omega)}_j e^{i[(\omega+\Omega)t - (\beta+K)z]}$$

$$+ S^{(-\Omega)}_{k\ell} E^{(\omega)}_j e^{i[(\omega-\Omega)t - (\beta-K)z]} + \text{c. c.}\Big] \qquad (41)$$

for the polarization wave arising from the nonlinear mixing of an electric field at frequency ω

$$\frac{1}{2} E_j^{(\omega)} e^{i(\omega t - \beta z)} + \text{c. c.} \tag{42}$$

and a sound strain wave (Eq. (40)) at frequency Ω.

To be specific, assume that the input optical field is a TM mode. The equation governing the evolution of the TE mode due to the coupling will be derived. The information relating to crystal symmetry and orientation will be abbreviated by defining

$$\epsilon^2 pS \equiv \epsilon_i \epsilon_j p_{ijk\ell} S_{k\ell} \ell_{jk} \ell_{iy} \tag{43}$$

which upon substitution into Eq. (41) yields

$$[P_{pert}(\vec{r}, t)]_y = \frac{\epsilon^2}{4\epsilon_0} p \left[S^{(\Omega)} E_x^{(\omega)} e^{i[(\omega + \Omega)t - (\beta_{TM} + K)z]} \right.$$

$$\left. + S^{(-\Omega)} E_x^{(\omega)} e^{i[(\omega - \Omega)t - (\beta_{TM} - K)z]} \right] + \text{c. c.}$$

The equation governing the field amplitudes is therefore

$$\frac{dA_m^+}{dz} e^{i(\omega_{TE} t - \beta_{TE})z} - \frac{dA_m^-}{dz} e^{i(\omega_{TE} t + \beta_{TE} z)}$$

$$= - \frac{i}{8} \int_{-\infty}^{\infty} \frac{\epsilon^2 p(x, z) S^{\Omega}(x)}{\epsilon(x) \quad \epsilon_0} \beta_\ell B_\ell H_y^{(\ell)}(x) E_y^{(m)}(x)\, dx$$

$$\left[e^{i[(\omega+\Omega)t - (\beta_{TM} + K)z} + e^{i[(\omega-\Omega)t - (\beta_{TM} - K)z]} \right]. \tag{44}$$

Each of the two terms on the right hand side of Eq. (44) represents a traveling polarization wave. Both input waves, i.e., $S^{(\Omega)}$ and $E_x^{(\omega)}$, were taken as traveling in the +z direction. Ordinarily, β_{TE} is close to, but slightly larger than, β_{TM}. In this case the coupling is via the first term on the right side of Eq. (44) and the wavelength of the sound wave is adjusted so that

$$\beta_{TE} = \beta_{TM} + K \tag{45}$$

and the resulting TE mode is shifted up in frequency to

$$\omega_{TE} = \omega + \Omega \quad .$$

Since the sign of β_{TE} and β_{TM} is the same, the coupling is codirectional. This is the case which we consider in detail below. Since $K/\beta = (c/V_s)(\Omega/\omega)$, where v_s is the sound velocity, it is possible for reasonable values of the sound frequency Ω to have $K \simeq 2\beta$. In this case the second term on the right side of Eq. (44) represents a polarization wave traveling in the -z direction with a phase velocity

$$-\frac{\omega}{K-\beta} \simeq \frac{-\omega}{\beta} \quad .$$

This wave is capable of coupling to the backward TE (or TM) mode. In this case

$$\beta_{TE} = \beta_{TM} - K < 0$$

$$\omega_{TE} = \omega + \Omega \quad . \tag{46}$$

Returning to the codirectional coupling case represented by Eq. (45), the equation determining the field amplitude A_m (Eq. (44)) can be simplified for the case of well confined wave-guide modes when the guiding layer is assumed uniformly acousto-optic and the strain field is taken as constant over the same region. Since $\beta_m^{TM} \simeq \beta_m^{TE}$ (ℓ = m) the term involving $A_m^{(-)}$ may be neglected. The coupling thus involves only forward TE and TM modes. The simplified equation is

$$dA_m^{(\omega+\Omega)}/dz = ikB_m^{(\omega)} e^{-i\Delta z} ,$$

and from Eqs. (2) and (4)

$$dB_m^{(\omega)}/dz + -ikA_m^{(\omega+\Omega)} e^{i\Delta z}$$

$$\Delta \equiv K - (\beta_{TE} - \beta_{TM}) \tag{47}$$

where it is assumed that $\omega >> \Omega$. The coupling constant, is given by

$$k = \frac{\pi p S^{(\Omega)} n_2^3}{2\lambda_o} \tag{48}$$

The solution of Eq. (47) is given by Eq. (6) and illustrated by Fig. 1. Complete power transfer can take place only when $\Delta = 0$ i. e., when

$$K = \beta_{TE} - \beta_{TM} \ . \tag{49}$$

Since $K = \Omega/v_s$, this condition can be fulfilled by adjusting the sound frequency Ω. Under phase matched conditions we have, according to Eq. (6)

$$|A_m^{(\omega+\Omega)}|^2 = |B_m(0)|^2 \sin^2 kz$$

$$|B_m^{(\omega)}|^2 = |B_m(0)|^2 \cos^2 kz \tag{50}$$

with complete power exchange in a distance

$$\ell = \pi/2k \ . \tag{51}$$

It is of interest to estimate the acoustic power needed to satisfy the switching condition Eq. (51). Solving Eq. (51) for the strain using Eq. (48) gives

$$S^2 = \frac{\lambda_o^2}{\ell^2 p^2 n^6} \ .$$

The corresponding acoustic intensity I (W/m^2) can be obtained using the relation $I = \rho v_s^3 S^2/2$ where ρ is the mass density. The result is

$$I_{switching} = \frac{\lambda_o^2 \rho v_s^3}{2\ell^2 p^2 n^6} = \frac{\lambda_o^2}{2\ell^2 M} \tag{52}$$

where $M \equiv n^6 p^2/\rho v_s^3$ is the acoustic figure of merit.[9]

In a GaAs crystal, as an example, using the data $M \simeq 10^{-13}$, $\ell = 5$ mm, and an optical wavelength $\lambda_o = 1\ \mu m$, results in

$$I_{switching} = 20\ W/cm^2 .$$

The corresponding strain amplitude is

$$S^{(\Omega)} \cong 2.3 \times 10^{-5}$$

where $\rho = 5.34\ g/cm^3$ and $v_s = 5.15 \times 10^3 m/sec$ was used.

5. Coupling by a Surface Corrugation

Consideration will now be given to perturbation caused by mechanical corrugation of the surface of a slab wave guide as is illustrated in Fig. 3. Using the relation

$$\vec{\rho} = [\epsilon(\vec{r}) - \epsilon_o]\vec{E}$$

the driving polarization is

$$\vec{P}_{pert} = \Delta\epsilon(\vec{r})\ E(\vec{r}, t) = \Delta n^2(\vec{r})\ \epsilon_o\ \vec{E}(r, t) \tag{53}$$

where $\epsilon(r) = n^2(r)\epsilon_o$ so that coupling is only between TE or TM modes but not from TE to TM. To be specific, consider the case of a TE mode of order m propagating in the +z direction in a smooth waveguide. At $z = 0$ it encounters a corrugated region, as shown in Fig. 3, extending to $z = L$.

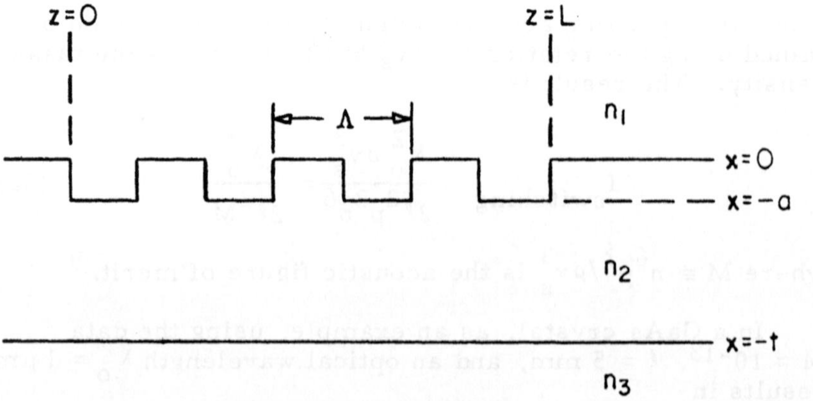

Fig. 3. Corrugated dielectric waveguide.

Using Eq. (17) for E_y and limiting the summation to a single term $\ell = m$ gives upon substitution in Eq. (53)

$$[P_{pert}(\vec{r}, t)]_y = \frac{\Delta n^2(\vec{r})\epsilon_o}{2}$$

$$\left[A_m^{(+)}(z)\, E_y^{(m)}(x)\, e^{i(\omega t - \beta_m z)} + c.c.\right] . \qquad (54)$$

Now anticipate that the period Λ will be chosen so that $2\pi/\Lambda \simeq 2\beta_m$ and the coupling will thus, take place predominantly between the forward and backward modes of order m. Substituting Eq. (54) in Eq. (19) and limiting the left side of the latter to the backward $A_m^{(-)}$ term gives

$$\frac{dA_m^{(-)}}{dz} = \frac{i\omega\epsilon_o}{4} A_m^{(+)} e^{-2i\beta_m z} \int_{-\infty}^{\infty} \Delta n^2(x, y)\, [E_y^{(m)}(x)]^2 \, dx \; . \tag{55}$$

From Fig. 3 the Fourier expansion of Δn^2 is

$$\Delta n^2(x, z) = \Delta n^2(x) \left[\frac{1}{2} + \frac{2}{\pi} \left(\sin \eta z + \frac{1}{3} \sin 3\eta z + \ldots\right)\right] \equiv \sum_{\ell} \Delta n_{\ell}^2 (x, z) \tag{56}$$

where

$$\Delta n^2(x) = \begin{cases} n_2^2 - n_1^2 & -a \le x \le 0 \\ 0 & \text{elsewhere} \end{cases} \tag{57}$$

$$\eta \equiv 2\pi/\Lambda \; .$$

Coupling from $A_m^{(+)}$ to $A_m^{(-)}$ occurs when, for some ℓ,

$$\ell\eta \equiv \ell \frac{2\pi}{\Lambda} \cong 2\beta_m \; .$$

In this case, keeping only the synchronous term, Eqs. (55) and (56) yield

$$\frac{dA_m^{(-)}}{dz} = \frac{\omega\epsilon_o}{4\pi\ell} A_m^{(+)} e^{i\Delta z} \int_{-\infty}^{\infty} \Delta n^2(x) \left[E_y^{(m)}(x)\right]^2 dx \tag{58}$$

$$\Delta \equiv \ell\eta - 2\beta_m \; . \tag{59}$$

The next task is to evaluate the integral in Eq. (58). Using Eq. (57), the integral can be written as

$$\int_{-\infty}^{\infty} \Delta n^2(x)\left[E_y^{(m)}(x)\right]^2 dx = \left(n_2^2 - n_1^2\right)$$

$$\int_{0}^{+a} \left[E_y^{(m)}(x)\right]^2 dx = \left(n_2^2 - n_1^2\right) A_m^2 \tag{60}$$

$$\int_{0}^{+a} \left[\cos(h_m x) + \frac{q_m}{h_m}\sin(h_m x)\right]^2 dx \ .$$

Although the integral in Eq. (60) can be calculated exactly using the TE field solutions, an especially simple result ensues if we consider that operation is sufficiently above the propagation cutoff, so that $q_m >> h_m$. Performing the integration and assuming $\sin ha \simeq ha$ results in

$$\left(n_2^2 - n_1^2\right)\int_{0}^{a}\left[E_y^{(m)}(x)\right]^2 dx = \frac{A_m^2 q_m^2 a^3}{3}\left[1 + \frac{3}{q_m^2 a^2} + \frac{3}{q_m a}\right] \tag{61}$$

In the well confined regime $q_m, p_m >> h_m$ so that

$$A_m^2 \rightarrow \frac{4h_m^2 \omega\mu}{\beta_m t q_m^2} \ . \tag{62}$$

Using $\beta_m \rightarrow n_2 k$, $h_m \rightarrow \pi/t$ and Eq. (62) in Eq. (61) leads to

$$\left(n_2^2 - n_1^2\right)\int_{0}^{+a}\left[E_y^{(m)}(x)\right]^2 dx = \left(n_2^2 - n_1^2\right)$$

$$\frac{4\pi^2\omega\mu}{3n_2 k}\left(\frac{a}{t}\right)^3\left[1 + \frac{3}{q_{ma}} + \frac{3}{q_m^2 a^2}\right]$$

which upon substitution into Eq. (58) gives

$$\frac{dA_m^{(-)}}{dz} = \kappa_\ell A_m^{(+)} e^{i\Delta z} \tag{63}$$

and using Eq. 10

$$\frac{dA_m^{(+)}}{dz} = \kappa_\ell A_m^{(-)} e^{-i\Delta z} \qquad \kappa_\ell = \frac{\pi k}{3\ell} \frac{\left(n_2^2 - n_1^2\right)}{n_2} \left(\frac{a}{t}\right)^3$$

$$\left[1 + \frac{3}{2\pi} \frac{\lambda_o/a}{\left(n_2^2 - n^2\right)^{1/2}} + \frac{3}{4\pi^2} \frac{(\lambda_o/a)^2}{\left(n_2^2 - n_1^2\right)}\right] \tag{64}$$

where ℓ is the order of the (corrugation function) harmonic responsible for the coupling.

The behavior of the incident and reflected waves $A_m^{(+)}$ and $A_m^{(-)}$ is given by Eq. (14) and is illustrated by Fig. 2 in which the reflected wave is A(z) and the incident wave B(z)). The exponential "decay" behavior occurs only for a narrow range of frequencies which satisfy, according to Eq. (13), the condition

$$\Delta \equiv \eta - 2\beta(\omega) \leqq 2\kappa \tag{65}$$

where $\eta = 2\pi/\Lambda$ and $\ell = 1$ only. This behavior is formally analogous to Bragg scattering of Bloch electron waves in a crystal from one edge of the Brillouin zone to the other by the crystal periodicity.[10] The latter phenomenon is responsible for the appearance of forbidden energy gaps. The behavior of the corresponding optical gap can be elucidated by considering the total propagation constant of, say, the incident wave B(z) of (12) ($A^+(z)$ in the notation of this section). From Eqs. (8) and (12) the propagation constant is

$$\beta'(\omega) = \beta(\omega) - \frac{\Delta}{2} \pm i\frac{S}{2} = \frac{\eta}{2} \pm i\frac{S}{2} \tag{66}$$

$$S = \sqrt{4\kappa^2 - \Delta^2} \, .$$

The imaginary part of β' is then given by

$$\operatorname{Im} \beta'(\omega) = \sqrt{\kappa^2 - (\beta - \eta/2)^2} \simeq \sqrt{\kappa^2 - \frac{n_{eff}^2}{c^2}(\omega - \omega_o)^2} \tag{67}$$

where ω_o, the midgap frequency, is defined by $\beta(\omega_o) = \eta/2$ and to get the second equality the unperturbed behavior of β was approximated by $\beta(\omega) \simeq (\omega/c)\, n_{eff}$. The height of the energy gap is thus the frequency region over which β' is complex. Using Eq. (67) it is given by

$$(\Delta\omega)_{gap} \equiv \omega_u - \omega_\ell = \frac{2\kappa c}{n_{eff}} \tag{68}$$

where ω_u and ω_ℓ are the upper and lower gap frequencies, respectively.

The behavior of β' at $\omega > \omega_u$ and $\omega < \omega_\ell$ is likewise derivable from Eq. (66). It is given for $\omega > \omega_u$, as an example, by

$$\beta' = \frac{\eta}{2} \pm \sqrt{\frac{n_{eff}^2}{c^2}(\omega - \omega_u)^2 + \frac{2 n_{eff} \kappa}{c}(\omega - \omega_u)} \; . \tag{69}$$

Equations (67) and (69) are valid for any coupling between the forward and backward modes of a waveguide which is describable by a set of equations, such as Eq. (63), regardless of the physical mechanism responsible for the coupling. As an illustration of the above ideas consider the ω-β' diagram in the vicinity of the optical gap using the case of coupling by a surface corrugation. Referring to Fig. 4 the following were used to plot ω-β'; $t = 3\ \mu m$, $\Lambda = 0.143\ \mu m$, $a = 0.3\ \mu m$, $n_2 = 3.6$, $n_3 = 3.4$, $n_1 = 1$ and $n_{eff} = n_2$. The midgap wavelength if $\lambda_o (= 2\pi c/\omega_o) = 1\ \mu m$. The plot corresponding to Eqs. (67) and (69) is dashed. The solid curves are the result of an "exact" analysis.[11].

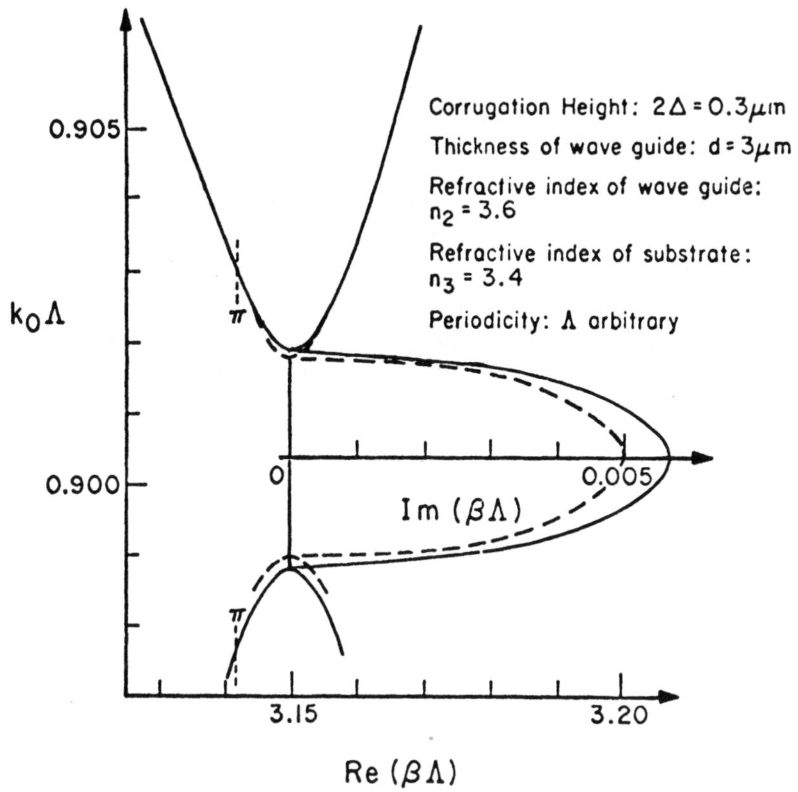

Fig. 4. Plot of the dispersion ($k\Lambda$ vs $\beta\Lambda$) diagram in the vicinity of the optical gap.

6. The Eigenmodes of a Perturbed Waveguide

Up to this point we adopted the point of view of a perturbation that couples the (otherwise coupled) TE and TM modes of a waveguide. This coupling, in the codirectional case, was shown on a number of occasions to lead to equations of the form

$$\frac{dA}{dz} = \kappa B e^{-i\Delta z}$$

$$\frac{dB}{dz} = -\kappa^* A e^{i\Delta z} . \qquad (70)$$

A and B are the normalized mode amplitudes and the corresponding field variables vary as

$$a(x, z, t) \propto A(z)\, E_y^{(m)}(x) e^{i(\omega_a t - \beta_a z)}$$

$$b(x, z, t) \propto B(z)\, H_y^{(\ell)}(x) e^{i(\omega_b t - \beta_b z)} \quad . \tag{71}$$

An alternative point of view is to find the eigenmodes of the perturbed system, i.e., those linear combinations of A(z) and B(z) which, except for a propagation factor exp ($i\gamma z$), are independent of z.

Define a column vector $\tilde{E}$ as

$$\tilde{E}(z) = \begin{vmatrix} B(z)\, e^{-i\beta_b z} \\ A(z)\, e^{-i\beta_a z} \end{vmatrix} \equiv \begin{vmatrix} E_1(z) \\ E_2(z) \end{vmatrix} \quad . \tag{72}$$

The evolution of $\tilde{E}(z)$ is obtained from Eq. (70) and is described by

$$\frac{d\tilde{E}}{dz} = \tilde{\tilde{C}}\, \tilde{E} \tag{73}$$

with

$$\tilde{\tilde{C}} = \begin{vmatrix} -i\beta_b & -\kappa^* \\ \kappa & -i\beta_a \end{vmatrix} \quad . \tag{74}$$

An eigenmode of the waveguide will have a solution of the form

$$\tilde{E}(z) = \tilde{E}(0) \exp(i\gamma z) \quad .$$

Substituting this form in Eq. (73) leads to two homogeneous equations for E_1 and E_2

$$-i(\beta_b + \gamma)E_1 - \kappa^* e_2 = 0$$

$$\kappa E_1 - i(\beta_a + \gamma)E_2 = 0 \quad . \tag{75}$$

The solution of the resulting determinantal equation is

$$\gamma_{1,2} = -\frac{(\beta_a + \beta_b)}{2} \pm \frac{1}{2}\sqrt{(\beta_a - \beta_b)^2 + 4\kappa^2} \tag{76}$$

$$= -\bar{\beta} \pm \frac{S}{2}$$

$$\bar{\beta} \equiv (\beta_a + \beta_b)/2 \quad , \quad S \equiv \sqrt{\Delta^2 + 4\kappa^2} \quad , \quad \Delta \equiv \beta_a - \beta_b \quad . \tag{77}$$

The corresponding eigenvectors are found from Eq. (75)

$$\tilde{E}_1 = \begin{vmatrix} \dfrac{2i\kappa^*}{\Delta + S} \\ 1 \end{vmatrix} e^{-i(\bar{\beta} - \frac{S}{2})z}$$

$$E_2 = \begin{vmatrix} \dfrac{2i\kappa^*}{\Delta - S} \\ 1 \end{vmatrix} e^{-i(\bar{\beta} + \frac{S}{2})z} \quad . \tag{78}$$

Note that $\tilde{E}_1 \cdot \tilde{E}_1^*$ and $\tilde{E}_2 \cdot \tilde{E}_2^*$ are the mode powers and that $\tilde{E}_1 \cdot \tilde{E}_2^* = 0$. The two components $2i\kappa^*/(\Delta \pm S)$ and 1 of each eigenvector represent the normalized amplitudes of the TE and TM components of each mode, so that the amount of admixture, i.e., the ratio of the powers in the two polarizations is $4\kappa^2/(\Delta \pm S)^2$. In the limit of $\kappa/\Delta \to 0$, $\tilde{E}_1$ and $\tilde{E}_2$ become

$$\tilde{E}_1 \to \begin{vmatrix} 0 \\ 1 \end{vmatrix} , \quad \text{and} \quad \tilde{E}_2 \to \begin{vmatrix} 1 \\ 0 \end{vmatrix}$$

to within a multiplicative constant, and the eigenvectors become the uncoupled TE and TM modes. Another important consequence is that when $\Delta = 0, S = 2\kappa$ and the admixture is 50%-50% regardless of κ. Even a more unconventional consequence is the fact that the two components E_1 and E_2 of each eigenvector do not, according to Eq. (71), correspond necessarily to fields of the same frequency. Thus, it is possible, as an example, when the perturbation is time periodic, to have an eigenmode traveling in a wave-guide with

$$E_x \propto (2i\kappa^* / (\Delta + S) \exp\left\{i\left[\omega_b t - (\bar{\beta} - \frac{S}{2})z\right]\right\} H_y^{(\ell)}(x)$$

while

$$E_y \propto \exp\left\{i\left[\omega_a t - (\bar{\beta} - \frac{S}{2})z\right]\right\} E_y^{(m)}(x) \quad . \tag{79}$$

This is the case in photoelastic coupling by a traveling sound wave. What makes these strange bedfellows into a mode is the fact that they travel with the same phase factor $\exp\ [-i(\beta - S/2)z]$. This phenomenon may be important in a laser oscillator which contains in its optical path a time modulated coupling section.

REFERENCES

1. Amnon Yariv, "On the Coupling Coefficients in the Coupled-Mode Theory, Proc. IRE 46, 1956 (1958).

2. Dietrich Marcuse, Light Transmission Optics (Van Nostrand Reinhold Company, New York, 1972).

3. Dietrich Marcuse, "Mode Conversion Caused by Surface Imperfections of a Dielectric Slab Waveguide," Bell Syst. Tech. J. 48, 3187 (1969).

4. M. L. Dakss, L. Kuhn, P. F. Heidrich, and B. A. Scott, "Grating Coupler for Efficient Excitation of Optical Guided Waves in Thin Films," Appl. Phys. Letters 16, 523 (1970).

5. Amnon Yariv, Introduction to Optical Electronics (Hold Rinehart and Winston, Inc., 1971), p. 190.

6. D. B. Anderson and J. T. Boyd, Appl. Phys. Lett. 19, 266 (1971).

7. Sasson Somekh and Amnon Yariv, "Phase Matching by Periodic Modulation of the Nonlinear Optical Properties," Opt. Commun. 6, 301 (1972).

8. L. Kuhn, M. L. Dakss, P. F. Heidrich, and B. A. Scott, "Deflection of Optical Guided Waves by a Surface Acoustic Wave," Appl. Phys. Lett. 17, 265 (1970).

9. R. W. Dixon, "The Photoelastic Properties of Selected Materials and Their Relevance to Acoustic Light Modulators and Scanners," J. Appl. Phys. 38, 5149 (1967).

10. C. Kittel, Introduction to Solid State Physics (John Wiley and Sons, Inc., 4th ed., Chapter 9).

11. K. Sakuda and A. Yariv, Optics Commun. 8, 1 (1973).

Chapter 11 – OPTICAL DIRECTIONAL COUPLERS

Sasson Somekh*

California Institute of Technology
Pasadena, California

1. Introduction

The optical directional coupler, analogous to the microwave element[1] of the same name, consists of parallel channel optical waveguides sufficiently closely spaced that energy is transferred from one to another. For this coupling to take place cumulatively over a substantial length, the light must propagate with the same phase velocity in each channel. The fraction of the power coupled per unit length is determined by the overlap of the modes in the separate channels. Thus, it depends on the separation of the guides, the mode penetration into the substrate, and the interaction length. Figure 1 is an example of a multichannel directional coupler. It shows a diagram of a large number of coupled waveguides (produced by proton implantation) and typical intensity profiles of the guided light. The incident light is focused into a single channel at $z = 0$, but is coupled into the adjacent guides as it propagates.

*Present address, Bell Telephone Laboratories, Murray Hill, New Jersey.

A theoretical analysis of the coupling between dielectric waveguides usually involves a formal solution of Maxwell's equations.[2] We shall attempt here to derive an expression for the coupling coefficient and its sign from an intuitive physical point of view without resorting to an elaborate mathematical solution. We shall also derive the power distribution as a function of propagation distance for two coupled guides and an infinite number of coupled guides.

Experimental results of a dual channel directional coupler and a multichannel coupler will be given. These channels were embedded at the surface of GaAs by proton implantation. A different type of directional coupler suitable for ridged waveguides is also described. We conclude with an account of the different applications of these devices in an integrated optics circuit, such as couplers, modulators, multiplexers and polarizers.

2. Coupled Mode Formalism

Let us consider the two waveguides shown in Fig. 2(a). The waveguides can have an arbitrary cross section; however, for the sake of simplicity, we consider only guides with rectangular cross sections. The electric field of a propagating mode in a guide is described by the complex amplitude A(z) which includes the phase term $e^{i\beta z}$, and by the mode $\vec{\mathcal{E}}(x, y)$ solved for in the absence of the other guide. Thus we have

$$\vec{E}(x, y, z) = A(z)\, \vec{\mathcal{E}}(x, y) \quad \cdot \qquad (1)$$

The mode profiles $\mathcal{E}(x, y)$ are always normalized to carry one unit of power; the power carried by $E_1(x, y, z)$ (where the subindex 1 denotes the guide number) at a given z is

$$P_1(z) = |A_1(z)|^2 = A_1(z)\, A_1^*(z) \cdot \qquad (2)$$

Employing the coupled mode formalism,[1] we write the general coupled mode equations for the amplitudes of the two modes

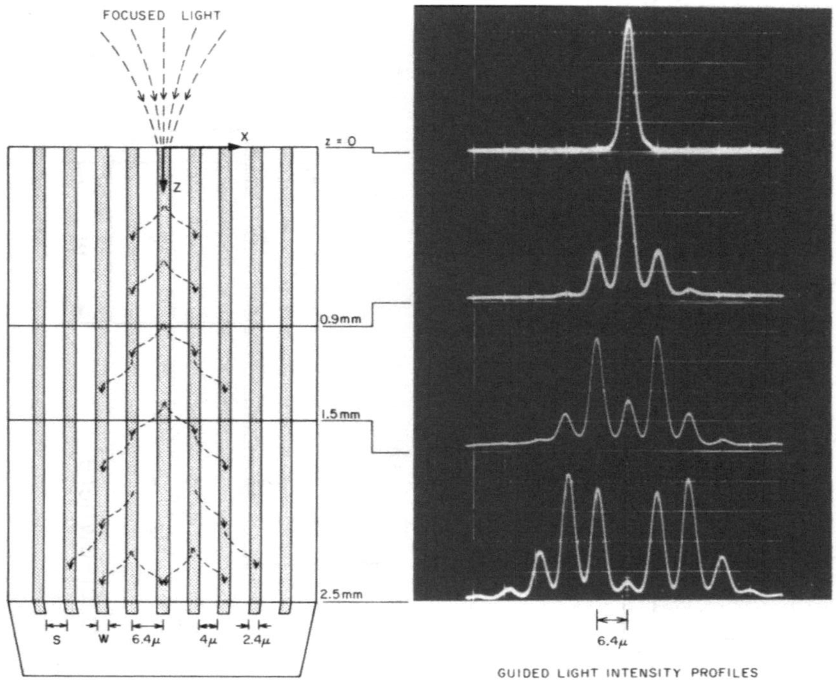

Fig. 1. Sketch of channel optical waveguide directional coupler showing flow of light energy into adjacent channels. To the right are photographs of guided light intensity profiles for various lengths. Intensity scale is arbitrary.

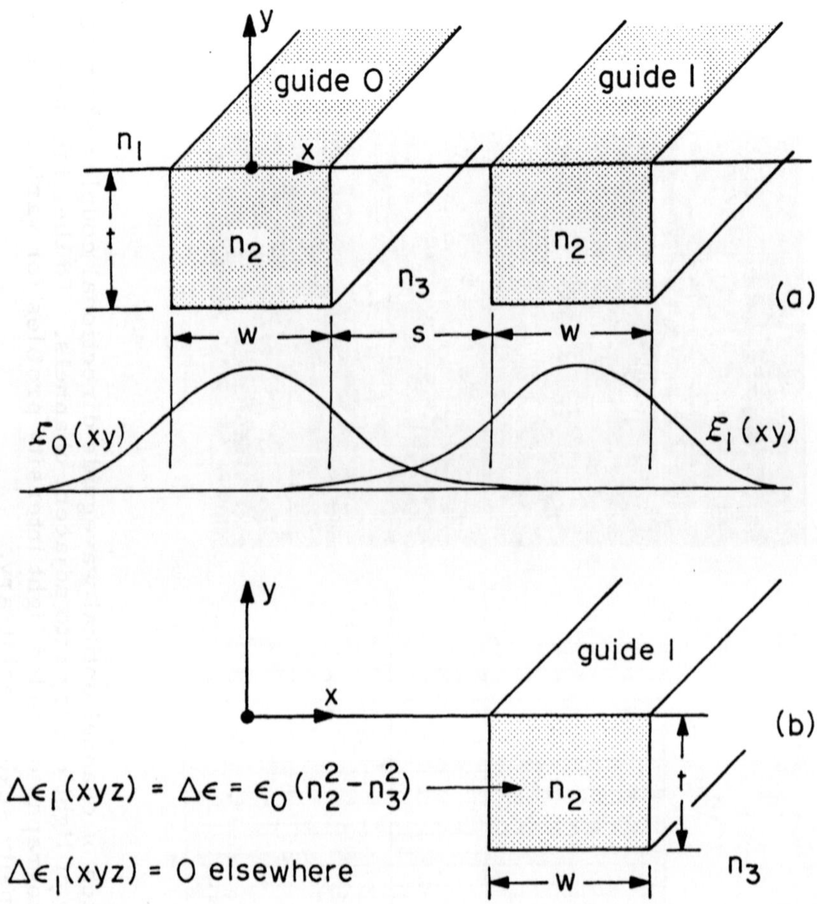

Fig. 2. (a) Sketch of two adjacent rectangular waveguides propagating modes.
(b) Definition of the dielectric discontinuity $\Delta\epsilon_1(xyz)$.

$$\frac{dA_0(z)}{dz} = -i\beta_0 A_0(z) + K_{01}A_1(z) \tag{3}$$

$$\frac{dA_1(z)}{dz} = i\beta_1 A_1(z) + K_{10}A_0(z) \cdot \tag{4}$$

The terms β_0 and β_1 are the propagation constants of the modes in the two guides. (An imaginary in β describes loss or gain that may exist in the guides.) K_{01} and K_{10} represent the coupling between the two modes, and a great deal of the theory in this chapter will be devoted to their derivation. Before treating that we consider a theoretical and an experimental example of the coupled mode behavior.

3. Dual Channel Directional Coupler-Theory and Experiment

Consider the guides shown in Fig. 2(a). We assume that the guides are identical and that both have an exponential loss constant of α. Thus the propagation constant of the guide is given by

$$\beta = \beta_r - i\frac{\alpha}{2} \cdot \tag{5}$$

As we shall see later the coupled mode equation for this case can be written as

$$\frac{dA_0(z)}{dz} = -i\beta\ A_0(z) - iKA_1(z) \tag{6}$$

$$\frac{dA_1(z)}{dz} = -i\beta\ A_1(z) - iKA_0(z) \tag{7}$$

where K, the coupling coefficient, is a real quantity. We also assume that at $z = 0$ light is coupled into guide 0 so that the boundary conditions for the problem are given by

$$A_0(0) = 1 \qquad A_1(0) = 0 \tag{8}$$

The solution is described by

$$A_0(z) = \cos Kz\, e^{i\beta z} \quad (9)$$

$$A_1(z) = -i \sin Kz\, e^{i\beta z} . \quad (10)$$

The power flow in the guides therefore is

$$P_0(z) = A_0(z)\, A_0^*(z) = \cos^2 Kz\, e^{-\alpha z} \quad (11)$$

$$P_1(z) = A_1(z)\, A_1^*(z) = \sin^2 Kz\, e^{-\alpha z} . \quad (12)$$

From Eqs. (11) and (12) we learn that the power will oscillate back and forth between the two guides as shown in Fig. 3. Because of the losses present in the guide, the power will also decay exponentially as a function of the propagation distance. The length necessary for a complete transfer of power from one guide to the other is given by

$$L = \frac{\pi/2}{K} . \quad (13)$$

Such a directional coupler that allows the transfer of light from one channel to another is one of the building blocks of optical integrated circuits. We shall describe below the first operation of such a device in the optical regime.

The dual channel directional coupler was fabricated by proton implantation in GaAs (Ref. 3) in a fashion described in detail in Ref. 19. The channel guides were formed by implanting the protons through a gold mask with the appropriate slit openings in it. Bombardment causes defect centers in the material, which trap the free carriers. This compensation of the free carriers eliminates their negative plasma contribution to the index of refraction, and thus increases the index for the samples used here ($N_{substrate} \approx 2 \times 10^{18}$) by $\Delta n \simeq 0.005$. The cross section of the gold mask on top of the GaAs substrate is shown in Fig. 4(a). This way the guides embedded at the surface have a cross

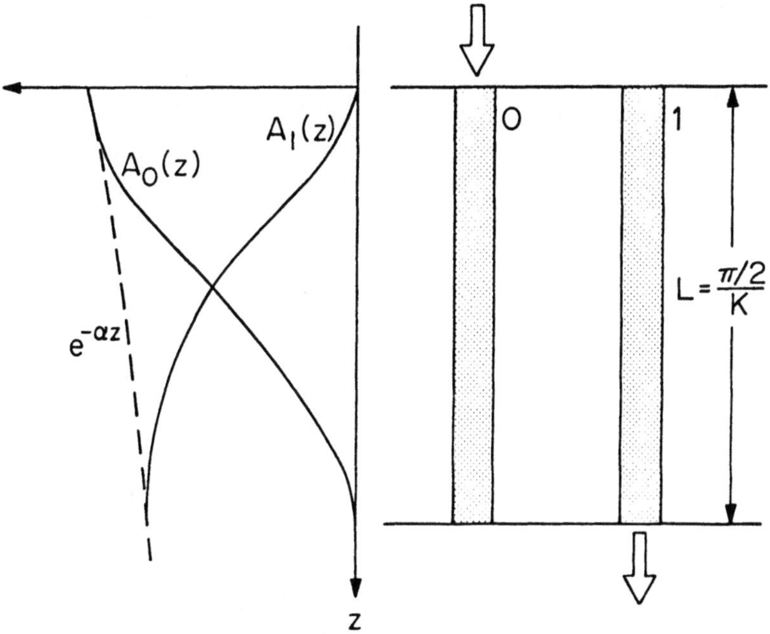

Fig. 3. Sketch of a dual optical directional coupler showing flow of light energy between the guides.

Fig. 4. Scanning electron microscope photograph of the gold mask on the GaAs substrate used in the fabrication of the (a) dual directional coupler, (b) multichannel directional coupler. The remains of the photoresist can be seen on top of the gold stripes.

section of about 3 x 3 μm, an index discontinuity of about $\Delta n \simeq 0.005$, and a separation of 3 μm between the guides.

Figure 5 describes three pairs of such directional couplers. On the left is the input plane of the device and the black dot indicates the point at which the input light was focused. On the right is a series of scans of the light distribution at the output plane taken by the setup described in Ref. (3). Let us confine our attention to the directional coupler at the center. The second row from the top shows that when the input light is coupled into the channel on the right, it emerges at the output plane from both channels. Similarly, the second row from the bottom indicates that the light also emerges from both channels when it is coupled into the left channel. Since the intensity of the emerging light from both channels is about equal, we can write for this sample,

$$K\ell \simeq \frac{\pi}{4}$$

$$\ell = 1\ \text{mm} \qquad \Rightarrow \qquad K \simeq 0.79\ \text{mm}^{-1}\ .$$

By increasing the length of the sample by about a factor of two, a complete transfer of power should be observed because

$$KL = \frac{\pi}{2}$$

This is shown in Fig. 6. The second row from the top describes a situation in which light coupled into the left channel emerges from the one on the left, and the second row from the bottom shows the transfer of power from left to right.

This device can be turned into a switch by controlling K, the coupling coefficient, with the application of an electric field through the electro-optic effect. This exciting possibility will be discussed in detail later.

Now we derive theoretical expressions for the value of the coupling coefficient.

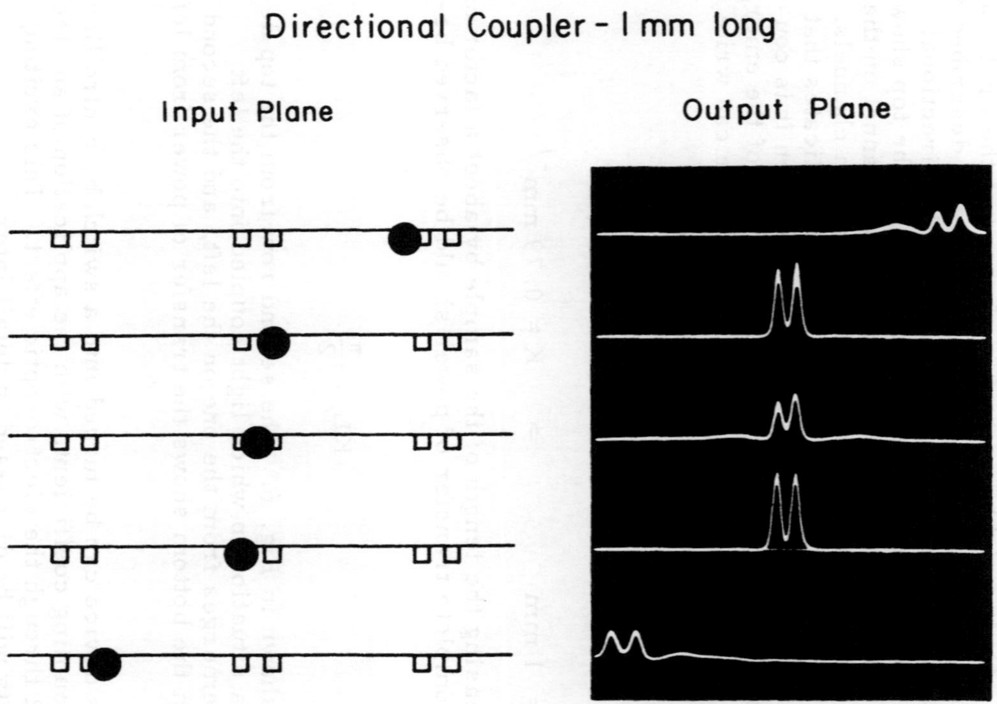

Fig. 5. Sketch of the input plane of a 1 mm long directional coupler. On the right are photographs of guided light intensity profiles at the output for various input conditions.

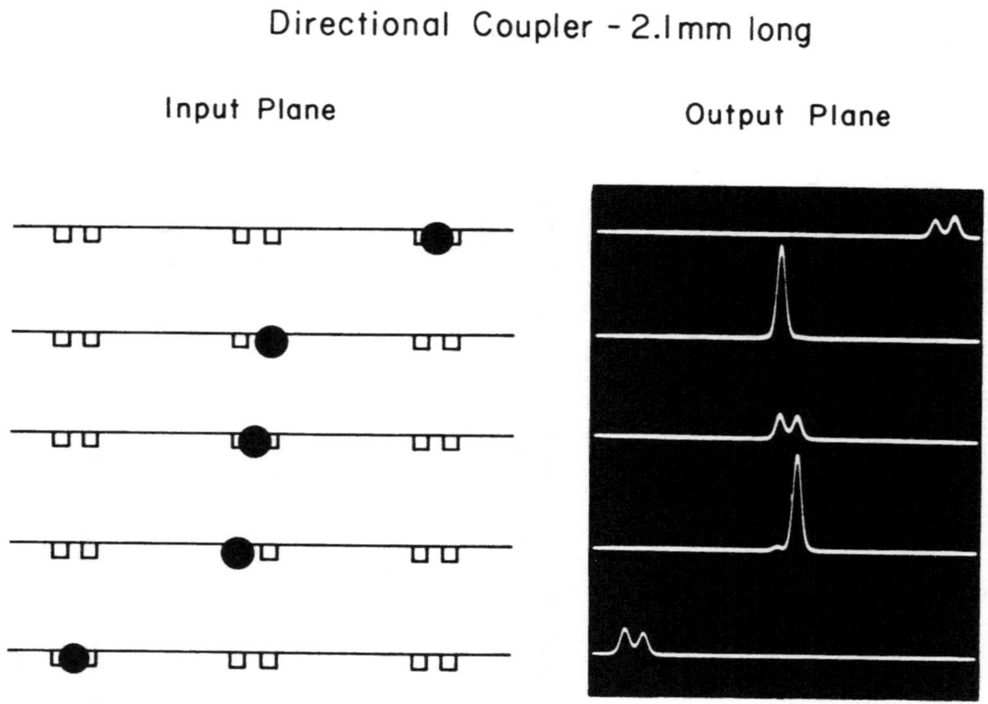

Fig. 6. Sketch of the input plane of a 2.1 mm long directional coupler. On the right are photographs of guided light intensity profiles at the output for various input conditions.

4. Derivation of the Coupling Coefficient

We consider again the coupled mode equations:

$$\frac{dA_0(z)}{dz} = -i\beta_0 A_0(z) + K_{01} A_1(z) \tag{14}$$

$$\frac{dA_1(z)}{dz} = i\beta_1 A_1(z) + K_{10} A_0(z) \cdot \tag{15}$$

We also recall that the electric field in a guide is given by

$$\vec{E}(x, y, z) = A(z)\, \vec{\mathcal{E}}\,(xy) \tag{16}$$

and that the power carried by $\vec{E}_1(x, y, z)$, for example, is

$$P_1(z) = |A_1(z)|^2 = A_1(z)\, A_1^*(z) \cdot \tag{17}$$

Next we obtain the rate of power growth in mode 1 because of the coupling from mode 0. (For this purpose we assume that β is real.) To do this we multiply Eq. (15) by $A_1^*(z)$ and the complex conjugate of Eq. (15) by $A_1(z)$. This gives

$$A_1^*(z)\, \frac{dA_1(z)}{dz} = -i\beta_1 A_1(z)\, A_1^*(z) + K_{10} A_0(z) A_1^*(z) \tag{18}$$

$$A_1(z) \frac{dA_1^*(z)}{dz} = i\beta_1 A_1^*(z)\, A_1(z) + K_{10}^* A_0^*(z)\, A_1(z) \cdot \tag{19}$$

Summing the two equations yields

$$\frac{d[A_1(z)\, A_1^*(z)]}{dz} = 2\,\mathrm{Re}\,[A_1(z)\, K_{10}^* A_0^*(z)] \cdot \tag{20}$$

Using Eq. (2) gives

$$\frac{dP_1(z)}{dz} = 2 \text{ Re } [A_1(z) \, K_{10}^* A_0^*(z)] \; . \tag{21}$$

We shall now obtain a different expression for the power growth from a physical point of view in the following manner. Guide 0 in the absence of guide 1 supports a mode with a mode profile given by $\mathcal{E}_0(xy)$. This mode, in spite of having a long tail in the x and y directions, will suffer no loss, and propagate with a constant power. When guide 1 is introduced, the tail of $\mathcal{E}_0(xy)$ will penetrate into guide 1 and generate additional polarization because of the higher index inside the guide. This perturbation in the polarization in turn will drive the mode in guide 1.

The net power per unit volume expended by an electric field on the electric polarization is given by the well known relation

$$\frac{\text{Power}}{\text{Volume}} = \overline{E \frac{\partial \rho}{\partial t}} \tag{22}$$

where E is the field, ρ is the polarization, and the horizontal bar denotes time averaging.

Thus, we can express the power generation, caused by perturbation in the polarization at a given point in guide 1, by changing the sign on the right of Eq. (22) yielding

$$\frac{dP_1}{dV} = \overline{-E_1(xyzt) \cdot \frac{\partial \rho_{10}(xyzt)}{\partial t}} \tag{23}$$

where ρ_{10} is the driving polarization caused by the mode in guide 0. This perturbation in the polarization can be approximated by multiplying the electric field of mode 0 by the disturbance of the dielectric constant caused by the introduction of guide 1.

$$\rho_{10}(xyz) = \vec{E}_0(xyz) \cdot \Delta\epsilon_1(xyz) \cdot \qquad (24)$$

Over the cross section of guide 1, $\Delta\epsilon_1(xyz)$ has a constant value, independent of z, of

$$\Delta\epsilon = \epsilon_2 - \epsilon_3 = \epsilon_0(n_2^2 - n_3^2)$$

and is zero anywhere else, as shown in Fig. 2(b). Using Eq. (3), Eq. (24) and time dependence of $e^{i\omega t}$, Eq. (23) becomes

$$\frac{dP_1}{dV} = \frac{}{-A_1(z)e^{i\omega t}\mathcal{E}_1(xy)\Delta\epsilon_1(xy)i\omega A_0(z)e^{i\omega t}\mathcal{E}_0(xy)} \cdot \qquad (25)$$

Performing the time averaging yields

$$\frac{dP_1}{dV} = 1/2\,\mathrm{Re}\left\{A_1(z)\mathcal{E}_1(xy)\left[-i\omega\Delta\epsilon_1(xy)\mathcal{E}_0(xy)A_0(z)\right]^*\right\} \cdot \qquad (26)$$

The growth of the total power P_1 carried by the mode in guide 1 can now be derived by simply integrating over x and y. We obtain

$$\frac{dP_1(z)}{dz} = 1/2\,\mathrm{Re}\left\{A_1(z)\left[i\omega\int\mathcal{E}_1(xy)\Delta\epsilon_1^*(xy)\mathcal{E}_0^*(xy)dxdy\right]A_0^*(z)\right\} \cdot \qquad (27)$$

This can also be written as

$$\frac{dP_1(z)}{dz} = 1/2\mathrm{Re}\Big\{A_1(z)\Big[-i\omega\int \mathcal{E}_1^*(xy)\Delta\epsilon_1(xy)\mathcal{E}_0(xy)dxdy\Big]^* A_0^*(z)\Big\} \; . \tag{28}$$

A comparison of Eqs. (21) and (28) immediately reveals that

$$K_{10} = \frac{-i\omega}{4}\int \mathcal{E}_1^*(xy)\Delta\epsilon_1(xy)\mathcal{E}_0(xy)dxdy \; . \tag{29}$$

Similarly one obtains

$$K_{01} = \frac{-i\omega}{4}\int \mathcal{E}_0^*(xy)\Delta\epsilon_0(xy)\mathcal{E}_1(xy)dxdy \tag{30}$$

where $\Delta\epsilon_0(xy)$ as opposed to $\Delta\epsilon_1(xy)$, is equal to $\Delta\epsilon$ over the cross section of guide 0, and is zero elsewhere.

From Eqs. (29) and (30) it is evident that as long as $\Delta\epsilon$ is real

$$K_{10} = -K_{01}^* \; . \tag{31}$$

As shown in Appendix I this is a general property of lossless codirectional coupling. It is interesting to note that this relation holds also for lossy propagation in which β is complex but $\Delta\epsilon$ is real. (Such a case may occur when the absorption in the guide and in the substrate are the same, or when the losses are caused by scattering.) In the cases we consider, $\Delta\epsilon$ is mainly real. In general, a complex $\Delta\epsilon$ will modify, for example, the solutions described in Fig. 3 for two modes, by causing the two modes to carry the same power when z is sufficiently large.[4]

Equations (29) and (30) are identical to those obtained by Marcuse[4] who solved the Maxwell equations using perturbation theory. They are independent of the guide geometry and can be applied to various cross section guides once the profile of the propagating mode is known.

5. Coupling Between Planar Guides

As a first example we apply our result to the planar waveguides shown in Fig. 7. The mode profile of a single planar waveguide can be derived analytically (see Chapter III). We consider a TE mode which has its electric field in the y direction, carries a unit power per unit length in the y direction, and is given by

$$\mathcal{E}_0(x) = \left(\frac{4\omega\mu}{\beta\left(w + \frac{2}{p_x}\right)}\right)^{1/2} \cos h_x x\,, \qquad |x| \le \frac{w}{2},\ -\infty \le y \le -\infty$$

$$\mathcal{E}_0(x) = \left(\frac{4\omega\mu}{\beta\left(w + \frac{2}{px}\right)}\right)^{1/2} \cos h_x t\, e^{-p_x\left(|x| - \frac{w}{2}\right)} \tag{32}$$

$$|x| \ge \frac{w}{2}, \quad -\infty \le y \le -\infty$$

where k_0 is the free space wave number and

$$h_x = \left(n_2^2 k_0^2 - \beta^2\right)^{1/2}$$

$$p_x = \left(\beta^2 - n_3^2 k_0^2\right)^{1/2} \tag{33}$$

In addition, the eigenvalue equation imposes the relation for the mth order mode (m+1 is the number of extrema)

$$h_x w = (m+1)\pi - 2\tan^{-1}\frac{h_x}{P_x}\,. \tag{34}$$

A similar profile describes the mode in guide number 1, the only difference being that the center of the mode is at x = w+s rather than x = 0.

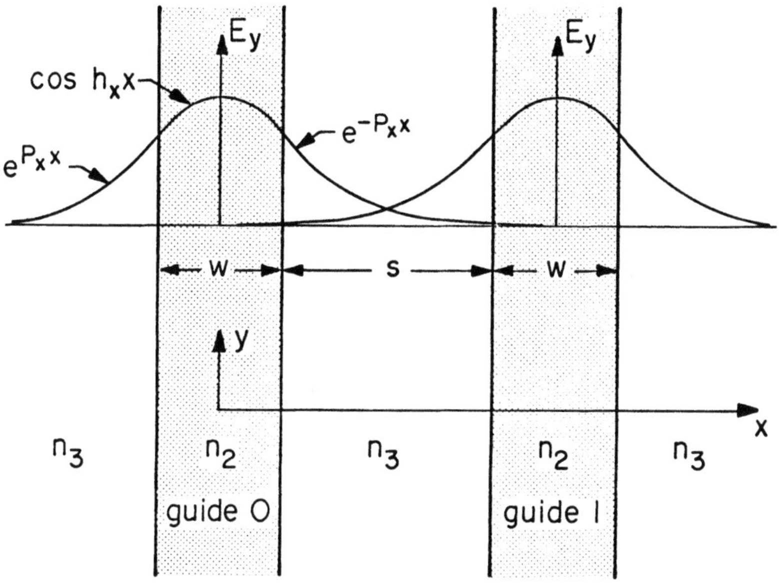

Fig. 7. Sketch of two adjacent planar guides and their propagating modes.

Knowing the mode profiles we can obtain the coupling coefficient between the guides with the help of Eq. (29). We note first that $\Delta\epsilon_1$ for this case is given by

$$\Delta\epsilon_1(xy) = \epsilon_2 - \epsilon_3 = \epsilon_0(n_2^2 - n_3^2) \quad \text{for}$$

$$s + \frac{w}{2} \le x \le s + \frac{3w}{2}, \tag{35}$$

$$-\infty \le y \le -\infty$$

$$\Delta\epsilon_1(xy) = 0 \qquad \text{elsewhere} .$$

Thus we obtain the following expression for the coupling coefficient:

$$K_{10} = K_{01} = iK = -i \frac{2h_x^2 P_x}{\beta\left(w + \frac{2}{p_x}\right)\left(h_x^2 + p_x^2\right)} e^{-p_x s} . \tag{36}$$

In the derivation of this expression, the y integration is performed over one unit length and we have assumed that

$$|p_x \cos h_x \frac{w}{2} + h_x \sin h_x \frac{w}{2}| >> |p_x \cosh_x \frac{w}{2} - h_x \sinh_x \frac{w}{2}| e^{-p_x w} \tag{37}$$

which is always true for well confined modes.

Equation (36) is the same as approximate expressions for the coupling coefficient between planar waveguides given in Ref. 4, 5, and 6.

In Ref. 5 a comparison of this expression to the exact solution of the problem is made and a good agreement is shown.

6. Coupling between Channel Guides

A rigorous solution for the mode profile of a rectangular cross section channel waveguide requires a computer.[7] However, once the mode profile is known, Eq. (29) can be applied for the calculation of the coupling coefficient. Marcatili[2] has shown that it is possible to introduce a drastic simplification in the derivation of the mode profile which enables one to get a simple solution. Consider the waveguide shown in Fig. 8. Most of the power propagates in the high index region and the fields decay exponentially in the x and y directions. Consequently, for a reasonably confined mode only a small part of the power propagates in the four shaded areas. This means that in matching the boundary conditions the field along the edges of the shaded area can be ignored.

The solution of the simplified problem yields modes that are essentially of the TEM kind and can be grouped in two families, $\mathcal{E}^{x}_{mn}$ and $\mathcal{E}^{y}_{mn}$. The first family has most of the electric field in the x direction and the magnetic field in the y direction, while the second family has most of the electric field in the y direction and the magnetic field in the x. Our notation is such that m+1 and n+1 indicate the number of extrema of the fields in the x and y directions. Furthermore, the solution of the problem indicates that mode profiles in the x and y directions are actually given by two equations, each of which is identical with the eigenvalue equations one obtains for a one-dimensional (planar) waveguide. For example, we consider an $\mathcal{E}^{y}$ mode in the guide shown in Fig. 8. The eigenvalue equation for the x direction is identical with that of a TE mode in a symmetric planar guide. It is given by

$$h_x w = (m+1)\pi - 2\tan^{-1}\frac{h_x}{p_x} \tag{38}$$

where

$$h_x^2 + P_x^2 = \left(n_2^2 - n_3^2\right)k_0^2 \ .$$

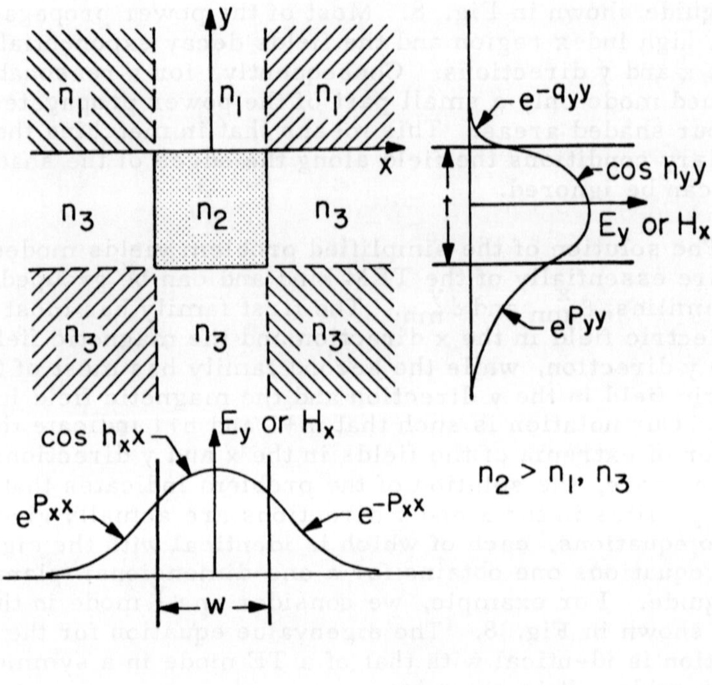

Fig. 8. Mode profile of a rectangular channel guide.

The eigenvalue equation for the y direction is identical with that of a TM mode in an asymmetric planar guide and is given by

$$h_y t = (n+1)\pi - \tan^{-1}\frac{n_3^2 h_y}{n_2^2 P_y} - \tan^{-1}\frac{n_1^2 h_y}{n_2^2 q_y} \tag{39}$$

where

$$h_y^2 + P_y^2 = \left(n_2^2 - n_3^2\right)k_0^2 \ ; \ h_y^2 + q_y^2 = \left(n_2^2 - n_1^2\right)k_0^2 \ . \tag{40}$$

The propagation constant along the z direction is therefore given by

$$\beta^2 = n_2^2 k_0^2 - h_x^2 - h_y^2 \ . \tag{41}$$

Therefore, we can approximate the propagating mode by regarding only the main components of the fields and using the mode profiles as given by Eqs. (39) and (40). The validity of this approximation can be determined by comparing the value of the expression

$$\frac{\beta^2 - n_3^2 k_0^2}{n_2^2 k_0^2 - n_3^2 k_0^2} = \frac{\left(n_2^2 - n_3^2\right)k_0^2 - h_x^2 - h_y^2}{\left(n_2^2 - n_3^2\right)k_0^2} \tag{42}$$

with that derived by computer calculation.[7] For

$$\frac{\beta^2 - n_3^2 k_0^2}{n_2^2 k_0^2 - n_3^2 k_0^2} \geq 0.3 \quad , \tag{43}$$

according to the figures in Ref. 2, the approximation is within a few percent of the exact value. Equation (43) also indicates what modes are supported by the guide, because only modes for which the expression is positive are above cutoff.

Figure 8 shows the mode profile of the $\mathcal{E}^y_{00}$ mode. This mode profile after the appropriate normalization can be used in conjunction with Eq. (29) to calculate the coupling coefficient.

We shall attempt to get a good estimate for the coupling coefficient without going through the above process. We note that the profile of the mode in the x direction is the same as that of the planar guide we have dealt with before. The major difference is that the planar guide and its propagating modes are independent of y. This basically means that $\Delta \epsilon_1 (x, y)$ (which is defined for the planar guide case by Eq. (35) has the same value for any value of y to which the modes of the guides extend. This would also be the case in a channel waveguide if the modes were very well confined in the y direction

$$\left(\frac{1}{q_y} + \frac{1}{p_y} << 1\right) .$$

In this case $\Delta\epsilon_1(x, y)$ is defined as

$$\Delta\epsilon_1(x, y) = \epsilon_2 - \epsilon_3 = \epsilon_0\left(n_2^2 - n_3^2\right) \text{ for}$$

$$s + \frac{w}{2} \le x \le s \frac{3w}{2} \quad ;$$

$$- t \le y \le 0$$

$$\Delta\epsilon_1(x, y) = 0 \quad \text{elsewhere} \quad . \tag{44}$$

Thus if the mode is well confined in the y direction it extends only from y = 0 to y = -t which is the region at which $\Delta\epsilon(xy)$ has a constant value. For this case can be used to express the coupling between the two channels. We can modify Eq. (36) to include coupling between modes that are not very well confined, but satisfy Eq. (43). This is done by multiplying Eq. (36) by a factor which is the ratio of the mode power propagating between y = 0 and y = -t (the region where $\Delta \epsilon_1 (x, y) \neq 0$) and the total mode power. This factor can be roughly approximated by

$$\frac{t}{(t + \sin^2 h_y t/p_y)} \tag{45}$$

where $1/p_y >> 1/q_y$ the penetration depths of the modes into the upper and lower boundaries satisfy. The expression for the coupling coefficient between two channel guides becomes

$$K_{10} = K_{01} = - iK$$
$$= -i \frac{2h_x^2 P_x t e^{-P_x s}}{(t + \sin^2 h_y t/p_y)(w + \frac{2}{p_x})(h_x^2 + p_x^2)} . \tag{46}$$

For a well confined mode

$$t >> \frac{1}{p_y} \text{ and } w >> \frac{2}{p_x}$$

and Eq. (46) reduces to

$$K_{10} = K_{01} = - iK = -i \frac{2h_x^2 p_x e^{-p_x s}}{\beta w(h_x^2 + p_x^2)} \tag{47}$$

which, aside from its negative sign, is the expression derived by Marcatili.[2]

In Marcatili's analysis the coupling coefficient is derived by solving for the propagation constants of two eigenmodes of the two-guide directional coupler, as shown in Fig. 9. The two are a symmetric mode $\mathcal{E}_s$ and an asymmetric mode $\mathcal{E}_a$, with propagation constants β_s and β_a. If at $z = 0$ light is coupled into the 0 guide only, the two modes are excited equally in such a way that they add on the 0 guide side and cancel over guide number 1 side. Since $\beta_s \neq \beta_a$ the two modes drift out of phase as they propagate along the z direction. At a distance L which is given by

$$\Delta \beta L = (\beta_s - \beta_s)L = \pi \tag{48}$$

the two modes are 180° out of phase, cancel over guide number 0 and add over guide number 1, as shown in Fig. 9. The net result is a complete transfer of power from guide 0 to guide 1. From a comparison of Eqs. (48) and (13) we find that $\Delta\beta$ is related to the coupling coefficient by

$$K = \frac{\Delta\beta}{2} \quad . \tag{49}$$

It is interesting to note that from this point of view it is possible to easily obtain an upper limit for the value of the coupling coefficient K. Clearly, this value reaches its upper limit as the two guides are brought closer together. Eventually they form one guide with twice the width. The symmetric mode turns into the lower order mode of the wide guide and the asymmetric mode turns into its higher order moder. The calculations of the propagation constants of the new modes becomes straightforward by using Eq. (41) and yields the upper limit for $\Delta\beta$.

Before comparing the theoretical expressions for the coupling coefficient with the experimental results, consider a geometry which allows an accurate measurement of the coupling coefficient.

7. Multichannel Directional Coupler-Coupling Coefficient Measurement

In Section 3 of this chapter we have considered a directional coupler made of two channels in which the power oscillates back and forth between the channels as a function of propagation distance. We consider now an infinite number of identical guides each coupled to its adjacent guides, as shown on the left of Fig. 1. When light is coupled into a central channel (defined as n = 0) it will leak to the sides. Because of the large number of channels, it is clear that the full power will not appear again in the central channel, but will constantly flow to the sides.

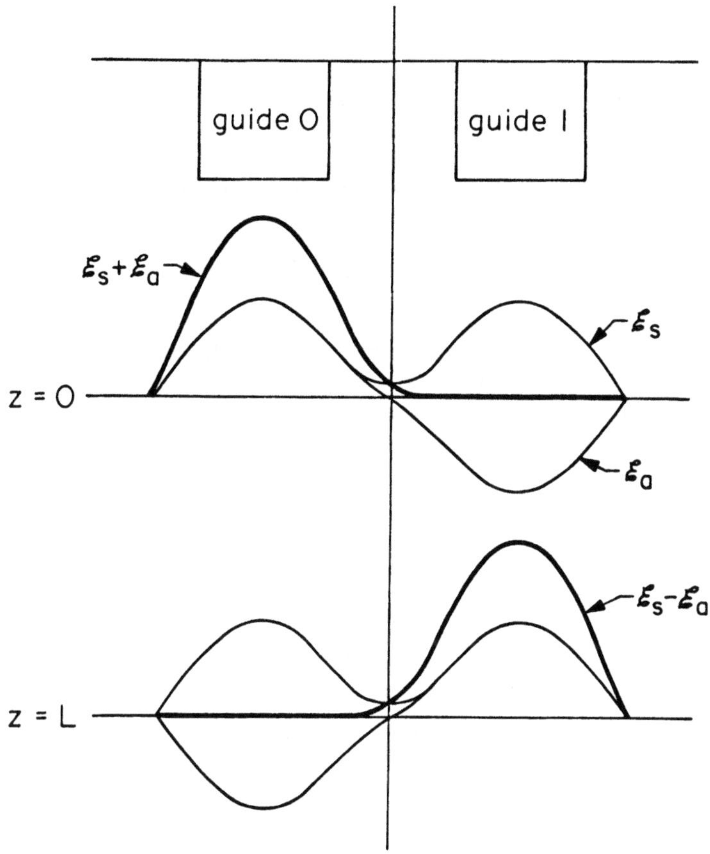

Fig. 9. The symmetric and antisymmetric eigenmodes of a dual directional coupler. Power is transferred from one channel to another because of not identical propagation constants.

The coupled mode equation for the nth channel is given by

$$\frac{dA_n(z)}{dz} = i\beta A_n(z) - iKA_{n-1}(z) - iKA_{n+1}(z) \qquad (50)$$

where β the propagation constant that includes the guide attenuation α

$$\beta = \beta_r - i\frac{\alpha}{2} , \qquad (51)$$

n is the guide number (n=0, ±1, ±2, ...), and K (a real number) is the coupling coefficient between two adjacent guides. (The coupling coefficient between nonadjacent guides is negligibly small.) When light is coupled into the 0 channel only, the boundary conditions become

$$A_0(0) = 1 \qquad A_{n\neq 0}(0) = 0 \cdot \qquad (52)$$

The solution of Eq. (50) is

$$A_n(z) = (-i)^n J_n(2Kz) e^{-i\beta z} \qquad (53)$$

and, therefore the power flow in the guides is

$$P_n(z) = A_n(z) A_n^+(z) = J_n^2(2Kz) e^{-\alpha z} \qquad (54)$$

where J_n represents the Bessel function of nth order.

A multichannel direction coupler [8] was fabricated by proton implantation through an appropriate gold mask on the surface of the GaAs substrate as shown in Fig. 4(b). The width of the embedded channels can be estimated by the width of the notches in the mask to be about 2.4 μm separated by 3.9 μm. Their depth is determined by the known penetration of the 300 keV protons in GaAs which is 3 μm. The measured index discontinuity is about 0.005 at 1.15 and about 10% larger at 1.06 μm.

Figure 1 describes the multichannel directional coupler with the input light coupled into a central guide and the output light scanned at different sample lengths. The argument of the Bessel functions at each of the scans was determined by fitting the square root of the normalized intensities of the light in the different channels, to a plot of the Bessel functions as shown in Fig. 10. The experiment was performed with light having a wavelength of $\lambda = 1.15\ \mu m$ (from a HeNe laser) and $\lambda = 1.06\ \mu m$ (a Nd:YAG laser). The coupling coefficient can be deduced readily and accurately now, simply by plotting the argument of the Bessel functions against the distance of propagation, and measuring the slope.

Using Eq. (53) the coupling coefficient is thus given by

$$K = \text{slope}/2 \cdot \tag{55}$$

A series of measurements at $1.15\ \mu m$ and $1.06\ \mu m$ is plotted in Fig. 11. By measuring the slopes we find that for the same directional coupler,

$$\begin{aligned} K_{1.15\ \mu m} &= 0.52 \pm 0.01\ \text{mm}^{-1} \\ K_{1.06\ \mu m} &= 0.30 \pm 0.01\ \text{mm}^{-1} \cdot \end{aligned} \tag{56}$$

It has also been found that different polarizations (E^x or E^y) of the input beam had no noticeable effect on K.

Only a qualitative comparison between theory and experiment can be made with these samples. The modes in these guides are not well confined modes, to which the theory of Section 6 pertains. In addition, the dielectric discontinuity may not be uniform across the cross section of the guide and the accuracy in measuring it is not high. However, the coupling coefficients calculated on the basis of the dimensions given earlier in (Eq. (46)) are close to the actual values. The theoretical results are (for both polarizations)

$$K_{1.15} = 0.32\ \text{mm}^{-1} \qquad K_{1.06} = 0.21\ \text{mm}^{-1} \tag{57}$$

Considering the limitations mentioned above, these results are in good agreement with Eq. (56).

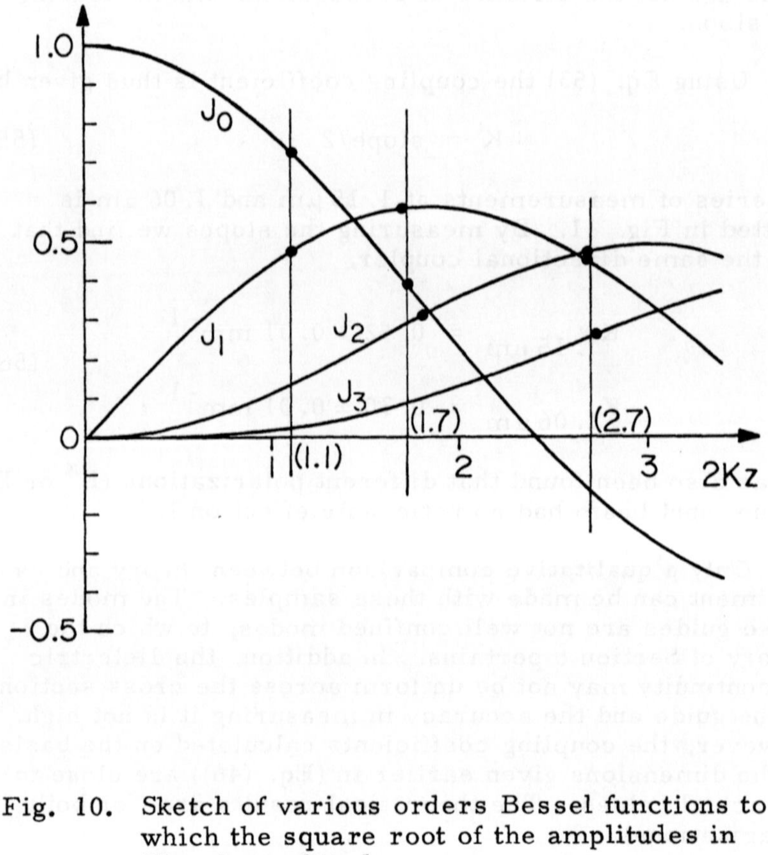

Fig. 10. Sketch of various orders Bessel functions to which the square root of the amplitudes in Fig. 1 are fitted.

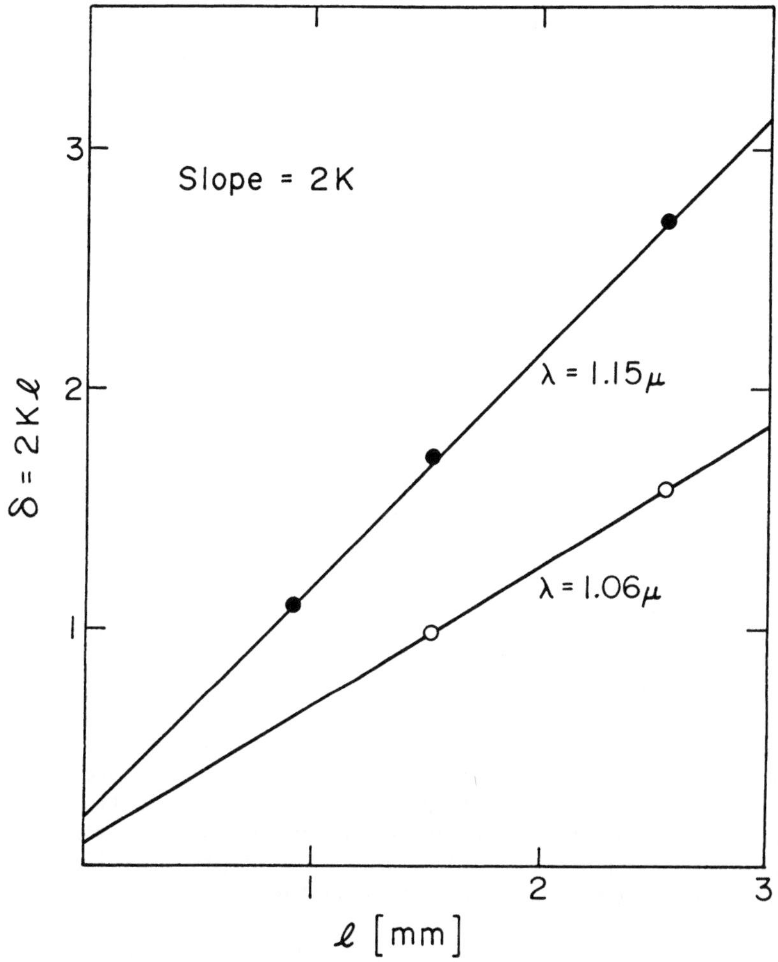

Fig. 11. Plot of the arguments of the Bessel functions against the distance of propagation.

The absorption of the propagating light is relatively high in these ion implanted waveguides. It amounts to about 5 cm^{-1} which is the same as the absorption of planar ion implanted guides.[3] It is caused by the absorption of the defect centers created by the implantation and by the free carrier absorption in the substrate. Other methods for fabrication of directional couplers will be discussed in Section 9.

8. The Coupling Coefficient Sign

Let us consider again the dual directional coupler shown in Fig. 3. The amplitude of the fields in the two guides is given by Eqs. (58) and (59) as

$$A_0(z) = \cos Kz\, e^{i\beta z} \tag{58}$$

$$A_1(z) = -i \sin Kz\, e^{i\beta z} . \tag{59}$$

From these equations we learn that a distinct phase difference exists between the two guides. We also find that the phase of the guide in which the power is increasing will always lag 90° behind the phase of the guide in which the power is decreasing. Formally, this is a direct correspondence of the sign of the coupling as given in Eq. (29). Physically, the reason for the time lag is the necessary phase relation between the polarization (caused by the field in guide zero), and the field in guide 1, if power is to be generated in guide 1. It is well known[9] that power dissipation in a dielectric occurs when the polarization lags the field. Therefore, in our case, for power generation in guide 1 the field there must lag the polarization, which is caused by the field of guide 0 and has its phase. This reasoning gives the same result as the above equations.

The phase relation between the two guides might be of importance in an optical circuit in which the phase of the light as well as its amplitude may be used to carry information. The following verifies that the phase difference is indeed a time lag.

Phase information can be acquired by interference. In this case the interference was performed in the fashion described in Fig. 12. Drawing (b) describes a geometry in

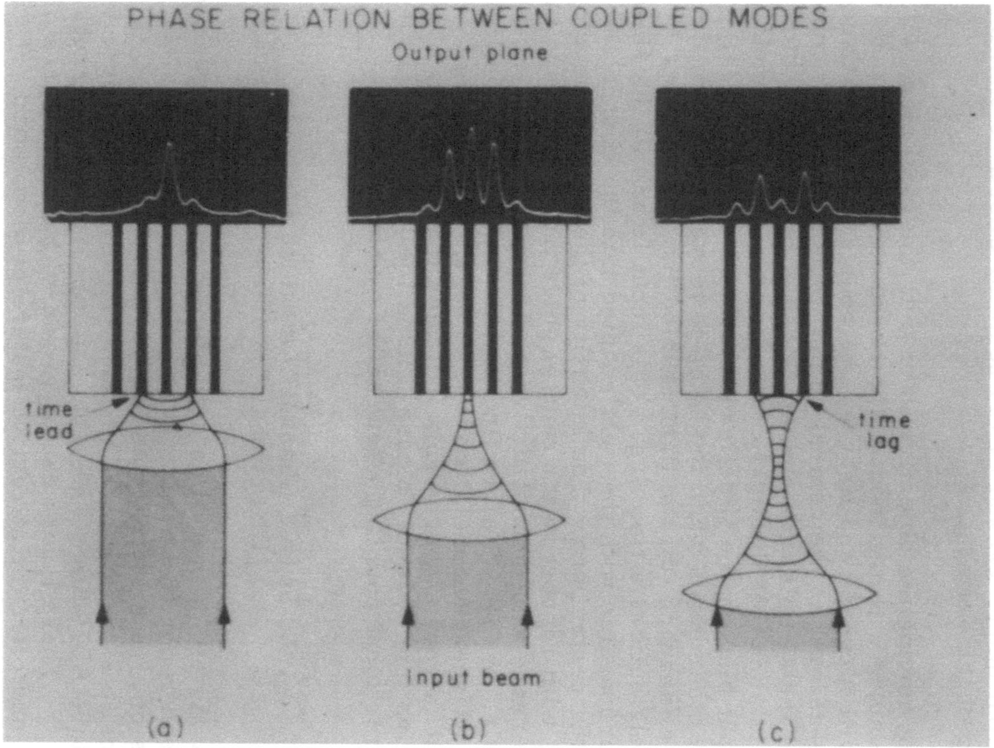

Fig. 12. Phase relation between coupled modes is obtained by interference. (See text for details.)

which the input beam is focused with an objective into the central channel of the multichannel directional coupler. The input face at the channels is in the focal plane of the objective and the spot size of the beam is small enough to couple into one channel only. The distribution of light at the output plane, caused by the coupling effect, is shown at the top of the drawing.

By moving the objective in or out, the focal plane will not coincide with the input plane of the sample, the spot size will increase and light will be coupled into the adjacent guides as well as the central guide. When the lens is moved toward the sample, the curvature of the phase fronts of the beam is such that the light coupled into the side channels at the input plane has a time lead (as shown in drawing (a)). On the other hand, when the beam is moved away from the sample (drawing (c)) the light coupled into the adjacent channels suffers a time lag. This input light will interfere, as it propagates with the light coupled into the channels from the central one. The intensity of the light in the output plane, with respect to the central guide indicates where this interference was a constructive or destructive one. For example, in Fig. 12(a) we find that the intensity of the light in the adjacent channels is close to zero, while in Fig. 12(c) it is very strong. This indicates that there is a constructive interference in case (c) and the light coupled via the directional coupling effect suffers a time lag.

9. Ridged Channel Waveguides and Directional Couplers

A variety of thin films suitable for optical waveguiding has been reported so far. These films include epitaxial layers of high resistivity on low resistivity GaAs (Ref. 10), composite structure of GaAs-GaAlAs (Ref. 11), composite structure ADP-KDP (Ref. 12), and single crystal garnet films.[13] These layers can be grown with a high degree of purity and offer the attractive feature of modulation.[10, 11, 14]

All these layers are planar guides which confine the light in one dimension only, and therefore, as such, are not compatible with the concept of optical circuitry. To form a channel waveguide in which the radiation is confined in the two dimensions perpendicular to the direction of propagation it is possible, as shown earlier, to use ion implanatation or diffusion.[15] It is interesting to investigate the possibility of fabricating a channel guide by a removal of the superfluous sections of the epilayer (as shown in Fig. 13). Because of the large dielectric discontinuity the scattering losses in this guide are very sensitive to the smoothness of its walls as well as its top surface. For example, a 3 μm wide waveguide in GaAs will suffer roughly a loss of 4 cm^{-1} when the roughness of the walls is 500 Å rms. This rough estimate[16] based on the Rayleigh criterion indicates that special care must be taken in the fabrication of these guides. An extreme case of rough walls and top surface is shown in Fig. 14(a). This guide (3 μm high, 7 μm wide) scattered away most of the propagating light in a sample less than 1 mm long.

Smoother channels were fabricated by exposing the photoresist with a better mask, and ion polishing the sample after the ion machining. Figure 14(b) shows a channel (1.4 μm high, 2 μm wide) with much smoother walls.

The number of modes supported by a ridged guide, as discussed in Section 6, can be controlled for a given dimension by choosing an appropriate guide substrate dielectric discontinuity or by covering the guide with an appropriate index material.

We have discussed directional couplers made of single mode waveguides embedded in the surface of GaAs. In that case the coupling between two adjacent guides is caused by the overlap of the propagating modes as shown in Fig. 15(a). Figure 15(b) shows single-mode ridged waveguides in which the mode is very well confined in the x direction (because of the large dielectric discontinuity) and two closely spaced guides will thus have a negligible coupling. To increase the coupling between the guides, only a partial removal of the epilayer between them was performed as shown in Fig. 15(c). Fig. 16 is an indication of the coupling. It shows the cross section of a high resistivity GaAs epilayer (about 8 μm thick) that was machined down to the substrate, to form a large number of isolated channels. Light coupled into one channel emerges at the other end of the sample from that channel only (bottom photograph). However, when the same epilayer is only partially machined (Fig. 16(b)) light coupled into one guide emerges from three guides.

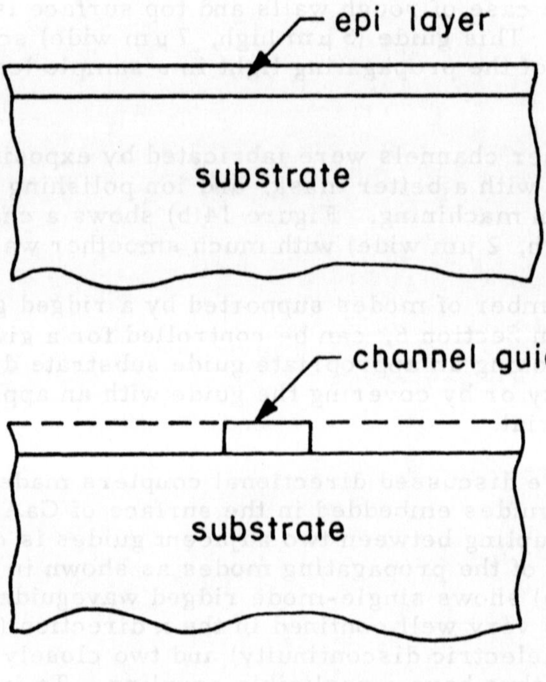

Fig. 13. Channel waveguide fabrication by removal of superfluous sections of an epilayer.

Fig. 14. (a) An extreme example of a guide with rough walls and top surface. (The guide is 3 μm high and 7 μm wide.)
(b) Good quality channel waveguide fabricated by using holographically prepared mask and ion polishing. (The guide is 1.4 μm high and 2 μm wide.)

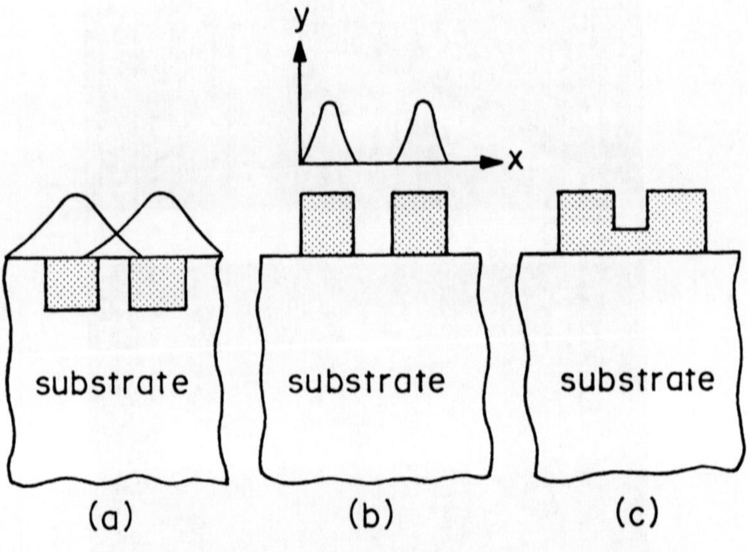

Fig. 15. Directional couplers fabricated in GaAs.
(a) Single mode embedded guides directional coupler.
(b) Two single mode ridged guides with no coupling.
(c) Ridged guides with increased degree of coupling.

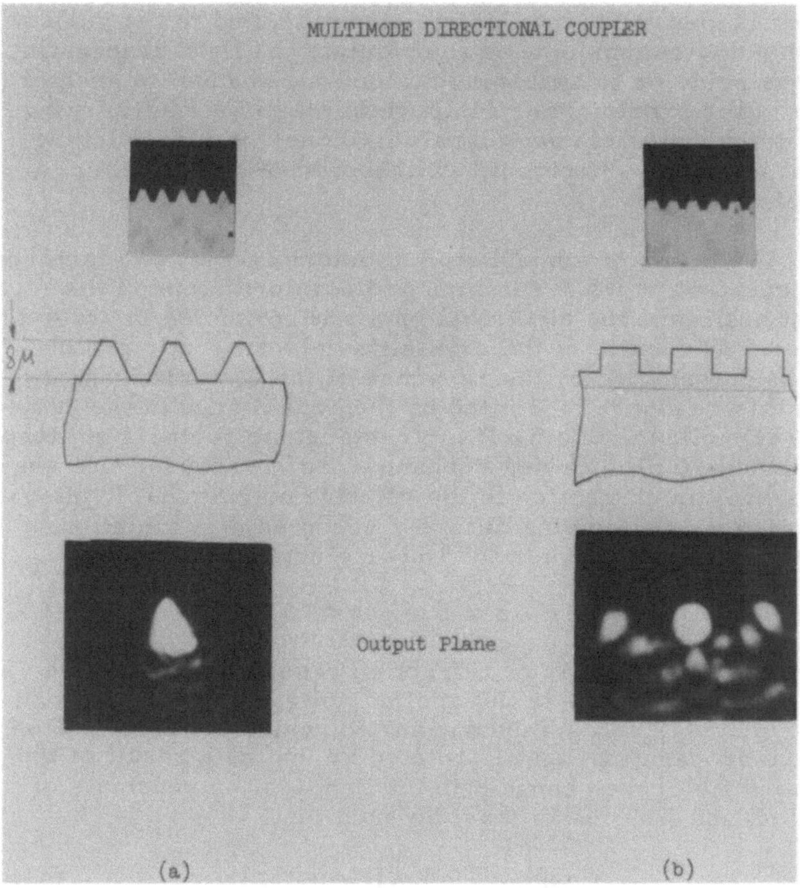

Fig. 16. Directional coupler fabricated by ion machining in a GaAs epilayer.
(a) Isolated channels. (b) Coupled channels.

10. Directional Coupler — Switch Modulator

As mentioned earlier, one of the most exciting applications of the directional coupler is the switch modulator in which the amount of light coupled from one channel to the other is controlled by electric signal applied to the guides. Such a device can be used to modulate the light propagating in one guide or to switch light from one channel to another. The latter property is an important mean for multiplexing (or demultiplexing) two signals into one, or for switching light in an opto-electronic switchboard of an optical communication network.

The merit of a modulator is determined by two factors: the speed at which it can impose the information on the light beam and the electrical power it consumes in order to do so. The speed of the crystalline electro-optic modulator is determined by the response of the electrical circuit, and this response is limited by the capacitance of the device. The capacitance of a 3 x 3 μm rectangular guide, 1 cm long, is about ≈10 pF allowing subnanosecond response. We shall thus limit our attention to the relative magnitude of voltage necessary for deriving different schemes of modulation. First we recall that for the linear electro-optic effect

$$\delta n \propto E \Rightarrow \delta n \propto V \tag{60}$$

where δn is the index of refraction caused by the applied field E and where V is the applied voltage. We also recall that for polarization modulation[9] (in which the polarization of the propagating light is rotated by 90° as a result of the applied field) the change β in the propagation constants of the TE and TM modes must be such that

$$(\delta\beta_{TE} - \delta\beta_{TM})L = \pi \tag{61}$$

where L is the length of the modulator.

For the sake of simplicity we assume

$$\delta\beta_{TM} = 0 \tag{62}$$

and that $\delta\beta_{TE}$ can be written as

$$\delta\beta_{TE} = \delta n \cdot k_0 \quad . \tag{63}$$

Using Eqs. (63) and (61) we can write that

$$\delta n = \frac{\pi}{k_0 L} \quad . \tag{64}$$

Our task now is to compare this δn which is required for a 100% modulation in the polarization modulator (as well as mode conversion modulator [17]) with the one needed for a directional coupler modulator.

Therefore, consider a typical example which is the directional coupler whose cross section is shown in Fig. 17. The material is assumed to be GaAs (n = 3.5). Because of the relatively large dielectric discontinuity ($\Delta n = 0.01$) the modes are well confined, yielding (with the use of Eq. 46 for the 3 x 3 μm guides) the following value for the coupling coefficient at $\lambda = 1.15\,\mu m$.

$$K = 0.155 \text{ mm}^{-1} \quad . \tag{65}$$

We choose the modulator length to be the length necessary for complete transfer of power. According to Eq. (13)

$$L = \frac{\pi/2}{K} \approx 10 \text{ mm} \quad . \tag{66}$$

Thus light entering the device at the left channel will emerge from the one on the right. In order to turn the device into a switch modulator Marcaliti[2] has suggested changing the coupling coefficient (by changing the dielectric discontinuity, through the electro-optic effect) to a new value K_1 so that

$$K_1 = \frac{\pi}{L} \quad . \tag{67}$$

According to Eq. (11), this will cause the light to be coupled back to the original channel resulting in a 100% switch

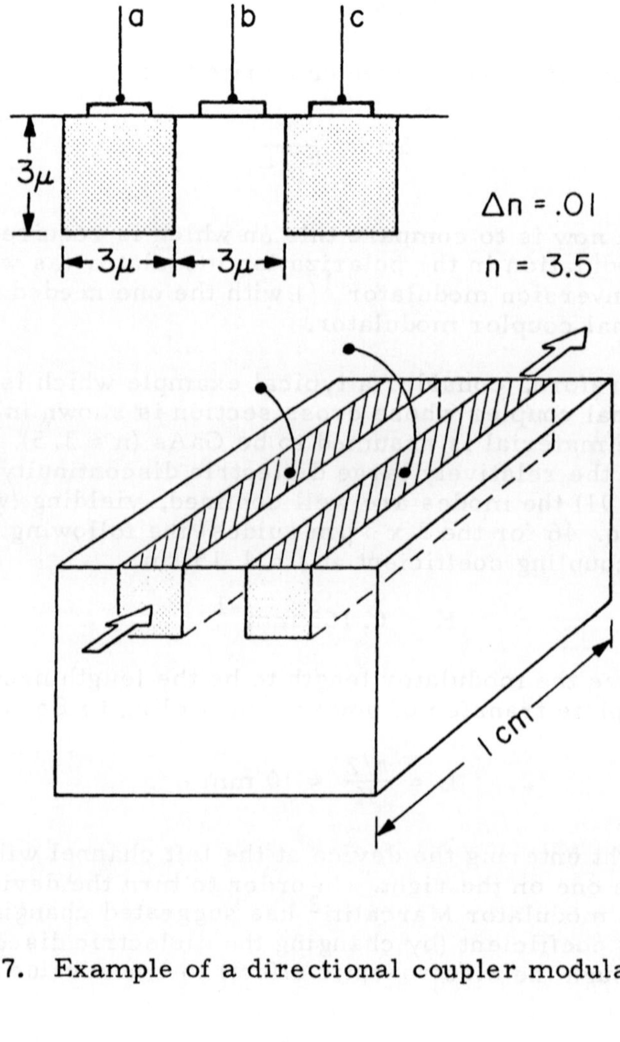

Fig. 17. Example of a directional coupler modulator.

modulation. The coupling coefficient can be varied by applying field to the section between the two guides (schematically via electrode b in Fig. 17), or alternatively by applying the field to the guides (electrodes a and c). To obtain the above value of K_1 the dielectric discontinuity of the guide must be reduced from $\Delta n = 0.01$ to $\Delta n = 0.007$. Thus the change in the index of refraction required for switching is

$$\delta n \approx 0.003 = 52 \frac{\pi}{k_0 L} \cdot \tag{68}$$

The conclusion drawn from this extreme example is disappointing. It indicates that the electrical power necessary to drive this directional coupler modulator is 2500 times larger than the driving power of a regular polarization modulator (Eq. (64)). A smaller δn can be achieved for a given length geometry by decreasing the dielectric discontinuity and increasing the separation between the guides. However, this reduces the confinement of the propagating modes.

To devise a new scheme that will allow directional coupling switching with less drive power, consider again the coupled mode, Eqs. (6) and (7), but this time allowing for different propagation constants β_0 and β_1 for the two guides.

$$\frac{dA_0(z)}{dz} = -i\beta_0 A_0(z) - iK A_1(z) \tag{69}$$

$$\frac{dA_1(z)}{dz} = -i\beta_1 A_1(z) - iK A_0(z) \cdot \tag{70}$$

Using the regular boundary conditions

$$A_0(0) = 1 \qquad A_1(0) = 0 \tag{71}$$

we obtain the solution

$$A_0(z) = \left[\cos\left(K^2 + \left(\frac{\Delta\beta}{2}\right)^2\right)^{1/2} z - i\,\frac{\Delta\beta/2}{\left(K^2+\left(\frac{\Delta\beta}{2}\right)^2\right)^{1/2}} \sin\left(K^2+\left(\frac{\Delta\beta}{2}\right)^2\right)^{1/2} z\right] \cdot e^{-i\left(\beta_0 - \frac{\Delta\beta}{2}\right)} \tag{72}$$

$$A_1(z) = -\left[\frac{-iK}{\left(K^2+\left(\frac{\Delta\beta}{2}\right)^2\right)^{1/2}} \sin\left(K^2+\left(\frac{\Delta\beta}{2}\right)^2\right)^{1/2} z\right] e^{-i\left(\beta_1+\frac{\Delta\beta}{2}\right)z}, \tag{73}$$

where $\Delta\beta$ is

$$\Delta\beta = \beta_0 - \beta_1 . \tag{74}$$

This solution becomes identical with Eqs. (11) and (12) when $\Delta\beta = 0$.

Going back to our previous example, where with two identical guides ($\Delta\beta = 0$) we had a complete transfer of power in a length L ($A_1(L) = 1$), it is possible by applying field to one of the guides only to destroy the identity between the guides so that $\Delta\beta \neq 0$. According to Eq. (73) if

$$\left(K^2 + \left(\frac{\Delta\beta}{2}\right)^2\right)^{1/2} L = \pi \tag{75}$$

we will get

$$A_1(L) = 0$$

$$A_0(L) = 1 .$$

This means that for such as $\Delta\beta$ there is no power transfer, and thus a 100% switch modulation is achieved. Equations

(75) and (76) give the value of $\Delta\beta$ needed for destroying the coupling between the guides

$$\Delta\beta L = \sqrt{3}\,\pi \tag{76}$$

and the necessary change δn is thus

$$\delta n \approx 1.73 \frac{\tau}{k_0 L} = 1.10^{-4} \tag{77}$$

This δn is much smaller than the previous one, Eq. (68), and is approaching the value required for polarization modulation. It is interesting to note that Eq. (77) applies to the ridge directional couplers described in Section 9 as well as the embedded ones.

A directional coupler can be made polarization sensitive by choosing an appropriate guiding plane (for example, 100 in GaAs) in which application of field causes a certain δn for TE modes but $\delta n = 0$ for TM mode (or vice versa). Thus in our example, before the application of the field, TE as well as TM modes are coupled from one guide to the other. However, with the application of the field, the coupling of the TE mode is destroyed while the coupling of the TM mode remains the same. Such a device with a dc electric field is capable of separating the two polarizations of an incoming optical signal into two different channels. As such, it can turn a polarization modulator (or a mode converter) into a directional coupler switch, as shown in Fig. 18(a). This composite device will have a low modulating power together with a directional coupler switch capability. Reversing the direction of propagation of the pulse (Fig. 18(b)) enables the devices to combine (multiplex) into one output channel, pulses with the appropriate polarizations from two input channels.

We can apply electric fields in a semiconductor [18] guide by back-biasing a metal semiconductor junction (Schottky barrier). The field is supported in a depletion region which is swept free of carriers to a depth d, given by $d = \epsilon E/eN$, where N is the free carrier concentration (assumed independent of depth) and E is the peak value of the electric field. In this depletion region the field falls off linearly with depth. The maximum depth to which a field

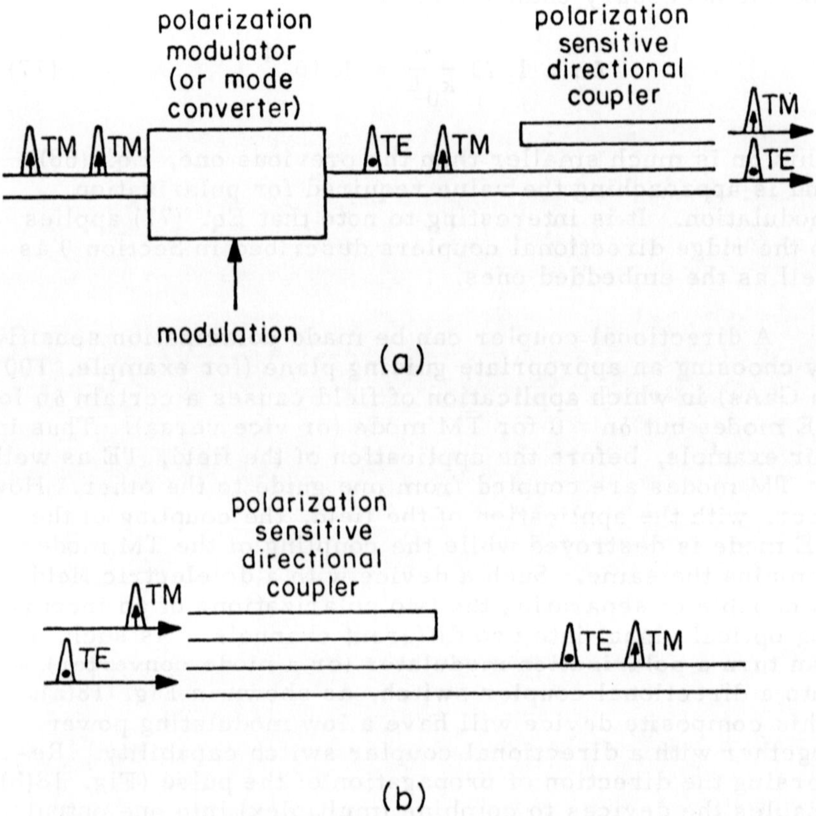

Fig. 18. (a) Combination of a phase modulator and a polarization sensitive directional coupler. (b) Polarization sensitive directional coupler as a multiplexer.

can be applied in such a junction is given by the field at which the material breaks down. In GaAs the breakdown field is roughly independent of doping level, and has a value $E_B \simeq 5 \times 10^5$ V/cm. This means that the maximum depletion width is inversely proportional to the carrier concentration. Numerically, the maximum depth d_m in microns is related to the carrier concentration N in cm^{-3} by

$$d_m = 3.5 \times 10^{16}/N \quad . \tag{78}$$

This indicates that in order to penetrate about 3 μm into the guiding layer, its concentration must be lower than

$$N < 10^{16} cm^{-3} \quad .$$

Unfortunately, the free carrier concentration of the proton implanted waveguides after the anneal (which is necessary for low optical attenuation) is about $10^{17} cm^{-3}$, thus preventing the penetration of the field into the bulk of the guide. Gallium arsenide epitaxial layers with low carrier concentration have been grown, [10] but are not yet readily available. A ridged directional coupler (Section 9) made of such an epitaxial layer may be a suitable candidate for a modulator.

In a GaAs guide in the (100) plane TM modes are unchanged by the applied field. However, TE modes experience a change in refractive index

$$\delta n = 1/2\, n^3\, r_{41} E \quad , \tag{79}$$

where r_{41} is the electro-optic coefficient and n is the index of refraction. Substituting the value of n_4 and r we get

$$\delta n = 3 \cdot 10^{-9} \cdot E \tag{80}$$

where E is the applied field in volts per centimeter.

To achieve a $\delta n = 10^{-4}$ which is necessary for the directional coupler switching, we need an average electric field of

$$E = 3.3 \cdot 10^4 \frac{V}{cm} = 3.3 \frac{V}{\mu m}$$

and voltage required for a 3 μm thick guide in this

$$V = 10\ V\ .$$

In the next section we shall discuss the use of this switch and a similar device for multiplexing and demultiplexing of optical pulses.

11. Light Multiplexing by Directional Coupling

The high optical frequency of the laser radiation enables it to carry large amounts of information. However, imposing a high rate of information on a laser beam is difficult. One way to avoid this difficulty is to modulate the information on a number of beams at a relatively moderate rate, and then to multiplex these beams into one beam as shown in Fig. 19(a). We shall assume for the following discussion that the laser radiation is in the form of very short pulses with a period T, and that they appear in the different channels with the appropriate shift in time so that they can be multiplexed with no overlap between them. Such pulses can be generated, for example, by mode locking[9] an Nd:YAG laser, and the appropriate shift in time between the channels can be achieved by coupling the pulses to the channel via optical fibers with different lengths. Another method is to have an array of injection lasers on the same substrate as part of an integrated optical circuit, where each one of these lasers is pulsed at a moderate rate but with an appropriate shift in time.

Let us examine the operation of the directional coupler switch described in the previous section as a two-channel multiplexer. Figure 19(b) shows the channels into which the two already modulated (pulse code modulation-PCM) beams are coupled. In order to combine the two beams into one that will propagate in the lower guide, we must see to it that when a pulse appears in the upper guide it is completely coupled into the lower guide, but at T/2 sec later when a pulse appears in the lower guide it is not coupled into the upper one but keeps propagating in the lower guide. This can be accomplished by choosing the coupling coefficient and the length of the two guides to be such that with no application of electric field a complete transfer of power

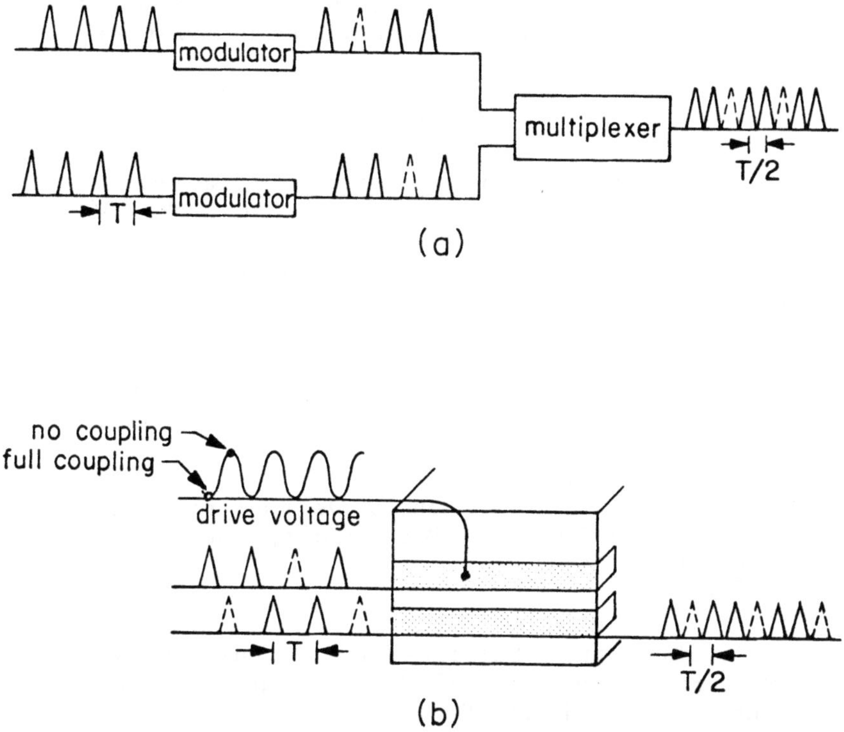

Fig. 19. (a) Multiplexing scheme for a higher rate of information modulation.
(b) Directional coupler multiplexer.

from one channel to the other is possible ($KL = \pi/2$). On the other hand, to prevent the coupling of pulses from the lower to the upper channel, we have to apply field with the appropriate strength for destroying the coupling whenever a pulse in the lower guide is due. This can be accomplished by driving the directional coupler with a sinusoidal wave whose period is exactly T as shown in Fig. 19(b). To perform demultiplexing the same configuration can be used, but the direction of propagation of the pulses is reversed.

REFERENCES

1. W. H. Louisell, Coupled Modes and Parametric Electronics (John Wiley and Sons, New York, 1960).

2. E. A. J. Marcatili, "Dielectric Rectangular Waveguide and Directional Coupler for Integrated Optics," Bell Syst. Tech. J. 48, 2071 (1969).

3. E. Garmire, H. Stoll, A. Yariv, and R. G. Hunsperger, "Optical Wave-Guiding in Proton-Implanted GaAs," Appl. Phys. Lett. 21, 87 (1972).

4. D. Marcuse, "The Coupling of Degenerate Modes in Two Parallel Dielectric Waveguides," Bell Syst. Tech. J. 50, 1791 (1971).

5. M. F. Bracey, A. L. Cullen, E. F. F. Gillespie, and J. A. Staniforth, "Surface-Wave Research in Sheffield," IRE Trans. Antennas and Propagation AP-7, Special Supplement, S219 (1959).

6. N. S. Kapany, Fiber Optics (Academic Press, New York, 1967).

7. J. E. Goell, "A Circular-Harmonic Computer Analysis of Rectangular Dielectric Waveguides," Bell Syst. Tech. J. 48, 2133 (1969).

8. S. Somekh, E. Garmire, A. Yariv, H. L. Garvin, and R. G. Hunsperger, "Channel Optical Waveguide Directional Coupler," Appl. Phys. Lett. 22, 46 (1973).

9. See for example, A. Yariv, Introduction to Optical Electronics (Holt, Rinehart and Winston, New York, 1971).

10. D. Hall, A. Yariv, and E. Garmire, "Optical Guiding and Electrooptic Modulation in GaAs Epitaxial Layers," Opt. Commun. 1, 403 (1970). Also, D. Hall, A. Yariv, and E. Garmire, "Observation of Propagation Cutoff and Its Control in Thin Optical Waveguides," Appl. Phys. Lett. 17, 127 (1970).

11. F. K. Reinhart and B. I. Miller, "Efficient GaAs-$Al_xGa_{1-x}As$ Double-Heterostructure Light Modulators," Appl. Phys. Lett. 20, 36 (1972).

12. V. Ramaswamy, "Epitaxial Electrooptic Mixed Crystal $(NH_4)_xK_{1-x}H_2PO_4$ Film Waveguide," Appl. Phys. Lett. 21, 183 (1972).

13. P. K. Tien, R. J. Martin, S. L. Blank, S. H. Wemple and L. J. Vernerin, "Optical Waveguides of Single-Crystal Garnet Films," Appl. Phys. Lett. 21, 207 (1972).

14. P. K. Tien, R. J. Martin, R. Wolfe, R. C. LeCraw, and S. L. Blank, "Switching and Modulation of Light in Magneto-Optic Waveguides of Garnet Films," Appl. Phys. Lett. 21, 394 (1972).

15. W. E. Martin and D. B. Hall, "Optical Waveguides by Diffusion in II-VI Compounds," Appl. Phys. Lett. 21, 325 (1972).

16. P. K. Tien, "Light Waves in Thin Films and Integrated Optics," Appl. Optics 10, 2395 (1971).

17. A. Yariv, "Coupled Mode Theory for Guided-Wave Optics," to be published in IEEE J. Quantum Electron. (1973).

18. E. Garmire, "Integrated Optics in Semiconductors," Presented at the Solid State Circuits Conference, NEREM 1972.

19. H. R. Garvin, E. Garmire, S. Somekh, H. Stoll and A. Yariv, "Ion Beam Micromachining of Integrated Optics Components," Appl. Optics 12, 455 (1973).

APPENDIX I: THE RELATIVE SIGNS OF THE COUPLING COEFFICIENT

Let us consider two lossless modes a and b, whose amplitudes A and B are related by

$$\frac{dA}{dz} = -i\beta_a A + k_{ab} B \qquad \text{(A-1)}$$

$$\frac{dB}{dz} = -i\beta_b B + k_{ba} A \ . \qquad \text{(A-2)}$$

We consider first the case where both modes carry power in the same direction. The conservation of total power at any given z is expressed by

$$\frac{d}{dz}(AA^* + BB^*) = 0 \qquad \text{(A-3)}$$

where

$$P_a = AA^* \quad P_b = BB^* \ . \qquad \text{(A-4)}$$

Because of the plus sign in Eq. (A-3), it is satisfied when

$$K_{ab} = -K_{ba}^* \ . \qquad \text{(A-5)}$$

Next we consider a case in which the two modes carry power in opposite directions. The conservation of power is described now by

$$\frac{d}{dz}(AA^* - BB^*) = 0 \ . \qquad \text{(A-6)}$$

Because of the minus sign the equation is satisfied only when

$$K_{ab} = K_{ba}^* \ . \qquad \text{(A-7)}$$

Chapter 12 – PERIODIC COUPLERS

J. H. Harris

Electrical Engineering Department

University of Washington

I. INTRODUCTION

When the optical properties of a portion of a planar waveguide are caused to vary in a periodic fashion the resulting structure may be used to couple energy from a collimated beam, e. g. a laser beam, into the waveguide. In principle, the device may be used to couple the light in a beam of any size into a film that need only be thick enough to support one guided mode. Because the coupling length, i. e., the beam size, may be arbitrarily large, the device belongs to the class of wide aperture couplers. The prism coupler also belongs to this class. As might be expected, the coupling length of practical structures is not arbitrary but is limited by tolerance requirements on the uniformity of the waveguide and the periodicities. These requirements not withstanding, couplers exhibiting high efficiency have been demonstrated with coupling lengths approaching one thousand times the thickness of the film into which the light is introduced.

Coupling between waves by periodic structures is a phenomenon central to many branches of physics, especially those that have some relation to the naturally occurring periodicities in crystalline materials. In reviewing the development of periodic couplers for integrated optics, it therefore seems judicious to discuss this application directly rather than attempt to correlate the very extensive literature of periodic media and periodic couplers. The first periodic coupler for integrated optics was demonstrated by Dakss et al.[1] Their structure, the grating coupler, consisted of a uniformly spaced array of bars of photoresistant material spaced 0.665 μm between centers and placed atop a glass optical waveguide (Fig. 1(a)). The bars were fabricated by first covering the guide with photoresist, then exposing a region of the resist with interfering beams of an argon laser, and finally etching all but the strongly exposed bar regions of the resist. They achieved approximately 40% coupling efficiency. Dalgoutte[2] has reported 70% efficiency with a grating coupler. A similar structure had been previously used by Teng and Stern[3] with the bars placed on a metal to couple into a surface plasma wave which is a lossy guided wave that propagates along the interface between a metal and a dielectric and exhibits evanescent penetration into the dielectric.

Succeeding periodic coupler configurations include the holographic coupler and the etched grating coupler. The former was first demonstrated by Kogelnick and Sosnowski[4] and consisted of a gelatin film atop a glass waveguide. The gelatin was exposed with interfering beams from a cadmium laser to provide an index variation of 0.25 μm period that was tilted with respect to the film as illustrated in Fig. 1(b). They reported a coupling efficiency of 71 to 74%. The etched grating coupler (Fig. 1(c)) consists of grooves placed directly in the guiding film and was first demonstrated by Ostrowsky and Jacques[5] using a photoresist guide that was exposed and etched to provide grooves only a portion of the way through the film.

In addition to these three coupler configurations, it also is possible to couple by electronically inducing periodic refractive index variations in the guiding film or adjacent regions. Acousto-optic, electro-optic, or similar methods may be employed, although they generally result in coupling lengths too long to yield high efficiency because the induced index variation is small. Gfeller and Pitt[6] have demonstrated coupling by launching a collinear acoustic wave along a film.

The theory of periodic couplers is concerned with both the basis of operation of the device[7, 8] and the design[8] of structures that provide high coupling efficiency. Efficiencies of 100% are achievable in principle. Analysis of the

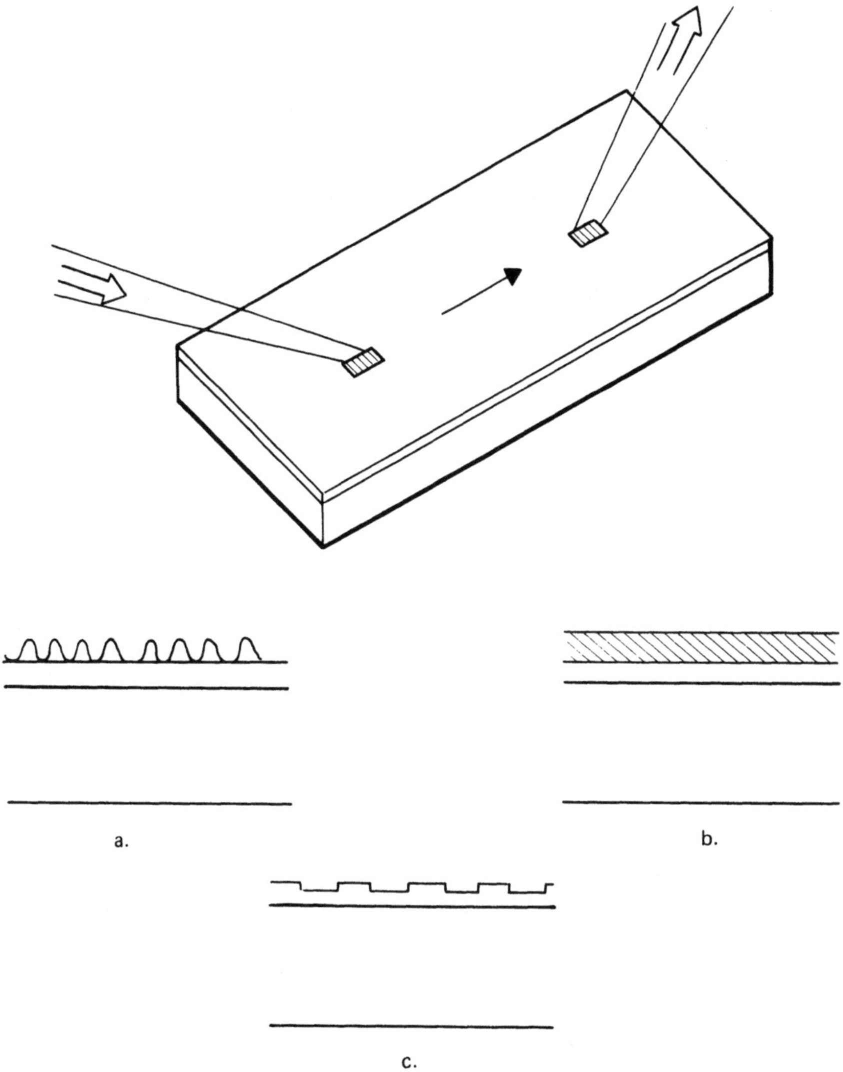

Figure 1. Input and output couplers on a planar optical waveguide. Inserts show cross section of substrate, film, and (a) grating coupler, (b) holographic coupler, and (c) etched grating coupler.

properties of a prescribed coupler may be effected by considering either the direct problem of determining the disposition of energy in an incident beam or the reciprocal problem of examining the radiation properties of the coupler when a guided wave is incident. Both of these approaches will be examined in this article. The latter approach affords a simple view of why the device functions as a coupler and will be briefly discussed for introductory purposes.

With reference to Fig. 2, suppose a guided wave propagating in the -x direction is to be incident on a periodic structure. The presence of the periodicities results in diffraction of the wave. To determine the diffracted fields the Maxwell equations

$$\nabla x \underline{E} = i\omega\mu\underline{H}$$

$$\nabla x \underline{H} = -i\omega\epsilon\underline{E} + \underline{J} \qquad (1)$$

are to be solved. Note that harmonic time dependence exp-iωt of the fields is assumed and suppressed. For the problem at hand the current $\underline{J}$ is zero, $\underline{E}$ and $\underline{H}$ may be represented at the sum of incident and diffracted fields in the form $\underline{E}^i + \underline{E}^d$, and the permittivity ϵ may be expressed

$$\epsilon = \overline{\epsilon} + (\epsilon - \overline{\epsilon}) = \overline{\epsilon} + \Delta\epsilon \cos \kappa x \qquad (2)$$

where $\overline{\epsilon}$ is the average of ϵ with respect to the x direction and $\Delta\epsilon$ is the amplitude of what for convenience is assumed to be a single harmonic representation for the changing index. Both $\overline{\epsilon}$ and $\Delta\epsilon$ are functions of z with $\Delta\epsilon$ nonzero only in the grating.

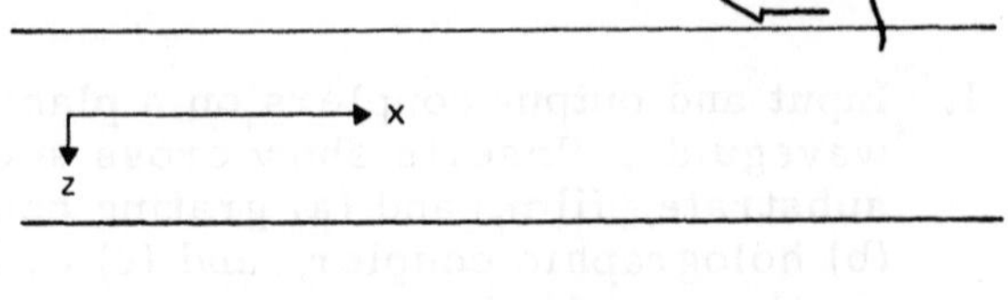

Figure 2. Guided wave incident on grating coupler.

If $\underline{E}^i$, $\underline{H}^i$ represents the field in the waveguide characterized by $\bar{\epsilon}$, substitution of eq. (2) into eq. (1) results in the following equations satisfied by the diffracted field

$$\nabla x \underline{E}^d = i\omega\mu \underline{H}^d$$

$$\nabla x \underline{H}^d = -i\omega\bar{\epsilon}\underline{E}^d - i\omega\Delta\epsilon \cos \kappa x(\underline{E}^i + \underline{E}^d) \ . \qquad (3)$$

The Born approximation for scattering consists in neglecting the very last term $\Delta \epsilon \underline{E}^d$ in eq. (3) with the result that eq. (3) assumes the form of eq. (1) in which an inhomogeneous or current source term

$$\underline{J} = -i\omega(\Delta\epsilon \cos \kappa x)\underline{E}^i \qquad (4)$$

is present. Since we are after a simple explanation of how the coupler works, the Born approximation suffices. Using an incident propagating wave of the form

$$\underline{E}^i(x, y, z) = \underline{E}^i(y, z) \exp ik_w x \qquad (5)$$

where k_w is the propagation constant, the current source takes the form

$$\underline{J} = i\omega(\Delta\epsilon /2)\underline{E}^i(y, z) \left[\exp i(k_w - \kappa)x + \exp i(k_w + \kappa)x\right] \ . \qquad (6)$$

Now we note that a traveling wave current source of the kind expressed in eq. (6) will radiate a plane wave into the regions above and below the guide so long as the wave number $k_w - \kappa$ is less than the wave numbers of the regions above and below the guide which may be respectively termed k_a and k_s for air and substrate. The direction of the radiation may be established with the aid of the diagram shown in Fig. 3. The plane wave in air, for example, must have propagation constant k_a and an x component $k_w - \kappa$ equal to the propagation constant of the current wave.

The current source expressed in eq. (6) is generated by the incident field in the coupling region so that the current source is not a uniform traveling wave, but rather is a source that begins at the input to the coupler and decays

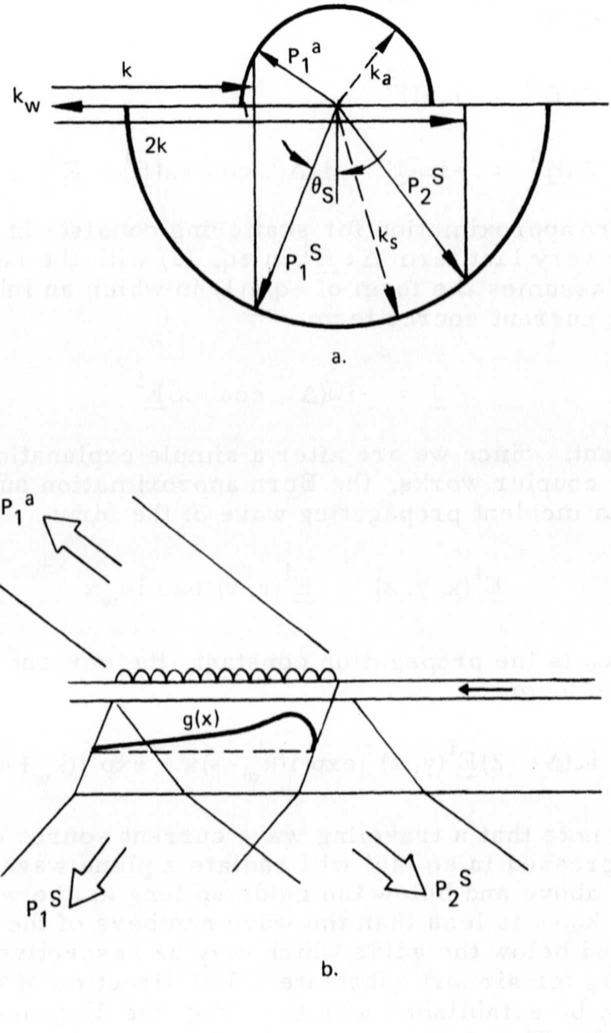

Figure 3. Angular direction of radiating beams in a coupler. (a) Semicircle diagram showing direction of P_1^s, P_1^a, and P_2^s as determined by wave numbers $k_w - \kappa$ and $k_w - 2\kappa$. (b) Associated beams emerging from the coupler.

in magnitude as energy radiates from the guide. The resulting radiated fields exhibit a similar behavior. As illustrated in Fig. 3(b), some components of the radiated field just below (or above) the guide have a spatial distribution g(x) that varies slowly relative to the optical wavelength if the energy emerges as a beam. The power in this beam we denote as P_1^s to indicate the first-order beam diffracted into the substrate. This power is some fraction of the power P_o that is incident on the coupler. As will be shown in Section III, the reciprocity principle indicates that the properties of the device acting as a coupler may be determined from the radiation properties discussed here. In particular, if a beam containing power P_{in} is incident on the structure and if a field component of this beam is characterized by the spatial distribution h(x), then the power P_c coupled into the device may be expressed

$$\frac{P_c}{P_{in}} = \frac{|\int g(x)h(x)dx|^2}{(\int|g|^2dx)(\int|h|^2dx)} \frac{P_1^s}{P_o} \tag{7}$$

Equation (7) may be interpreted as follows: the coupling efficiency of a periodic coupler is limited first to the relative power P_1^s/P_o radiated by the structure in a prescribed direction when the structure acts as a transmitter. In addition, the efficiency is enhanced by matching the general shapes of the radiated and incident beams. When $g(x) = h^*(x)$ and $P_1^s = P_o$ the coupling efficiency is 100%. Design of periodic couplers therefore consists in selecting the various device parameters so that these equalities are approached as closely as possible.

Three aspects of periodic couplers will be discussed in this article. Section II consists of an analysis of the interesting problem of the disposition of scattered energy when a beam is incident on a uniform coupler. It is found that at certain angles of incidence there is a strongly excited wave in the guiding film that contains a substantial portion of the incident power. Termination of the periodic structure prevents radiation of this wave and results in a coupler. In Section III the coupling problem is treated from the reciprocal point of view in a manner similar to that discussed above. This analysis results in approximate design equations for couplers. The concluding paragraph deals with some physical interpretations and suggestions for useful designs.

II. DIRECT ANALYSIS OF BEAM COUPLING

When a beam is incident on a periodic coupler at an angle for which the first or higher diffraction orders is phase matched to the guided wave, the energy in and near the guiding film reaches substantial values that are not achieved at other angles. Thus, although coupling is associated with generation of a diffraction order, the relationship between the efficiencies of coupling and diffraction is indirect and the two can be inversely related. In this chapter we review the solution to the problem of diffraction by a coupler. The analysis is treated in four steps that involve an introduction to the spectral representation of beams, the solution of the prism coupler problem, the analysis of propagation in periodic structures, and finally, a discussion of the problem at hand.

II. 1 Spectral Representation of Electromagnetic Waves

To introduce the concept of the angular or plane wave spectral representation of optical fields, let us consider the solution of the problem of determining the fields to the right of the plane z = 0 when monochromatic fields in this plane are prescribed. The region z > 0 is assumed to be free space and, in particular, contains no sources of optical energy nor any reflectors so that all energy enters this region from the left (Fig. 4).

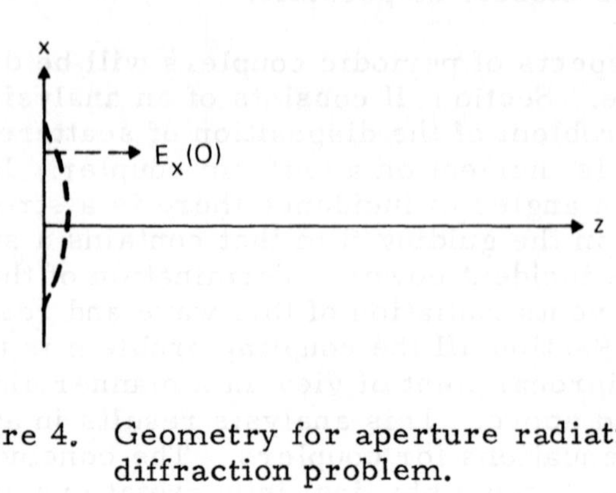

Figure 4. Geometry for aperture radiation or diffraction problem.

From Maxwell's equations one readily derives the fact that each of the six components of $\underline{E}$ and $\underline{H}$ satisfies the wave equation, e.g.

$$\left(\nabla^2 + k^2\right)E_x = 0 \quad . \tag{8}$$

If $E_x|_{z=0} = E_x(0)$ is prescribed, then $E_x(z)$ may be found by solving eq. (8) and this may be accomplished in a number of ways. Perhaps most familiar is the Kirchhoff-Huygens integral

$$E_x(z) = -\frac{1}{2\pi}\frac{\partial}{\partial z}\int\int_{-\infty}^{\infty} E_x(0)\left(e^{ikR}/R\right)_{z=0} dx'dy' \tag{9}$$

which is an expression of Huygens principal to the effect that the field is a superposition of spherical waves (exp ikR/R) generated at each point of the aperture. By examining Maxwell's equations it may be established that only two of the six field components can be independently specified, e.g., $E_x(0)$ and $E_y(0)$. The other field components can be obtained from the integral expressions (eq. (9)) for $E_x(z)$ and $E_y(z)$ using a Maxwell divergence and curl equation, for example.

The spectral representation provides an alternative solution to the aperture radiation problem. The two-dimensional Fourier transform of the field in any plane z is the spectral representation and may be defined

$$\mathcal{E}_x(z) = \frac{1}{2\pi}\int\int_{-\infty}^{\infty} E_x(z)\, e^{-i\underline{k}_\rho \cdot \underline{\rho}} d\underline{\rho}$$

$$E_x(z) = \frac{1}{2\pi}\int\int_{-\infty}^{\infty} \mathcal{E}_x(z)\, e^{i\underline{k}_\rho \cdot \underline{\rho}} d\underline{k}_\rho \tag{10}$$

where $\underline{k}_\rho$ is the two dimensional transform variable $\underline{k}_\rho = k_x\hat{x} + k_y\hat{y}$. Other terms in eq. (10) are $\underline{\rho} = x\hat{x} + y\hat{y}$, $d\underline{\rho} = dxdy$, $d\underline{k}_\rho = dk_x dk_y$. If the solution of the radiation problem is expressed as the second equality in eq. (10) and if $E_x(z)$ is placed in eq. (8), it is found that the spectrum satisfies the one-dimensional wave equation

$$\frac{d^2}{dz^2} E_x + \left(k^2 - k_\rho^{\ 2}\right) E_x = 0 . \qquad (11)$$

The solution of this equation when waves propagate in only the positive z direction is

$$E_x(z) = E_x(0) \exp i\, uz$$

$$u = \left(k^2 - k_\rho^{\ 2}\right)^{1/2} . \qquad (12)$$

The spectral representation of the field in any plane z is thus the product of the representation in the aperture plane and the function exp i uz. When the two-dimensional Fourier transform can be computed, the spectral repre-

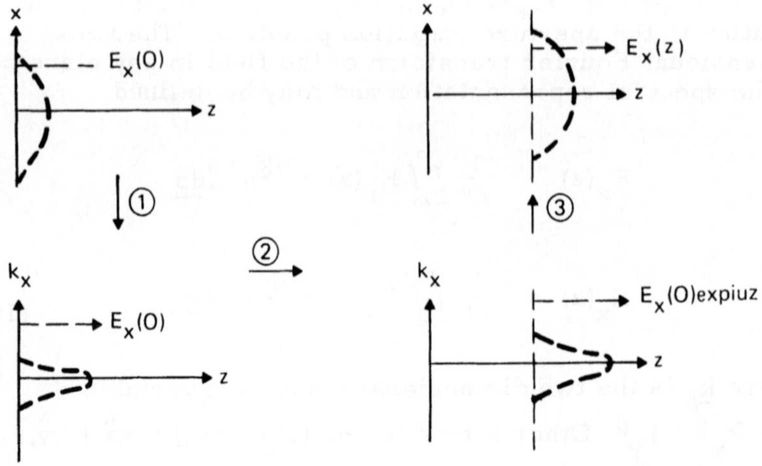

Figure 5. Three-step procedure for solving the diffraction problem using the spectral representation.

sentation is seen to lead to the straightforward solution technique illustrated in Fig. 5.

The spectral technique provides the easiest method for dealing with multilayer problems of interest in couplers. To illustrate the method let us consider the radiation of a gaussian beam of width σ produced by the aperture distribution

$$E_x(0) = \frac{1}{\sigma^2} \exp - \frac{\rho^2}{2\sigma^2} . \tag{13}$$

The two dimensional transform of eq. (13) is found as the integral in the first equality of eq. (10). Integration can be accomplished in cylindrical coordinates by completing squares in the exponential with the result that $E_x(0)$ is found to be gaussian in the transform variables, i. e.,

$$E_x(0) = \exp - \frac{k_\rho^2 \sigma^2}{2} . \tag{14}$$

Note that whereas the aperture field is effectively limited in space a distance of about 3σ, at which point the amplitude is 0.001 the peak value, the angular spectrum is limited to about $3/\sigma$.

Solution of the radiation problem consists in calculating the inverse transform

$$E_x(z) = \frac{1}{2\pi} \int\int_{-\infty}^{\infty} \exp \left[- \frac{k_\rho^2 \sigma^2}{2} + i\left(k_\rho \cdot \rho + uz\right)\right] d\underline{k}_\rho . \tag{15}$$

Because the gaussian is small for $k_\rho > 3/\sigma$, when σ is large compared with the wavelength λ so that $3/\sigma \ll 2\pi/\lambda = k$ the function u that appears in eq. (15) may be approximated by the first two terms of the series

$$u = \left(k^2 - k_\rho^2\right)^{1/2} = k\left[1 - \frac{1}{2}\frac{k_\rho^2}{k^2} - \frac{1}{8}\left(\frac{k_\rho^2}{k^2}\right)^2\right] + \ldots . \tag{16}$$

When this approximation is made the argument of the exponential in eq. (15) is quadratic in k_x and k_y, the integral is similar in form to the transform of a gaussian and so may be computed to be

$$E_x(z) \simeq \left(\frac{1}{\sigma'^2}\right) \exp - \frac{\rho^2}{2\sigma'^2} \qquad \sigma'^2 = \sigma^2 + \frac{iz}{k} \ . \tag{17}$$

The error in eq. (17) is negligible until the difference

$$uz - kz\left(1 - \frac{k_\rho^2}{2k^2}\right)$$

approaches some appreciable fraction of π. Using the third term in eq. (16) to estimate the difference at $k_\rho = 3/\sigma$ we note that (17) is good in the range $z < 0\ (\sigma^4/\lambda^3)$ which for a mm beam at 1 μm wavelength extends from the aperture plane to a distance of order 10^6 m. By expressing the inverse of σ'^2 in terms of real and imaginary parts, one may verify that eq. (17) is an expression for an expanding beam of the kind illustrated in Fig. 6.

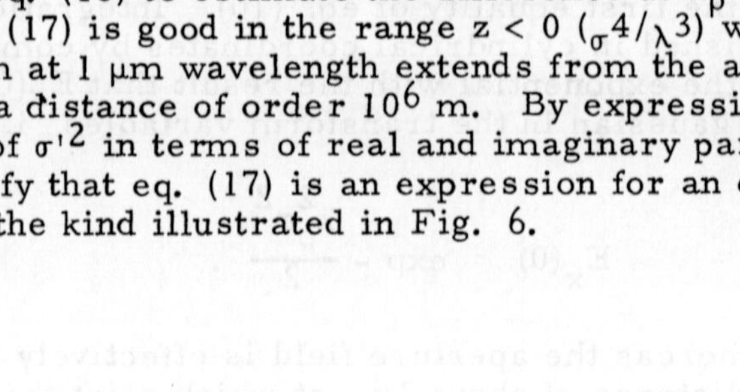

Figure 6. Expanding beam obtained from a gaussian aperture distribution.

The approximate inversion of the transform expressed in eq. (15) was facilitated by the narrow spectral character of the beam. The band limitation of wide beams also plays a role in the analysis of couplers. We will show that over a restricted band the spectrum of the field at a plane z in or near the guiding film in a coupler may usually be represented as*

$$E_x(z') = \frac{\beta}{k_\rho - (k_w' + i\alpha)} E_x^i(0) \qquad (18)$$

where $E_x^i(0)$ is the incident spectrum and β, k_w' and α are constants. The latter two constants are positive real and β is of order $(\alpha/t')^{1/2}$ where t' is the effective thickness of the guiding film. Equation (18) holds over a region of k_ρ near k_w'.

To show how eq. (18) describes coupling, consider a beam to be incident on the coupler at an angle θ as shown in Fig. 7. The beam may be characterized

$$E_x^i(0) = \frac{1}{\sigma_x' \sigma_y} \exp\left[-\frac{x^2}{2\sigma_x'^2} - \frac{y^2}{2\sigma_y^2} + i k_x^i x\right]$$

$$E_x^i(0) = \exp\left[-\left(k_x - k_x^i\right)^2 \frac{\sigma_x'^2}{2} - \frac{k_y^2 \sigma_y^2}{2}\right] \qquad (19)$$

where $\sigma_x' = \sigma_x/\cos\theta$ and $k_x^i = k \sin\theta$ are the beam width and phase constant, respectively, measured along the x axis. To invert eq. (18) and investigate the field we take $k_\rho \simeq k_x$ in eq. (18), since our interest lies in the spectral region near $k_x = k_x^i$, $k_y = 0$. The resulting inverse transform of eq. (18) is

$$E_x(z') = \sqrt{2\pi}\, i\beta \int_0^\infty \exp i\left[k_w' + i\alpha\right]x'\, E_x^i(x-x', y, 0)\, dx' \ . \qquad (20)$$

---

*Tamir and Bertoni[7] have previously discussed the beam reflection problem using a similar formulation.

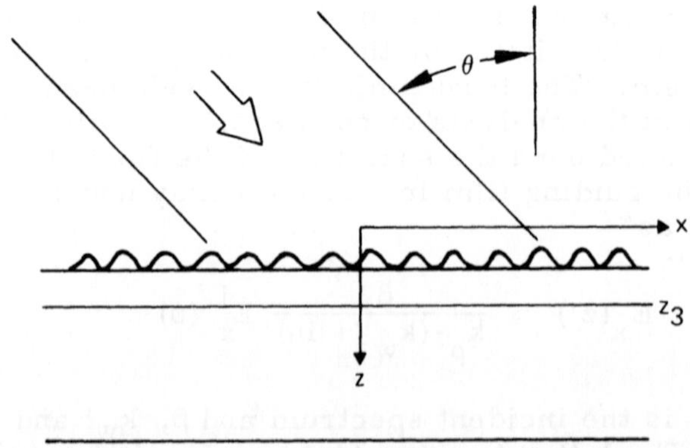

Figure 7. A beam incident on an unterminated coupler.

The integral in eq. (20) is obtained by convolving the inverse transform of the terms of eq. (18)(with $k_\rho \simeq k_x$). Note that the complex pole in eq. (18) contributes a leaky wave.

The field build-up in the film occurs at phase match when $k'_w x'$ and $-k^i_x x'$ cancel in eq. (20) so that integration is over the real function $\exp[-\alpha x' - (x-x')^2/2\sigma'^2_x]$. For useful couplers α is approximately $1/2\sigma'_x$ and eq. (20) is found to be of the order

$$E_x(z')_{max} = 0\left[\frac{\beta}{\alpha}\, E^i_x\,(0, y, 0)\right] = 0\left[\left(\frac{\sigma'_x}{t'}\right)^{1/2} E^i_x\,(0, y, 0)\right]. \tag{21}$$

Note the substantial increase in the field magnitude at some point in the film, sufficient to account for the fact that the power of the beam of width σ'_x is in the film of thickness t'. The position in x at which the maximum field occurs corresponds to best overlap of the exponential and the gaussian and is displaced a distance approximately σ'_x from the center of the beam. Formal integration of eq. (20) yields the function

$$E_x(z') = i\,\frac{\beta}{2}\exp\left[-\frac{x^2}{2\sigma'^2_x} + i\,k^i_x x - \left(k'_w + i\alpha - k^i_x - i\,\frac{x}{\sigma'^2_x}\right)^2 \frac{\sigma'^2_x}{2}\right]$$

$$\operatorname{erfc}\left[-i\left(k'_w + i\alpha - k^i_x - i\,\frac{x}{\sigma'^2_x}\right)\frac{\sigma'_x}{\sqrt{2}}\right] \frac{e^{-y^2/2\sigma_y^2}}{\sigma_y} \qquad (22)$$

Figure 8 shows a sketch of this function for some representative values of α near $(2\sigma'_x)^{-1}$ where useful coupling occurs. The aperture field represented by eq (22) is bell shaped and gives rise to a reflected beam that is spatially displaced from the incident beam by a distance $1/2\alpha$.

II. 2 The Prism Coupler

To show that the behavior of the spectrum of the field in a coupler is of the form indicated in eq. (18) we commence with the solution of the prism coupler[9] problem. The problem consists in determining transmitted and reflected fields when a beam is incident on a multilayer structure as in Fig. 9. It is useful to begin by specifying two independent polarizations for the representation of the incident beam that remain independent in the presence of multilayers. The number of equations that have to be manipulated is halved if the beam is represented by fields in which $\underline{E}$ is transverse to the layer interface (TE) plus fields in which $\underline{H}$ is transverse to the layer interfaces (TM).

Employing potential functions $\underline{F} = F\hat{z}$ and $\underline{A} = A\hat{z}$ for the TE and TM representation, respectively, where F and A are two independent scalar functions the fields may be expressed in terms of these functions as

$$\underline{E} = \nabla \times \underline{F} + \frac{\mu}{-i\omega\epsilon}\,\nabla \times (\nabla \times \underline{A})$$

$$\underline{B} = \nabla \times \underline{A} + \frac{1}{i\omega\mu}\,\nabla \times (\nabla \times \underline{F}) \quad . \qquad (23)$$

Figure 8. Normalized electric field inside the guiding film of a coupler (e.g., z_3 of Figure 7) versus distance along the film. The gaussing beam expressed in (19), centered at x=0, is incident at the angle for phase match. The function $f_y(x)$ is the ratio $f_y(x) = E_y(x, 0, z_3) / \left[(\sigma_{x'}^{1/2}\beta/2^{5/4}\alpha^{1/2})\, E^i_y(0, 0, 0)\right]$ with $E_y(x, o, z_3)$ expressed in (22) and $E^i_y(0, 0, 0)$ obtained from (19). Parameter α has values near optimum (see eq. (100)).

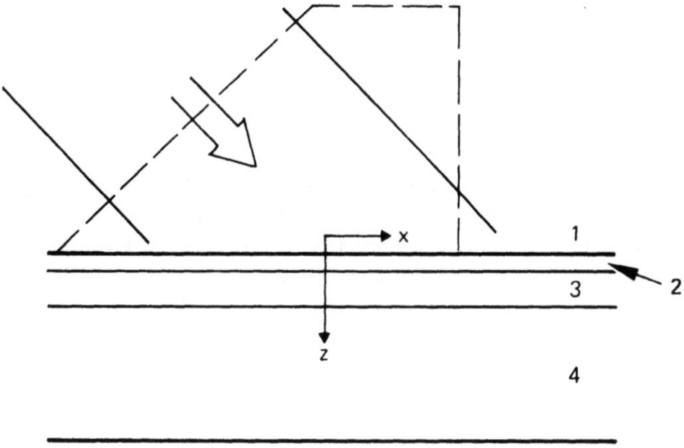

Figure 9. Undetermined prism coupler. The guiding film, region 3, and the prism, region 1, have high refractive index while the substrate, 4, and tunneling layer, 2, have low refractive index. Dotted line shows terminated prism location.

The second addends in eq. (23) are obtained from Maxwell's equations. The spectral relations equivalent to eq. (23) obtained by differentiation are

$$\underline{E} = i\left(k_y\hat{x} - k_x\hat{y}\right)F - \frac{\mu}{i\omega\epsilon}\left(k_\rho^2\hat{z} - i\underline{k}_\rho\frac{d}{dz}\right)A$$

$$\underline{B} = i\left(k_y\hat{x} - k_x\hat{y}\right)A + \frac{1}{i\omega\mu}\left(k_\rho^2\hat{z} + i\underline{k}_\rho\frac{d}{dz}\right)F. \qquad (24)$$

When eq. (23) is substituted into Maxwell's equations, A and F are found to be solutions of the wave equation. Their transforms therefore satisfy the one dimensional wave equation

$$\frac{d^2}{dz^2}A + \left(k^2 - k_\rho^2\right)A = \frac{d^2}{dz^2}A + u^2A = 0\ . \qquad (25)$$

As discussed in the previous section, solutions of eq. (25) for waves propagating in the z direction are of the form

$$A = a^+ \exp i\,uz \qquad F = f^+ \exp i\,uz \tag{26}$$

Using eq. (26), eq. (24) reduces to a set of linear equations that may be employed to establish a^+ and f^+ from aperture fields that are otherwise expressed in terms of tangential electric or magnetic fields. When $E_x(0)$ and $E_y(0)$ are prescribed, for example, eq. (24) yields

$$a^+ = \left(\frac{i\omega\epsilon}{\mu}\right)\left(k_x E_x + k_y E_y\right)/uk_\rho^2 \qquad f^+ = -i\left(k_y E_x - k_x E_y\right)/k_\rho^2 \tag{27}$$

Turning now to the multilayer problem, the more general solution to eq. (25) is a sum of two waves of the form

$$A_j = a_j^+ \exp iu_j z + a_j^- \exp -iu_j z = A_j^+ + A_j^- \tag{28}$$

where the minus sign indicates a reflected wave or one traveling in the negative z direction. The subscript j is used to indicate the layer in which eq. (28) represents the potential function. Note in particular that $u_j = (k_j^2 - k_\rho^2)^{1/2}$ and k_j is the wave number of region j. To determine the reflected and transmitted fields the $a_j^\pm$ must be found. This is accomplished by matching tangential $\underline{E}$ and H at each boundary. Equation (24) indicates that equivalently A and F must be continuous at each interface and also that

$$\frac{i}{i\epsilon}\frac{d}{dz}A = \frac{u}{\epsilon}A^+ - \frac{u}{\epsilon}A^- \quad \text{and} \quad \frac{1}{i}\frac{d}{dz}F = uF^+ - uF^-$$

must be continuous. Employing the symbol $K_j = u_j/\epsilon_j$ for TM fields or $K_j = u_j$ for TE fields, the continuity equations for both polarizations can be summarized by the matrix equation shown in Table I in which the symbol $b_j^\pm$ is used to represent either $a_j^\pm$ or $f_j^\pm$.

Table I. The matrix equation for solving multilayer (4 region) transmission and reflection problems. Symbols are e±12 = $\exp \pm i u_1 z_2$, $K_2^{TE} = u_2$. $K_2^{TM} = u_2/\epsilon_2$, $u_2 = (k_2^2 - k_\rho^2)^{1/2}$. The determinant of the square matrix is $D_{1,4}$ and the determinant of the submatrix within the dotted line is $D_{1,3}$. Blank spaces are zeros.

$$
\left[\begin{array}{cccc|cc}
e\text{-}11 & -e21 & -e\text{-}21 & & & \\
-u_1 e\text{-}11 & -u_2 e21 & u_2 e\text{-}21 & & & \\
 & e22 & e\text{-}22 & -e32 & -e\text{-}32 & \\
 & u_2 e22 & -u_2 e\text{-}22 & -u_3 e32 & u_3 e\text{-}32 & \\
\hline
 & & & e33 & e\text{-}33 & -e43 \\
 & & & u_3 e33 & -u_3 e\text{-}33 & -u_4 e43
\end{array}\right]
\left[\begin{array}{c}
b_1^- \\ b_2^+ \\ b_2 \\ b_3^+ \\ b_3^- \\ b_4^+
\end{array}\right]
=
\left[\begin{array}{c}
-b_1^+ e11 \\ -u_1 b_1^+ e11 \\ 0 \\ 0 \\ 0 \\ 0
\end{array}\right]
$$

Solution of the linear equations of Table I for any of the coefficients $b_j^{\pm}$ may be expressed as the quotient of determinants obtained from the matrix shown there. The transmitted and reflected waves, for example, may be expressed

$$b_4^+ = \frac{K_1 K_2 K_3}{D_{1,4}} b_1^+ \qquad b_1^- = -\frac{'D_{1,4}}{D_{1,4}} b_1^+ \tag{29}$$

where $D_{1,4}$ is the determinant of the matrix and is termed the system function for regions 1 through 4. The numerator of eq. (29) is found by vector substitution in the matrix to be

$$'D_{1,4} \pm D_{1,4}\left(-u_1, u_2, u_3 u_4\right) . \tag{30}$$

By expanding the determinant of the matrix of Table I about the lower right corner we find that the system function can be expressed

$$D_{1,4} = e^{iu_4 z_3}\left(K_3 + K_4\right) e^{-iu_3 z_3} D_{1,3} - e^{iu_4 z_3}\left(-K_3 + K_4\right)$$
$$e^{iu_3 z_3} D'_{1,3} \tag{31}$$

where $D_{1,3}$ is the determinant submatrix shown within the dotted line and $D'_{1,3}(u_1, u_2 - u_3)$. Had there been only three regions in the problem, $D_{1,3}$ would have appeared in eq. eq. (29) instead of $D_{1,4}$. Had there been only regions 3 and 4, we would find the denominator to be

$$D_{3,4} = e^{iu_4 z_3}\left(K_3 + K_4\right) e^{-iu_3 z_3} . \tag{32}$$

Using this notation, eq. (31) may be expressed

$$D_{1,4} = D_{1,3} D_{3,4} - D'_{1,3}\,'D_{3,4} . \tag{33}$$

The partition relation (33) can be generalized to the case in which any number of layers are present and any region is singled out for partition, e.g.,

$$D_{1,N} = D_{1,j} D_{j,N} \quad D'_{1,j}\,{}'D_{j,N} \quad . \tag{34}$$

The generalization of eq. (32) is

$$D_{j,\,j+1} = e^{iu_{j+1} z_j} \left(K_j + K_{j+1} \right) e^{-iu_j z_j} \tag{35}$$

and that of eq. (29) is

$$b_N^+ = \frac{K_1 K_2 \ldots K_{N-1}}{D_{1,N}} b_1^+$$

$$b_1^- = - \frac{{}'D_{1,N}}{D_{1,N}} b_1^+ \quad . \tag{36}$$

The solution of the linear equations for any $b_j^{\pm}$ has the same denominator $D_{1,N}$.

To show that the prism coupler exhibits the behavior of eq. (18), let us consider eq. (29) for a 4 region structure in which

$$k_2, k_4 < k_\rho < k_1, k_3 \tag{37}$$

The regions 2, 3, and 4 constitute a waveguide while region 1 is a high index region used to permit entry of the beam. Employing eqs. (34) and (35) and combining complex conjugates we may express the system function as

$$D_{1,4} = D_{1,2} D_{2,4} - D'_{1,2}\,{}'D_{2,4} \tag{38}$$

where

$$D_{2,4} = -2i|K_2+K_3||K_3+K_y| \sin\left(u_3t_3 - \tan^{-1}\frac{K_2}{iK_3}\right.$$

$$\left. -\tan^{-1}\frac{K_4}{iK_3}\right) \exp\left(-|u_4|z_2 + |u_2|z_2\right)$$

$$D_{1,2} = |K_1+K_2| \exp\left(-|u_2|z_1 + i\tan^{-1}\frac{K_1}{iK_2} - iu_1z_1\right)$$

$$'D_{2,4} = -2i|K_2+K_3||K_3+K_4| \sin\left(u_3t_3 + \tan^{-1}\frac{K_2}{iK_3}\right.$$

$$\left. -\tan^{-1}\frac{K_4}{iK_3}\right) \exp\left(-|u_4|z_2 - |u_2|z_3\right)$$

$$D'_{1,2} = -|K_1+K_2| \exp\left(|u_2|z_1 - i\tan^{-1}\frac{K_1}{iK_2} - iu_1z_1\right). \tag{39}$$

The function $D_{2,4}$ has zeros at values of k_ρ that set the argument of the sine to zero, i. e., when

$$u_2t_2 - \tan^{-1}\frac{K_2}{iK_3} - \tan^{-1}\frac{K_4}{iK_3} = m\pi . \tag{40}$$

Equation (40) is the eigenvalue equation for guided waves on a film of wave number k_3 surrounded by media with wave numbers k_2 and k_4. Let us denote this zero as $k_\rho = k_w$.

In the vicinity of k_w which is a zero of eq. (40) the sine function may be approximated by a term linear in k_ρ. Differentiating we find

$$D_{2,4} \simeq d_{2,4}\left(k_\rho - k_w\right) \tag{41}$$

where

$$d_{2,4} = -2i\,|K_2 + K_3|\,|K_3 + K_4|(-1)^m \left[\frac{k_w}{u_3}\left(t_3 + \frac{\eta}{|u_2|} + \frac{\eta}{|u_4|}\right)\right]$$

$$\exp\left(-|u_4|\,z_3 + |u_2|\,z_2\right) \tag{42}$$

and $\eta = 1$ for TE waves and $\eta = (k_3^2/k_w^2)$ for TM waves. Using eq. (41) in eq. (29) the transmission coefficient in the vicinity of k_w becomes

$$b_4^+ = \frac{(2K_1)(2K_2)(2K_3)/D_{1,2}d_{2,4}}{k_\rho - \left(k_w + D'_{1,2}{}'D_{2,4}/D_{1,2}d_{2,4}\right)}\, b_1^+ \quad . \tag{43}$$

The various terms in eq. (43) are all slowly varying with k_ρ with the exception of k_ρ itself so that when evaluated at $k_\rho = k_w$, eq. (43) has the form of eq. (18). The constants relating eq. (18) to eq. (43) are

$$k'_w + i\alpha = k_w + D'_{1,2}{}'D_{2,4}/D_{1,2}d_{2,4}$$

$$\beta = \frac{(2K_1)\,(2K_2)\,(2K_3)}{D_{1,2}\,d_{2,4}} \quad . \tag{44}$$

The right hand term in eq. (44) is complex and provides some small change in the propagation constant of the guided wave due to the presence of the prism. Examination of eqs. (39) and (42) shows that α, the attenuation constant of the leaky wave, is of the order

$$\alpha = I_m\left[\frac{D'_{1,2}{}'D_{2,4}}{D_{1,2}\,d_{2,4}}\right] = 0\left(t_3^{-1}\exp -2|u_2|\,t_2\right) \tag{45}$$

where t_2 is the thickness of the low index tunneling region and t_3 the thickness of the film. The constant β is seen to be of the form $(\alpha/t_3)^{1/2}$. The solution (43) thus gives rise to a coupled wave in accordance with the discussion of the previous section.

As a final comment we note from eqs. (29) and (39) that over a limited spectral band the reflection coefficient may be expressed

$$\frac{b_1^-}{b_1^+} = -\frac{'D_{1,4}}{D_{1,4}} = \frac{D_{1,4}^*}{D_{1,4}} = \frac{k_\rho - (k_w - i\alpha)}{k_\rho - (k_w + i\alpha)} = 1 + \frac{2i\,\alpha}{k_\rho - (k_w + i\alpha)} . \tag{46}$$

The form of eq. (46) indicates that there are two reflected beams, one of which is displaced along the surfaces of the coupler by approximately $1/2\alpha$ and represents reradiation of energy coupled into the guide.

II. 3 Fields in the Periodic Medium

To show that periodic couplers exhibit the behavior indicated in eq. (18) we examine the 4 region configuration shown in Fig. 10. Layer 2 is a periodic medium characterized by a relative permittivity that can be represented by the Fourier expansion

$$\epsilon = \bar{\epsilon}\left(1 + \sum_{p=-\infty}^{\infty} \epsilon^{(p)} \exp ip\underline{\kappa} \cdot \underline{\rho}'\right) \tag{47}$$

where $\underline{\kappa}$ is $(2\pi/\Lambda)\hat{q} = \kappa_x\hat{x} + \kappa_z\hat{z}$ defines the direction $\hat{q}$ and wavelength Λ of the variation in permittivity and $\rho' = x\hat{x} + z\hat{z}$ is the space vector. Examples of grating structures are shown in Fig. 11. The expansion of the square etched grating on Fig. 11(a) has coefficient

$$\epsilon^{(p)} = -i\left(\frac{\epsilon_a - \epsilon_b}{\epsilon_a + \epsilon_b}\right)\frac{\sin^2 p\pi/2}{p\pi/2} \tag{48}$$

where

$$\bar{\epsilon} = \frac{\epsilon_a + \epsilon_b}{2}$$

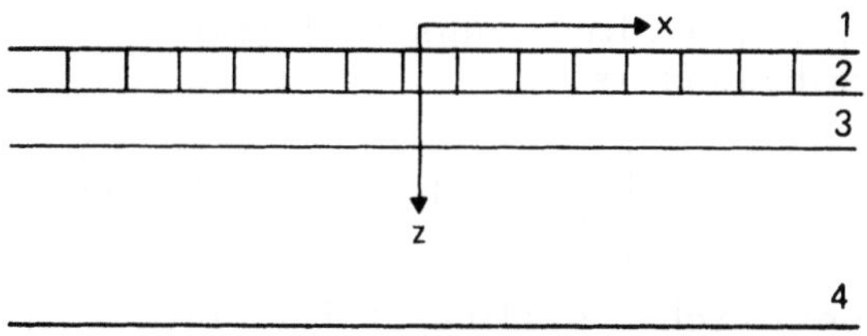

Figure 10. Geometry for periodic coupler.

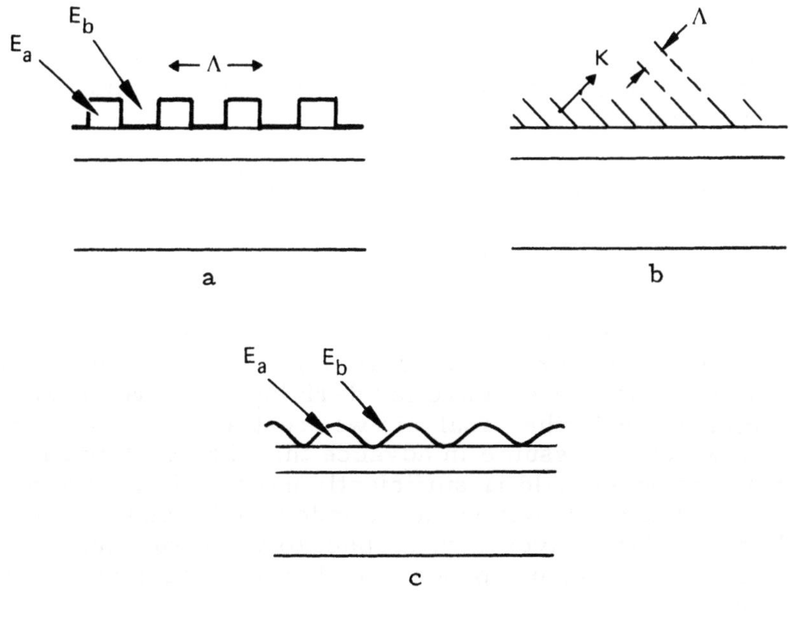

Figure 11. Example of periodic regions for a coupler.

and $\kappa_z = 0$. For a tilted sinusoidal grating the permittivity might be expressed as

$$\epsilon = \bar{\epsilon}\left[1 + \delta\left(e^{i\underline{\kappa}\cdot\underline{\rho}'} + e^{-i\underline{\kappa}\cdot\underline{\rho}'}\right)\right]. \tag{49}$$

The sinusoidal etched grating of Fig. 11(c) can be represented by eq. (47) by permitting the coefficients to vary with z, e.g.,

$$\bar{\epsilon} = \left[\left(\epsilon_a - \epsilon_b\right) \sin^{-1} z/t + \pi\epsilon_b\right]/2\pi \quad . \tag{50}$$

When the coefficients $\bar{\epsilon}$ and $\epsilon^{(p)}$ vary with z, approximate solutions to the coupling problem can be obtained by treating the grating system as a series of layers. We will not specifically consider such structures further, however, and we treat $\bar{\epsilon}$ and $\epsilon^{(p)}$ as constants.

Within the layer with periodic index variation the electric and magnetic fields satisfy the wave equations (for isotropic materials)

$$\left(\nabla^2 + k^2\right)\underline{E} = \nabla\left(\frac{\nabla}{\epsilon}\cdot \underline{E}\right)$$

$$\left(\nabla^2 + k^2\right)\underline{H} = -\frac{\nabla}{\epsilon} \times \nabla \times \underline{H} \tag{51}$$

where k^2 contains the factor ϵ of eq. (47). The expressions in eq. (51) are obtained from Maxwell's equations by following the customary procedure for deriving the wave equation. To simplify the further analysis we treat a quasi-plane wave case in which we assume in advance that the spectrum in k_y of the incident field is sufficiently narrow that coupling between TE and TM waves due to index variations can be neglected. Derivatives with respect to y are suppressed in eq. (51) so that two independent scalar wave equations are obtained, viz.,

$$\left(\nabla'^2 + k^2\right)E_y = 0$$

$$\left(\nabla'^2 + k^2 - \frac{1}{\epsilon}\nabla\epsilon\cdot\nabla'\right)H_y = 0 \quad . \tag{52}$$

The procedure for solving both equations in (52) is the same. For simplicity we consider only the TE case described by the first equation in (52). Solutions of the equation

$$(\nabla'^2 + \bar{k}^2) E_y = -\bar{k}^2 \left(\sum_{p=-\infty}^{\infty} \epsilon^{(p)} e^{ip\kappa \cdot \rho'} \right) E_y \tag{53}$$

in which $\bar{k}^2 = k_o^2 \bar{\epsilon}$ can be expressed in the form

$$E_y = \sum_{m=-\infty}^{\infty} C_{m\pm} \exp i\underline{k}_{m\pm} \cdot \underline{\rho}' \sum_{n=-\infty}^{\infty} h_m^n \exp in\underline{\kappa} \cdot \rho' \tag{54}$$

where

$$\underline{k}_{m\pm} = k_x \hat{x} + k_{z_{m\pm}} \hat{z} \ . \tag{55}$$

Equation (54) represents a sum of waves, one for each value of m± having propagation constant $k_{m\pm}$. The plus or minus refers to positive or negative going waves in z. Each wave (say m± = m'+) is composed of a summation of space harmonics designated by index n. The values of h_m^n are determined by the periodic medium with h_m^m taken as unity. The propagation constants $k_{z_{m\pm}}$ are also determined by the medium while the C_m are found by imposing boundary conditions. The parameter k_x is the same as the transform variable discussed previously and eq. (54) should be thought of as the representation of the spectrum of E_y. Figure 12 is an illustration of the waves in eq. (54).

When eq. (54) is placed in eq. (53) and terms that multiply like powers of exp $in\kappa_x x$ are equated, there results a set of linear homogeneous equations that determine h_m^n as eigenvectors (one set for each value of m) and $k_{z_{m\pm}}$ as eigenvalues. Numerical values are generally obtained by truncating in the manner shown in Table II. Insight into the representation (54) can be gained by obtaining approximate values for h_m^n and $k_{z_{m\pm}}$ based on the

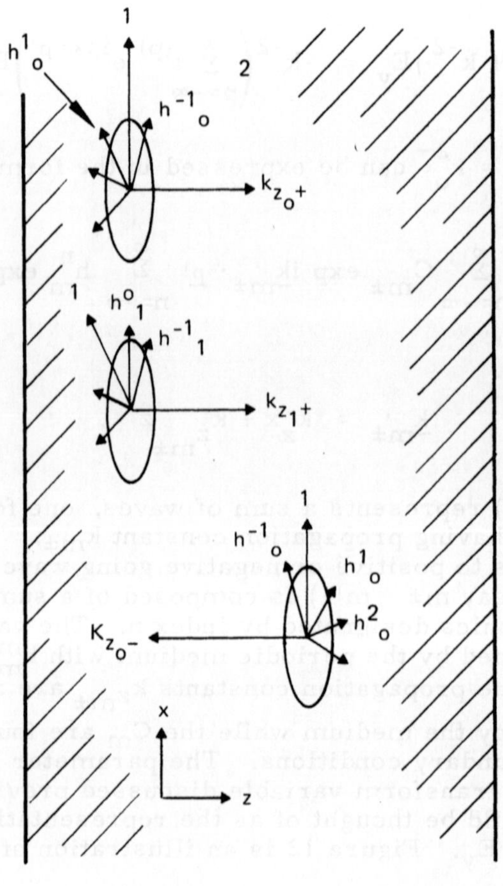

Figure 12. Examples of waves that propagate in a periodic medium. Each wave has a propagation constant in the z direction $k_{z_{m\pm}}$ and is composed of a sum of terms $h^n_m \exp i[k_x x + n\kappa_x x + n\kappa_z z]$.

Table II. Truncated eigenvector equation for establishing the wave function of a periodic medium. When the determinant of the above matrix is set to zero, 10 eigenvalues labeled $k_{z_{m\pm}}$ are obtained. These have associated eigenvectors $h^{n}_{m\pm}$.

$$
\begin{bmatrix}
k_z^2 - u(-2)^2 & -\bar{k}^2\,\epsilon(1) & -\bar{k}^2\,\epsilon(2) & -\bar{k}^2\,\epsilon(3) & -\bar{k}^2\,\epsilon(4) \\
-\bar{k}^2\,\epsilon(-1) & k_z^2 - u(-1)^2 & -\bar{k}^2\,\epsilon(1) & -\bar{k}^2\,\epsilon(2) & -\bar{k}^2\,\epsilon(3) \\
-\bar{k}^2\,\epsilon(-2) & -\bar{k}^2\,\epsilon(-1) & k_z^2 - u(o)^2 & -\bar{k}^2\,\epsilon(1) & -\bar{k}^2\,\epsilon(2) \\
-\bar{k}^2\,\epsilon(-3) & -\bar{k}^2\,\epsilon(-2) & -\bar{k}^2\,\epsilon(-1) & k_z^2 - u(1)^2 & -\bar{k}^2\,\epsilon(1) \\
-\bar{k}^2\,\epsilon(-4) & -\bar{k}^2\,\epsilon(-3) & -\bar{k}^2\,\epsilon(-2) & -\bar{k}^2\,\epsilon(-1) & k_z^2 - u(2)^2
\end{bmatrix}
\begin{bmatrix}
C_{-2} \\ C_{-1} \\ C_o \\ C_1 \\ C_2
\end{bmatrix}
= 0
$$

assumption that the $\epsilon^{(p)}$ are small compared with unity. For photoinduced index changes this assumption is clearly viable. Even in etched gratings, however, this assumption is useful for understanding the nature of the solution.

When the $\epsilon^{(p)}$ are small, the determinant of the matrix in Table II is approximately the product of the diagonal terms. With m and n truncated to values between ±M, the 2(2M + 1) eigenvalues of the determinant can therefore be expressed

$$k_{z_{m\pm}} = -m\kappa_z \pm \left(u^{(m)} + \Delta^{(m)}\right) \tag{56}$$

where

$$u^{(m)} = \left[k^2 - \left(k_x + m\kappa_x\right)^2\right]^{1/2} \tag{57}$$

is the propagation constant of the mth harmonic wave in the absence of index variations. The correction term $\Delta^{(m)}$ results from the index variations and is of order $\epsilon^{(p)2}$. The special case in which $-m\kappa_z \pm u^{(m)}$ are equal for two different values of m and prescribed plus or minus sign, which is termed the Bragg case, results in $\Delta^{(m)}$ proportional to $\epsilon^{(p)}$.

The eigenvector h_m^n ($h_m^m = 1$) for a prescribed value of $k_{z_{m\pm}}$ is generally small compared to unity and may be obtained from the nth row of the matrix in Table II by retaining only two terms as

$$h_m^n = \frac{\bar{k}^2 \epsilon^{m-n}}{\left(k_{z_m} + n\kappa_z\right)^2 - u^{(n)2}} + \delta_m^n . \tag{58}$$

The correction δ_m^n is of order $\epsilon^{(p)2}$. In the event of a Bragg degeneracy between two orders m and n, the denominator of the first term in eq. (58) is of order $\epsilon^{(p)}$ and h_m^n is of order unity. Periodic couplers have wavelengths comparable to that of light. This results in generally large values for the denominator of the first term in eq. (58) and permits rapid truncation of the series in eq. (54).

II. 4 Fields in the Coupler

In each region j of the multilayer structure the spectrum of the fields may be represented by a summation of the kind shown in eq. (54). In the homogeneous regions, however, we have

$$h^n_{jm} = 0 \qquad h^m_{j_m} = 1 \qquad k_{jz_{m\pm}} = -m\kappa_z \pm u_j^{(m)} \; . \tag{59}$$

Values of the coefficients $C_{j_{m\pm}}$ are obtained by imposing the boundary conditions that E_y and $d/idz\ E_y$ be continuous. Using the symbol $v^n_{j,m\pm}$ to represent

$$v^n_{m,m\pm} = n\kappa_z + k_{jz_{m\pm}} \tag{60}$$

so that when m and n are the same

$$v^m_{j,m\pm} = u_j^{(m)} + \Delta_j^{(m)} \sim u_j^{(m)} = \left[k^2 - \left(k_x + m\kappa_x\right)^2 \right]^{1/2} \tag{61}$$

the boundary conditions may be matched in the manner shown in Table III. The table is arranged so that equations matching a prescribed order n are written sequentially at each interface.

Solutions for wave amplitudes $C_{j_{m\pm}}$ may be expressed as a quotient of determinants obtained from the matrix in Table III. The matrix has been divided into submatrices with diagonal submatrices denoted $D^{(n)}_{1,N}$ in analogy with the system function for homogeneous regions. $D^{(n)}_{1,N}$ is the same as $D_{1,N}$ when $k_\rho = k_x + m\kappa_x$ except that $v^n_{j,n\pm}$ replaces $u_j^{(n)}$. Since all terms involving h^n_{jm} appear outside the diagonal submatrices, the solution for transmission

Table III. The matrix equation for the periodic coupler. Subscripts m and n indicating which submatrix an element appears in are not shown. The notational system may be explained by example. Denoting q_{ij} as a general element of the matrix obtained from row i and column j.

$$q_{3,4} = (e-32)^{n=-1}_{m=-1} = \exp -iu_3^{(-1)} z_2$$

$$q_{4,3} = \left[-v_2^- \hat{e}-22\right]^{n=-1}_{m=-1} = -\left(v^{-1}_{2,-1^-}\ \exp\right) iv^{-1}_{2,-1^-}\ z_2$$

$$q_{1,8} = \left[-h^+\hat{e}-21\right]^{-1}_{0} = -h^{-1}_{0^+}\ \exp iv^{-1}_{2,0^+}\ z_1$$

$$q_{10,3} = \left[-v_2^- h^- \hat{e}-22\right]^{0}_{-1} = \left(v^0_{2,-1^-}\right)\left(h^0_{-1^-}\right) \exp -iv^0_{2,-1^-}\ z_2$$

Parameters $u_j^{(n)}$, $v^n_{j,m\pm}$, and $h^n_{j,m\pm}$ are defined in the text.

	m=1						m=0					
n=1	$e-11$	$-\hat{e}21$	$-\hat{e}-21$					$-h^+\hat{e}21$	$-h^-\hat{e}-21$			
	$-u_1e-11$	$-v_2^+\hat{e}21$	$v_2^-\hat{e}21$					$-v_2^+h^+\hat{e}21$	$v_2h^-\hat{e}21$			
		$\hat{e}22$	$\hat{e}-22$	$-e32$	$-e-32$			$h^+\hat{e}22$	$h^-\hat{e}-22$			
		$v_2^+\hat{e}22$	$-v_2^-\hat{e}-22$	$-u_3e32$	u_3e-32			$v_2^+h^+\hat{e}22$	$-v_2^-h^-\hat{e}-22$			
				$e33$	$e-33$	$-e43$						
				u_3e33	$-u_3e-33$	$-u_4e43$						
n=0		$-h^+\hat{e}21$	$-h^-\hat{e}-21$				$e-11$	$-\hat{e}21$	$-\hat{e}-21$			
		$-v_2^+h^+\hat{e}21$	$v_2^-h^-\hat{e}21$				$-u_1e-11$	$-v_2^+\hat{e}21$	$v_2^-\hat{e}-21$			
		$h^+\hat{e}22$	$h^-\hat{e}-22$					$\hat{e}22$	$\hat{e}-22$	$-e32$	$-e-32$	
		$v_2^+h^+\hat{e}22$	$-v_2^-h^-\hat{e}-22$					$v_2^+\hat{e}22$	$-v_2^-\hat{e}-22$	$-u_3e32$	u_3e-32	
										$e33$	$e-33$	$-e43$
										u_3e33	$-u_3e-33$	$-u_4e43$
n=1		$-h\hat{e}2$	$-h^-\hat{e}-21$					$-h^+\hat{e}21$	$-h^-\hat{e}-21$			
		$-v_2^+h^+\hat{e}21$	$v_2^-h^-\hat{e}21$					$-v^+h^+\hat{e}21$	$v_2^-h^-\hat{e}21$			
		$h^+\hat{e}22$	$h^-\hat{e}-22$					$h^+\hat{e}^{22}$	$h^-\hat{e}-22$			
		$v_2^+h^+\hat{e}^{22}$	$-v_2^-h^-\hat{e}-22$					$v_2^+h^+\hat{e}22$	$-v_2^-h^-\hat{e}-22$			

$$
\begin{array}{c}
m=1
\end{array}
$$

$$
\left[\begin{array}{cccccc}
 & -h^{+}\hat{e}21 & -h^{-}\hat{e}\text{-}21 & & & \\
 & -v_2^{+}h^{+}\hat{e}21 & v_2^{-}h^{-}\hat{e}21 & & & \\
 & h^{+}\hat{e}22 & h^{-}\hat{e}\text{-}22 & & & \\
 & v_2^{+}h^{+}\hat{e}22 & -v_2^{-}h^{-}\hat{e}\text{-}22 & & & \\
 & & & & & \\
 & & & & & \\
\hline
 & -h^{+}\hat{e}21 & -h^{-}\hat{e}\text{-}21 & & & \\
 & -v_2^{+}h^{+}\hat{e}21 & v_2^{-}h^{-}\hat{e}21 & & & \\
 & h^{+}\hat{e}22 & h^{-}\hat{e}\text{-}22 & & & \\
 & v_2^{+}h^{+}\hat{e}22 & -v_2^{-}h^{-}\hat{e}\text{-}22 & & & \\
 & & & & & \\
 & & & & & \\
\hline
e\text{-}11 & -\hat{e}21 & -\hat{e}\text{-}21 & & & \\
-u_1e\text{-}11 & -v_2^{+}\hat{e}21 & v_2^{-}\hat{e}\text{-}21 & & & \\
 & \hat{e}22 & \hat{e}\text{-}22 & -e32 & -e\text{-}32 & \\
 & v_2^{+}\hat{e}22 & -v_2^{-}\hat{e}\text{-}22 & -u_3e32 & u_3e\text{-}32 & \\
 & & & e33 & e\text{-}33 & -e43 \\
 & & & u_3e33 & -u_3e\text{-}33 & -u_4e43
\end{array}\right]
\left[\begin{array}{c}
C_{1,-1^-} \\ C_{2,-1^+} \\ C_{2,-1^-} \\ C_{3,-1^+} \\ C_{3,-1^-} \\ C_{4,-1^+} \\
C_{1,0^-} \\ C_{2,0^+} \\ C_{2,0^-} \\ C_{3,0^+} \\ C_{3,0^-} \\ C_{4,0^+} \\
C_{1,1^-} \\ C_{2,1^+} \\ C_{2,1^-} \\ C_{3,1^+} \\ C_{3,1^-} \\ C_{4,1^+}
\end{array}\right]
=
\left[\begin{array}{c}
\\ \\ \\ \\ \\ \\ \\ \\
C_{1,0^+}\,e11 \\
-u_1C_{1,0^+}\,e11 \\
\\ \\ \\ \\ \\ \\ \\ \\
\end{array}\right]
$$

and reflection coefficients of the zeroth diffraction order, for example, may be expressed in the form

$$\frac{C_{1,0-}}{C_{1,0+}} = \frac{-'D_{1,4}^{(0)} D_{1,4}^{(1)} D_{1,4}^{(-1)} \cdots D_{1,4}^{(-M)} + n_{1,0} - (h^2)}{D_{1,4}^{(0)} D_{1,4}^{(1)} D_{1,4}^{(-1)} \cdots D_{1,4}^{(-M)} + d(h^2)}$$

$$\frac{C_{4,0+}}{C_{1,0+}} = \frac{\left(2u_1^{(0)}\right)\left(2v_{2,0}^{(0)}\right)\left(2u_3^{(0)}\right) D_{1,4}^{(1)} D_{1,4}^{(-1)} \cdots D_{1,4}^{(-M)} + n_{4,0+}(h^2)}{D_{1,4}^{(0)} D_{1,4}^{(1)} D_{1,4}^{(-1)} \cdots D_{1,4}^{(-M)} + d(h^2)} . \tag{62}$$

As indicated, the correction terms n and d are of order h^2 and therefore, except for Bragg related waves, are of order $\epsilon^{(p)2}$. The correction terms can be expressed as sums of products of submatrices.

Certain aspects of the nature of periodic couplers are evident in eq. (62). When the wavelength of the periodicities is of the order of the optical wavelength, most of the $u_j^{(n)}$ are imaginary and the corresponding function $D_{1,4}^{(n)}$ is large $\{0(\exp|u_j^{(n)}|t_j)\}$. The leading terms in eq. (62) dominate and the reflection and transmission coefficients are similar to those of the homogeneous medium with average coefficients as expressed in eq. (29), i.e.,

$$\frac{C_{1,0-}}{C_{0+}} \sim -\frac{'D_{1,4}^{(0)}}{D_{1,4}^{(0)}}$$

$$\frac{C_{4,0+}}{C_{1,0+}} \sim \frac{\left(2u_1^{(0)}\right)\left(2v_{2,0}^{(0)}\right)\left(2u_3^{(0)}\right)}{D_{1,4}^{(0)}} . \tag{63}$$

One exception to eq. (63) occurs at a Bragg angle where n

and d of eq. (62) may be of the same order as the leading terms in eq. (62). The Bragg case is not generally of interest in connection with couplers. Another exception occurs when the system function for one of the diffraction orders has a zero, i. e., when

$$D^{(n)}_{1,4}(k_x) = 0 \quad . \tag{64}$$

Since

$$D^{(n)}_{1,4}(k_x) \sim D^{(0)}_{1,4}\left(k_x + n\kappa_x\right) \tag{65}$$

eq. (64) will have zeros as long as the stratified medium of average index properties can support a guided wave. For beam incidence angles near a zero of $D^{(n)}_{1,4}$, eq. (62) will exhibit dramatic changes in amplitude.[1,2,4]

If we examine a cross transmission coefficient such as $C_{4,n+}$, which expresses the amplitude of the nth harmonic at the surface of the guiding film opposite the periodic region, the matrix in Table III shows that $C_{4,n+}$ can be expressed

$$\frac{C_{4,n+}}{C_{1,0+}} = \frac{n_{4,n+}(h)}{D^{(0)}_{1,4}\, D^{(-1)}_{1,4} \cdots D^{(-M)}_{1,4} + d(h^2)} \quad . \tag{66}$$

Note that the product of diagonal matrices such as appear in the numerators in eq. (62) do not appear in eq. (66). Furthermore, if we denote the zero of eq. (64) as $k_x = k_w$ and approximate $D^{(n)}_{1,4}$ by a linear term in the vicinity of k_w in the manner of eqs. (41) and (42), viz.,

$$D^{(n)}_{1,4} \sim D^{(n)}_{1,2}\, d^{(n)}_{2,4}\left(k_x + n\kappa_x - k_w\right) \;, \tag{67}$$

then eq. (66) is of the form of eq. (18). If the first diffraction order (n = 1) is the one that couples, the parameters of eq. (18) are

$$\alpha = \text{Im}\,\frac{d(h^2)}{\left(D_{1,4}^{(0)}\,D_{1,4}^{(-1)}\cdots D_{1,4}^{(-M)}\right)D_{1,2}^{(1)}\,d_{2,4}^{(1)}}$$

$$\beta = \frac{n_{4,1+}^{(h)}}{\left(D_{1,4}^{(0)}\,D_{1,4}^{(-1)}\cdots D_{1,4}^{(-M)}\right)D_{1,2}^{(1)}\,d_{2,4}^{(1)}} \quad . \tag{68}$$

The term $d_{2,4}^{(n)}$ is expressed in eq. (42) and is proportional to the effective thickness of the guide so that

$$\alpha = 0\left(h^2/t_3\right) \quad \beta = 0\left(h/t_3\right) = 0\,\sqrt{\alpha/t_3} \quad . \tag{69}$$

The parameters α and β thus have the general characteristics required of a coupler.

We have succeeded in demonstrating that periodic structures can exhibit coupling behavior similar to that of the prism coupler. Computation of the parameter in eq. (68) can proceed using the techniques outlined, but these data have not appeared in the literature as yet. To design a periodic coupler, the parameter values must be established. In Section III we approach that task using approximate methods.

III. RECIPROCAL ANALYSIS OF PERIODIC COUPLERS

We turn our attention to the analysis of periodic couplers using the approximate technique introduced in Section I. With the aid of the reciprocity theorem we consider structures in which α may vary along the input surface.

III. 1 The Equation for Coupling Efficiency

With reference to Fig. 13 we assume that the periodic coupler produces a series of collimated beams that radiate into the substrate and superstrate. The power contained in these beams is labeled P_n^s or P_n^a where n represents the diffraction order of the beam and s and a stand for substrate or air (or other superstrate material). When a guided wave of total power P_o is incident on the coupler this power is divided into a part that is reflected P_r, a part transmitted beyond the coupler, P_t, a part scattered at the input in a generally nondirective fashion P_s, and the remainder appears in the collimated beams. Power conservation in the coupler may thus be expressed

$$P_o = P_s + P_t + P_r + \sum_n \left(P_n^s + P_n^a\right) \; . \qquad (70)$$

The efficiency with which a beam can be coupled into a device will be shown to be limited to the ratio $P_n^{s(a)}/P_o$.

The reciprocity theorem can be used to relate the coupling efficiency to the ratio P_n^s/P_o. Identifying $\underline{E}_r$ and $\underline{H}_r$ as the fields of the (reciprocal) radiation problem and $\underline{E}$ and $\underline{H}$ as the fields produced when a beam is incident on a coupler, the reciprocity theorem may be expressed

$$\int \left(\underline{E} \times \underline{H}_r - \underline{E}_r \times \underline{H}\right) \cdot d\underline{s} = 0 \qquad (71)$$

where the integral is over any closed surface without regard for index discontinuities.[10] The surface of interest for periodic couplers is shown in Fig. 13. The theorem

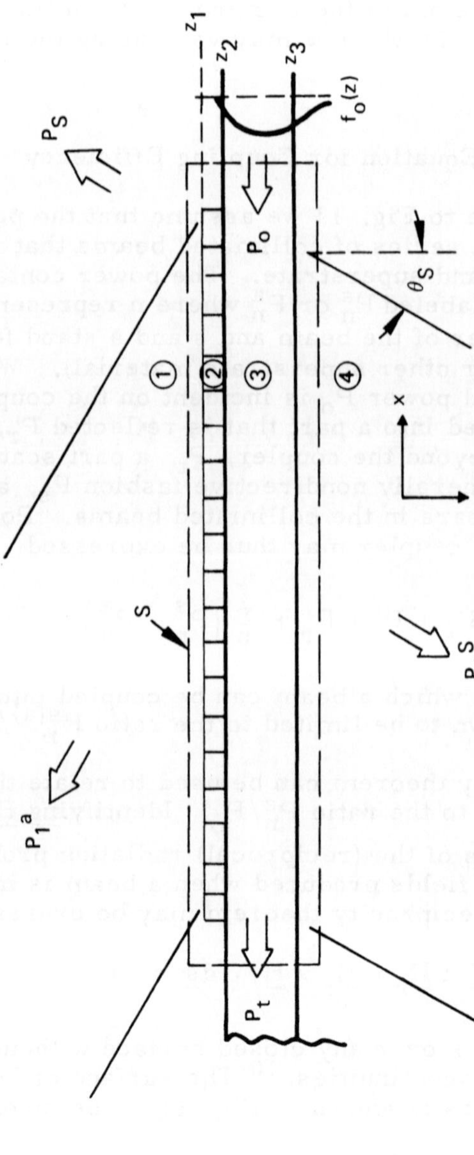

Figure 13. Surface S for applying the reciprocity theorem to the coupling problem. Region 1 is air, region 2 is the periodic medium, region 3 is the waveguide, and region 4 is the substrate.

follows from volume integration of the following relation obtained from Maxwell's equations:

$$\underline{H}\cdot[\nabla x\underline{E}_r - i\omega u\underline{H}_r] - \underline{H}_r\cdot[\nabla x\underline{E} - i\omega\mu\underline{H}] - \underline{E}_r\cdot[\nabla x\underline{H} + i\omega\epsilon\underline{E}]$$

$$+ \underline{E}\cdot[\nabla x\underline{H}_r + i\omega\epsilon\underline{E}_r] = \nabla\cdot[\underline{E}\, x\,\underline{H}_r - \underline{H}\, x\,\underline{E}_r] = 0 \quad . \tag{72}$$

To apply eq. (71), let us identify some fields in both the reciprocal (radiation) and coupling problems. We assume that all fields are gaussian in the y direction with a beamwidth wide enough to permit the problem to be treated as a two-dimensional one by suppressing the y dependence. We also treat only the TE case in detail. The TM case differs from the TE case, but not significantly to our approximation.

For the reciprocal problem the wave of total power P_o incident on the coupler is

$$E_{y_r} = \left(2\omega\mu P_o/k_w\right)^{1/2} f(z) \exp i\, k_w x$$

$$H_{z_r} = \left(k_w/\omega\mu\right)E_y \tag{73}$$

where f(z) is the orthonormal wave function of the optical waveguide. The function is

$$f(z) = \sqrt{\frac{2}{t'}}\begin{cases} \left[\dfrac{u_3^w}{\left(k_3^2 - k_2^2\right)^{1/2}}\right]\exp - iu_2^w\left(z-z_2\right) & z < z_2 \\ \cos\left[u_3{}^w\left(z-z_2\right) - \tan^{-1}u_2^w/iu_3^w\right] & z_2 > z > z_3 \\ \left[\dfrac{u_3^w}{\left(k_3^2 - k_4^2\right)^{1/2}}\right]\exp iu_4^w\left(z-z_3\right) & z > z_3 \end{cases} \tag{74}$$

where the effective guide thickness t' is

$$t' = t_3 + \frac{1}{|u_2^{w}|} + \frac{1}{|u_4^{w}|} \tag{75}$$

and

$$u_j^{w} = u_j \,|\, k_\rho = k_w = \left(k_j^2 - k_w^2\right)^{1/2} .$$

The wave number k_w in eq. (73) is the propagation constant of the guided wave. Since $\int_{-\infty}^{\infty} |f|^2 dz = 1$, the power in the incident wave is $1/2 \int E_y H^*_z dz = P_o$.

The field of one of the radiated beams at a point in a plane just below the film may be expressed

$$E^s_{y_{rn}} = \left(2\omega\mu P_n{}^s / k_s \cos\theta_s\right)^{1/2} g_n^s(x) \exp i\left(k_s \sin\theta\right) x$$

$$H^s_{x_{rn}} \simeq \left(k_s \cos\theta_s / \omega\mu\right) E^s_{y_{rn}} \tag{76}$$

where $g_n^s(x)$ is an aperture function whose square integral is unity ($\int |g|^2 dx = 1$). We will specify g(x) shortly but it should be noted that we have employed plane wave relations to obtain H_x from E_y so that we have tacitly assumed that g(x) is slowly varying with wavelength. The beam emerges at an angle θ_s with respect to the normal.

The field in the incident beam at the same plane at w which the field of eq. (76) appears is

$$E^i_y = \left(2\omega\mu P_i / k_s \cos\theta^{\,i}\right)^{1/2} h(x) \exp - i\left(k_s \sin\theta^{\,i}\right) x$$

$$H^i_x = -\left(k_s \cos\theta^{\,i} / \omega\mu\right) E^i_y \tag{77}$$

where h(x) is presumably a gaussian and $\int_{-\infty}^{\infty} |h|^2 dx = 1$. Under phase match conditions the angles θ_s and θ^i are equal. The final field of interest is that coupled into the waveguide mode which may be expressed

$$E_y^c = \left(2\omega\mu P_c/k_w\right)^{1/2} f(z) \exp -ik_w x$$

$$H_z^c = -\left(k_w/\omega\mu\right) E_y^c \tag{78}$$

where f(z) and k_w are the same parameters employed in eqs. (73) and (74).

We now proceed to apply the reciprocity theorem. When the gaussian beam is incident on a coupler it will generate many fields in addition to the power coupled into the waveguide which we can assume is less than the total power in the beam. All of these fields result in null contributions to the integral (71) because only waves in the coupling and reciprocal problems that travel in opposite directions through the surface S give nonzero values to the integral. Only the waves (73) and (78) and the waves (76) and (77) contribute to the integral. Integration of eq. (71) is thus found to yield the expression

$$\frac{P_c}{P_i} = \left| \int_{-\infty}^{\infty} g(x) h(x) dx \right|^2 \frac{P_n^s}{P_o} \quad . \tag{79}$$

In eq. (79) we have assumed the phase match condition so that the factor exp $ik_s (\sin\theta_s - \sin\theta^i)x$ that would otherwise appear under the integral sign has been suppressed.

III. 2 Aperture Fields

To evaluate the coupling efficiency we must determine $g_n^s(x)$ and P_n^s in eq. (79). To this end we assume that the wave that propagates in the coupler itself can be expressed as a slowly attenuating (with respect to wavelength) wave of the form

$$E_y^{(0)} = \left(\frac{2\omega\mu P'_o}{\bar{k}_w}\right)^{1/2} f_o(z) \exp\left(i\bar{k}_w x - \int_o^x \alpha\, dx\right) . \quad (80)$$

In eq. (80), $f_o(z)$ and $\bar{k}_w$ are the orthonormal wave function and guided wave propagation constant, respectively, of the waveguide composed of layers of average permittivity. The function $f_o(z)$ may be taken to be that expressed in eq. (74). The power P'_o in eq. (80) is the power coupled into the wave that excites the beam P_n^s. We may write

$$P'_o = \gamma P_o \quad (81)$$

and note that γ is unity only in the event that scattered and reflected powers P_s and P_r are zero and no other guided wave in the coupler region is excited. The attenuation of (80) is shown as an integral (sum) over an attenuation constant α that may in general be a function of x.

Coexistent with the wave (80) are harmonic waves that provide radiating as well as nonradiating beams. These waves are excited by eq. (80) and when we neglect coupling back into the guide may be taken as being approximately proportional to eq. (80) in the form

$$E_y^{(n)} = f_n(z) \exp\left[i\left(\bar{k}_w + n\kappa_x\right) x - \int_o^x \alpha dx\right] . \quad (82)$$

Equation (82) represents a first order solution to the wave equation

$$\left[\frac{\partial^2}{\partial x^2} + \frac{\partial^2}{\partial z^2} + \bar{k}^2\right]\left[\sum_{-\infty}^{\infty} E_y^{(n)}\right] = -\bar{k}^2\left[\sum_{-\infty}^{\infty} \epsilon^{(p)} \exp ip\underline{\kappa}\cdot\underline{\rho}'\right]\left[\sum_{-\infty}^{\infty} E_y^{(n)}\right] \quad (83)$$

under the assumption that α is small compared with $(k_w - n\kappa_x)$. Using the zeroth order term (80) on the right of eq. (83), assuming α to be small, suppressing $\exp - \int_0^x \alpha dx$, and equating appropriate harmonic terms in eq. (83), the approximate equation satisfied by $f_n(z)$ is found to be

$$\left[\frac{\partial^2}{\partial z^2} + u_j^{(n)^2}\right] f_n(z) = -\bar{k}^2 \epsilon^{(n)} \left(\exp in\kappa_z z\right)\left(\frac{2\omega\mu P_o}{k_w}\right)^{1/2} f_o(z) \tag{84}$$

where

$$u_j^{(n)} = \left[\bar{k}_j^2 - \left(k_x + n\kappa_x\right)^2\right]^{1/2} . \tag{85}$$

The solution to equation (84) may be expressed in terms of the Green's function $G(z, z')$ which itself is a solution of the equation

$$\left(\frac{\partial^2}{\partial z^2} + u_j^{(n)^2}\right) G(z, z') = -\delta(z-z') . \tag{86}$$

The exact form of $G(z, z')$ will be expressed shortly. For the moment we note as a point of reference that if the $u_j^{(n)}$ were approximately the same in each layer, i.e., if reflections from the surfaces could be neglected, then $G(z, z')$ would be

$$G(z, z') = \frac{-1}{2iu^{(n)}} \begin{cases} e^{iu^{(n)}(z-z')} & z > z' \\ e^{-iu^{(n)}(z-z')} & z < z' . \end{cases} \tag{87}$$

Utilizing the Green's function, integration of eq. (84) yields

$$f_n\left(z_{1(3)}\right) = \bar{k}_1^{-2}\,\epsilon^{(n)}\left(\frac{2\omega\mu P'_o}{\bar{k}_w}\right)^{1/2}\int_{z_1}^{z_2} f_o(z')$$

$$e^{in\kappa_z z'}\,G\left(z_{1(3)},\,z'\right)dz' \quad . \tag{88}$$

Comparing eqs. (82), (88), and (76), the normal amplitude distribution may be expressed

$$g_n^{s(a)}(x) = \bar{k}^2\left(\frac{k_{s(a)}\cos\theta_{s(a)}P'_o}{k_w p_n^{s(a)}}\right)^{1/2}$$

$$\left[\epsilon^{(n)}e^{-\int_0^x \alpha dx}\int_{z_1}^{z_2} f_o(z')e^{in\kappa_z z'}\,G_n\left(z_{3(1)},\,z'\right)dz'\right] \tag{89}$$

where, to ensure that the integral (89) yields unity,

$$P_n^{s(a)} = \bar{k}_1^{4}\left(\frac{k_{s(a)}\cos\theta_{s(a)}P'_o}{k_w}\right)\int_0^\infty \Big|\epsilon^{(n)}e^{-\int_0^x \alpha dx}$$

$$\int_{z_1}^{z_3} f_o(z')e^{in\kappa_z z'}\,G\left(z_{3(1)},\,z'\right)dz'\Big|^2 dx \quad . \tag{90}$$

Implicit in eq. (89) is the fact that some of the parameters under the integral sign may be slowly varying functions of x. The attenuation constant α is most readily obtained as the ratio of power loss per unit length to power transmitted,

i. e.,

$$\alpha = \sum_n \left(E^s_{y_{r_n}} H^{s*}_{x_{r_n}} + E^a_{y_{r_n}} H^{a*}_{x_{r_n}} \right) / P'_o \exp -2 \int_o^x \alpha dx$$

$$= \sum \frac{1}{2} \overline{k}_1^{-4} \left(\frac{k_{s(a)} \cos \theta_{s(a)}}{k_w} \right) \left| \epsilon^{(n)} \int_{z_1}^{z_2} f_o(z)' \exp in \kappa_z z' \right.$$

$$\left. G_n \left(z_{3(1)}, z' \right) dz' \right|^2 \qquad (91)$$

where the summation is over the radiating beams.

Equations (89) through (91) and (79) constitute an approximate solution to the radiation problem for periodic couplers. To analyze a prescribed structure one proceeds by first determining α using eq. (91). Then $P_n^{s(a)}/P'_o$ and $g_n^{s(a)}(x)$ are computed using eqs. (89) and (90) and are inserted in eq. (76) to express the radiated beams. To obtain the coupling efficiency the parameter γ of eq. (81) must be estimated and then $P_n^{s(o)}/P'_o$ and $g_n^{s(a)}(x)$ are inserted in eq. (79).

The Green's function for a source in region (3) of a four-region structure may be found by assuming waves above and below z' to be of the form

$$G = \begin{cases} \tau^+ A e^{iu_4^{(n)} z} & z > z_3 \\ A \left[e^{iu_2^{(n)}(z-z')} + \rho^+ e^{-iu_2^{(n)}(z-z')} \right] & z' < z < z_2 \\ B \left[e^{-iu_2^{(n)}(z-z')} + \rho^- e^{iu_2^{(n)}(z-z')} \right] & z_1 < z < z' \\ \tau^- B e^{-iu_1^{(n)} z} & z < z_1 \end{cases} \qquad (92)$$

where from the discussion of Section II we note

$$\rho^{+} = -{}'D^{(n)}_{2,4}/D^{(n)}_{2,4} \qquad \rho^{-} = -D'^{(n)}_{1,2}/D^{(n)}_{1,2}$$

$$\tau^{+} = \frac{\left(2u_2^{(n)}\right)\left(2u_3^{(n)}\right)}{D^{(n)}_{2,4}} \qquad \tau^{-} = \left(2u_2^{(n)}\right)/D^{(n)}_{1,3} \tag{93}$$

At z' the matching equations are

$$A(1 + \rho^{+}) - B(1 + \rho^{-}) = 0$$

$$A(1 - \rho^{+}) + B(1 - \rho^{-}) = \frac{-1}{iu_2^{(n)}} \tag{94}$$

so that

$$A = (1 + \rho^{-})/2iu^{(n)}_{2}(1 - \rho^{-}\rho^{+})$$

$$B = -(1 + \rho^{+})/2iu_2^{(n)}(1 - \rho^{-}\rho^{+}) \tag{95}$$

IV. DESIGN CONSIDERATIONS

Design of a periodic coupler consists in maximizing the efficiency expression (79). The factor P_n^s/P_o represents the relative power in a guided wave that is transferred to a single radiating beam by the coupler. To maximize this ratio one must minimize the power in undesired beam, and in scattered and uncoupled fields.

The number of beams depends on how many integer values of n satisfy the equations

$$\left|\bar{k}_w - n\kappa_x\right| < k_a \qquad \left|\bar{k}_w - n\kappa_x\right| < k_s \qquad (96)$$

where $\kappa_x = 2\pi/\Lambda_x$ and Λ_x is the period of the index variations measured along the guide. The wavenumber of the guided wave in a structure with permittivity averaged in x is $\bar{k}_w$. A backward wave coupler with a single radiating beam and therefore unit value for P_n^s/P_o results when κ_x satisfies the equation

$$\bar{k}_w + k_s > \kappa_x > \bar{k}_w + k_a \ . \qquad (97)$$

The drawback in this coupler is that it requires a period Λ_x on the order of $\lambda_o/2$ that necessitates beam interference or scanning electron microscope fabrication techniques. The radiating beam furthermore appears in the substrate unless a different high index prism is clamped above the guide.

Since the amplitude of the higher harmonics of the permittivity distribution reduce fairly rapidly with increasing order n, it is also possible to fabricate an efficient structure by having the first order beam radiate exclusively into the substrate. This requires,

$$k_w - k_a > \kappa_x > k_w - k_s \qquad (98)$$

and can involve long wavelength structures on the order of $\lambda/(n_{film} - n_{substrate})$ (where n is the refractive index) that can be fabricated using photolithographic techniques.

The introduction of undesirable beams that reduce efficiency also occurs if there is more than one excited guided mode in the coupler region, i. e., if $\bar{k}_w$ in eq. (96) has more than one value. In accordance with our approximate analysis of the previous section, the number of guided modes may be established by treating the variable permittivity region as a uniform layer of average permittivity. The transition region between the waveguide and the coupler can present a rapid change in guiding properties and can result in undesirable modes being excited in a coupler capable of supporting multiple modes. There will also be some scattered power that reduces the ratio P_n^s/P_o and therefore the efficiency. To minimize scattering effects in transition regions the average permittivity of the periodic layer should remain as close to that of the guiding region as possible. This implies wide etched regions for the grating coupler and narrow grooves in the etched grating coupler.

Suppression of undesirable beams can also be achieved by what may be viewed analytically as the introduction of a large assymetry in the overlap integral (88) when it is evaluated above or below the waveguide. The imposition of a totally reflecting layer either above or below, for example, can eliminate beams into one of the regions. With such a reflector the Green's function that appears in eq. (88) and is presented in eqs. (92) through (95) exhibits a small value of transmission coefficient τ^+ or τ^-.

Beams launched into one of the regions can also be suppressed using the holographic technique. With the approximate Green's function in eq. (87), the integral (90) shows that the relative power in the beams in the two directions are approximately related by

$$\frac{P_{-1}^{(a)}}{P_{-1}^{(s)}} \simeq \frac{\left[\left|u_3^{\,w}\right| - i\left(\kappa_z - u_3\right)\right|^2\right]}{\left|u_3^{\,w}\right| - i\left(\kappa_z + u_3\right)\right|^2} \quad . \tag{99}$$

When the evanescent tail of the guided wave into the periodic region is large ($|u_3^w|$ is small) the ratio (99) can be quite small. Calculations of the ratio $P_{-1}^{(a)}/P_{-1}^{(s)}$ made by Winn, et al.[8] for the holographic coupler show the relative powers to be 5% into air and 95% into the substrate. This ratio may be compared with the grating coupler which gave 55% in air. The principal difficulty with holographic couplers is that a tilted index variation is readily provided only in special materials that have a generally low index and yield small index variations. When a high index waveguide is employed the field penetration as expressed by f_o in eq. (74) is small. The combination of small index variations and low penetration can result in excessively long structures.

The other factor that enters into the equation for the coupling expression (79) is the aperture integral $\left|\int_0^\infty g(x)h(x)\right|^2$ where h(x) is the normalized aperture field of the incident beam and g(x) is the function expressed in eq. (89). Properties of this integral have been extensively investigated in connection with the prism coupler.[9] When the wave attenuation α (x) is constant in x, i. e., when the coupler is uniform and when h(x) is gaussian as shown in eq. (19), the maximum value of $\left|\int_0^x g(x)\, h(x)dx\right|^2$ is found by numerical integration to be 0. 80 and to occur when the parameter α is

$$\alpha_{max} = (0.96)\left(\frac{1}{2\sigma'_x}\right) . \tag{100}$$

The factor $2\sigma'_x$ is the beamwidth of the incident gaussian to the 1/e points in power.

When α (x) is permitted to vary, but the relative coupling to each beam remains constant, the form of g(x) is

$$g(x) = \sqrt{2\alpha}\, \exp - \int_0^x \alpha dx . \tag{101}$$

The maximum value of $\left|\int_0^\infty g(h)h(x)dx\right|^2$ is unity and occurs for real h(x) when

$$\alpha(x) = \frac{1}{2}\,\frac{h(x)}{1 - \int_0^x h^2 dx} \,. \qquad (102)$$

The solution (102) is found by equating the square of (101) to $h^2(x)$ and noting that

$$\alpha(x) = -\frac{1}{2}\frac{d}{dx}\,\ell n\left[1 - \int_0^x h^2 dx\right] . \qquad (103)$$

The key to maximizing the aperture integral factor of the coupling efficiency is to select parameter values so that α satisfies eq. (100) or (102) when the width of the incident beam is fixed or so that α provides a reasonable coupling length to which the incident beam may be adjusted. From a practical standpoint the problem is to avoid too small a value of α that results in an excessive coupling length.

Examples of some computed values of the aperture size provided by grating couplers are shown in Fig. 14. These were computed using the Green's function (92 through 95). To appreciate the elements that appear in α let us obtain an approximate expression for α using (87), $\kappa_z = 0$, and assuming $t_3 = z_3 - z_2$ is large compared with the decay distance of the guided wave into the grating $1/u_3^w = 1/\left(\bar{k}_3^2 - \bar{k}_w^2\right)^{1/2}$.

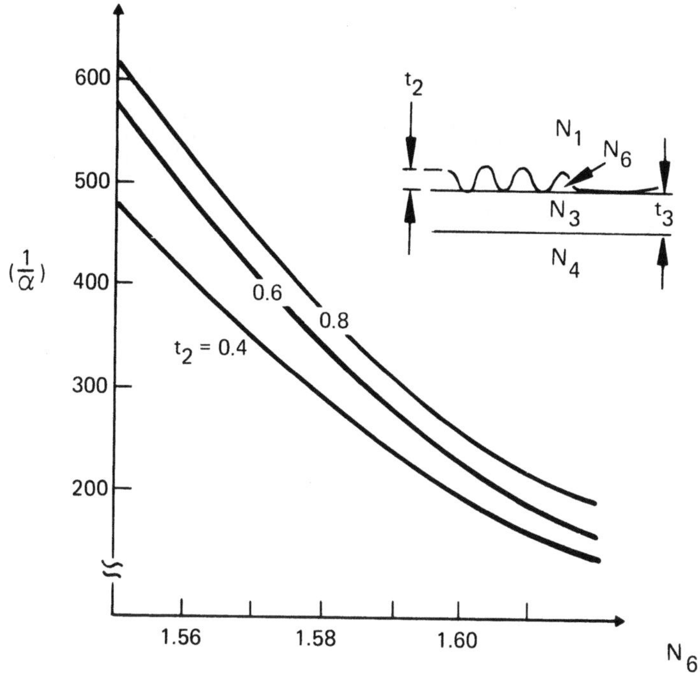

Figure 14. Sample approximate calculation of beamwidth $(1/\alpha)$ expressed in microns for optimum coupling to a grating coupler. The grating surface is a sinusoid and the grating material has index n_6. Other index values are $n_1 = 1.0$, $n_3 = 1.73$, $n_4 = 1.52$. The guide thickness is $t_3 = 0.76$ μm and $\lambda_o = 0.6328$ μm, $\kappa_x = 0.95\ k_o$. The calculation is after Winn, et al.[8] using data from the experiment of Dakss, et al.[1] The calculation is based on a somewhat different approximation than that discussed in Section III.

With f_o obtained from (74), the approximate expression for α is

$$\alpha = \frac{1}{t'} \, \epsilon^{(-1)\,2} \left[\frac{\left(k_3^2 - \bar{k}_w^2\right)}{\left(k_3^2 - \bar{k}_2^2\right)} \right] \frac{k_o^4 \, \bar{\epsilon}_2^2}{\left(2u_2^{(-1)}\right)^2 \left| \, |u_2^w| + iu_2^{(-1)} \right|^2} \left[\frac{u_1^{(-1)} + u_4^{(-1)}}{\bar{k}_w} \right] \quad (104)$$

where

$$u_j^{(-1)} = \left[\bar{k}_j^2 - \left(\bar{k}_w - \kappa_x \right)^2 \right]^{1/2}$$

$$u_j^w = \left[\bar{k}_j^2 - \bar{k}_w^2 \right]^{1/2}$$

$$t' = t + \frac{1}{|u_2^w|} + \frac{1}{|u_4^w|} \; . \quad (105)$$

Equation (104) shows that α is inversely proportional to the effective thickness of the guided wave in the coupler (t'). The proportionality constant involves the square of the amplitude of the first harmonic of the permittivity distribution $\epsilon^{(-1)2}$ and a measure $\left| \left(k_2^2 - \bar{k}_w^2\right) / k_2^2 - \bar{k}_3^2 \right|$ of how tightly the guided wave is bound to the film. These factors are generally much less than unity and result in a coupling length $1/\alpha$ that is large compared with t'. When these factors are too small the coupling length is excessive.

Perhaps the most important design concept to consider is the relationship between the physical model and calculations of the coupler parameter α. The calculations are sensitive to variations in the geometrical properties of the periodic medium as well as other geometric and index parameters of the structure. It would be inappropriate to expect a close match between calculated and experimental values of α unless great care was exercised in characterizing the physical properties of the coupler and unless an analysis for the specific structure more accurate than that presented in Section III was performed.

ACKNOWLEDGMENT

I am grateful for the help of Gordon Mitchel, Lawrence Stots, and Clare Khasha in preparing this article.

REFERENCES

1. M. L. Dakss, L. Kuhn, P. F. Heidrich, and B. A. Scott, Appl. Phys. Lett. 16, 523 (1970).

2. D. G. Dalgoutte, Opt. Comm. 8, 124 (1973).

3. Y. Y. Teng and E. A. Stern, Phys. Rev. Lett. 19, 511 (1967).

4. H. Kogelnik and T. P. Sosnowski, Bell Syst. Tech. J. 49, 1602 (1970).

5. D. B. Ostrowsky and A. Jacques, Appl. Phys. Lett. 18, 556 (1971).

6. F. R. Gfeller and C. W. Pitt, Co-linear Acousto-Optic Deflection in Thin Films, Dept. Electronics and Elect. Engrg. Rept., U. College, London (Sept. 1972).

7. A. Hessel and A. A. Oliner, Appl. Opt. 4, 1275 (1965).

T. Tamir and H. L. Bertoni, J. Opt. Soc. Am. 61, 1397 (1971).

S. T. Peng, T. Tamir, and H. L. Bertoni, Elect. L Lett. 9, 150 (1973).

C. Elachi and C. Yeh, J. Appl. Phys. 44, 3145 (1973).

L. L. Hope, Opt. Comm. 5, 179 (1972).

F. W. Dabby, A. Kestenbaum, and U. C. Paek, Opt. Comm. 6, 125 (1972).

M. Neviere, R. Petit, and M. Cadilhac, Opt. Comm. 8, 113 (1973).

8. J. H. Harris, R. K. Winn, and D. G. Dalgoutte, Appl. Opt. 11, 2234 (1972).

K. Ogawa, W. S. Chang, B. L. Sopori, and F. J. Rosenbaum, IEEE Trans. QE-9, 29 (1973).

9. L. V. Ioganse n, Soviet Phys. -Tech. Phys. 11, 1529 (1967).

J. H. Harris, R. Shubert and J. N. Polky, J. Opt. Soc. Am. 60, 1007 (1970).

J. E. Midwinter, IEEE Trans. QE-6, 583 (1970).

P. K. Tien and R. Ulrich, J. Opt. Soc. Am. 60, 1325 (1970).

R. Ulrich, J. Opt. Soc. Am. 60, 1337 (1970).

J. H. Harris and R. Shubert, IEEE Trans. MTT-19, 269 (1971).

R. Ulrich, J. Opt. Soc. Am. 61, 1467 (1971).

T. Tamir and H. L. Bertoni, J. Opt. Soc. Am. 61, 1397 (1971).

10. D. S. Jones, The Theory of Electromagnetism, pp. 63-65, Macmillan (1964).

Chapter 13 — MODULATION

J. H. Harris

Electrical Engineering Department

University of Washington

1. INTRODUCTION

A modulator is a device with an identifiable output optical wave whose amplitude and/or phase can be varied with the application of a modulating signal. There are a number of ways (see for example schematics of Fig. 1(a)-1(e)) in which modulation can be achieved. One method is to control the output of the optical source as occurs when the current in an injection laser is varied. Another is to alter the absorption properties of the material through optical bleaching or shifting an absorption band-edge by applying an electric field (Franz-Keldysh effect). The use of liquid crystals[1] or mechanical techniques represent other possibilities. Some of these modulation methods provide viable options for communication systems. Direct modulation of a laser in particular may well evolve as the method of choice.

Those techniques that have been more directly associated with the application of integrated optics to communications involve mode conversion or mode interference methods as induced by electro-, magneto-, or acousto-optic mechanisms. Mode conversion methods involve the gradual transfer of energy from a prescribed wave in a planar or linear (energy confined in all but the direction of propagation) guide to other waves that are removed at the

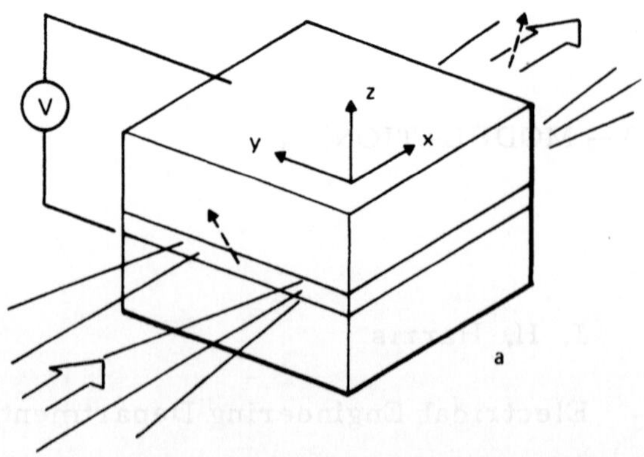

Figure 1. Examples of thin film modulators. (a) A reverse biased junction modulator. The incident wave (double arrow) is polarized in the direction of the dotted arrow so as to excite both TE_z and TM_z waves with approximately equal amplitude. As the phase difference between the waves increases, the polarization rotates.

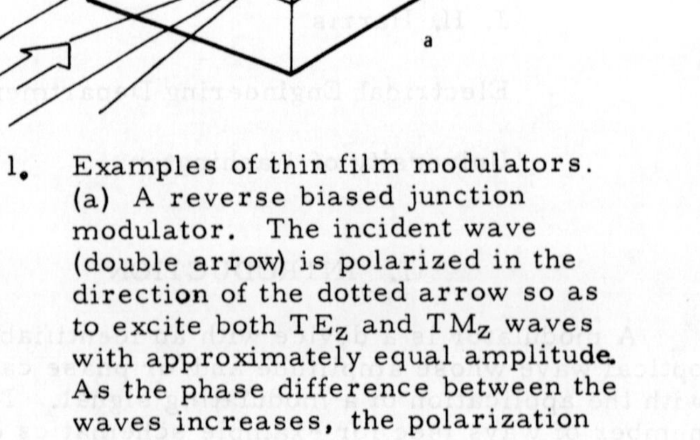

Figure 1(b). A collinear acousto-optic modulator. Excitation of a collinear acoustic surface wave (curved arrow) converts energy from a TE_0 optical (double arrow) to a TE_2 wave (dotted double arrow). Approximate mode profiles are shown at the side of the waveguide.

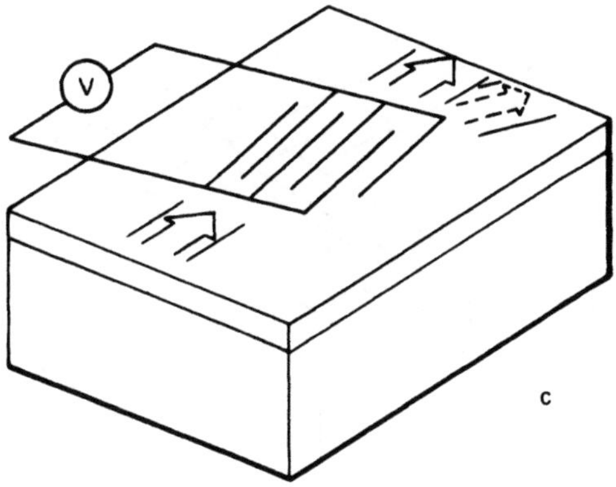

Figure 1(c). An electro-optic diffraction modulator. A voltage applied to the interdigitated electrodes produces a periodic electric field in the waveguide. Energy in a plane guided wave (double arrow) is converted to another guided wave(s) (dotted double arrow) propagating in a different direction.

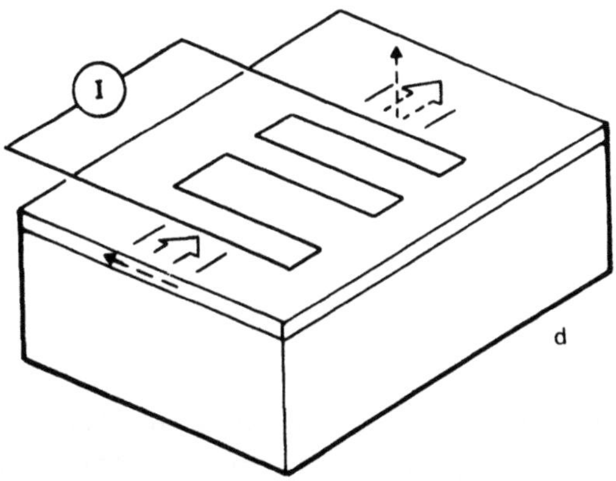

Figure 1(d). Phase matched magneto-optic modulator. A magnetic field periodic in the direction of propagation is produced when current flows in the wire snaked about the propagation path. Energy in a plane guided incident wave (double arrow) polarized in the direction of the dotted arrow is converted to the orthogonal polarization.

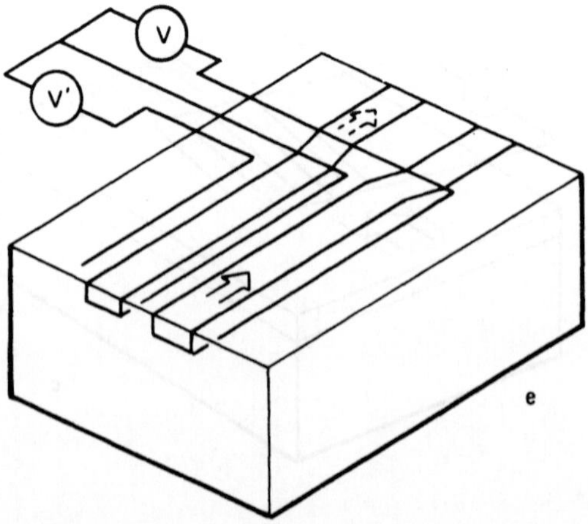

Figure 1(e). Proposed linear waveguide electro-optic switch. Energy launched in one waveguide (double arrow) couples to an adjacent guide (dotted double arrow). Voltage applied to the electrodes changes the coupling rate and energy is returned to the first guide.

output. Mode interference methods involve the excitation of two waves whose phase difference at the output can be varied. An analyzer such as a polarizer (in the case that the two modes yield a polarization dependent on the phase shift) provides amplitude modulation.

Mode interference methods have been demonstrated in waveguides that are formed in reverse biased p-n junctions of III-V compounds gallium phosphide (GaP) and gallium arsenide (GaAs),[2] in the waveguiding region of aluminum gallium arsenide ($Al_xGa_{1-x}As$) heterojunction structures used otherwise as injection lasers,[3] and in diffused waveguides formed in II-VI compounds (cadmium diffused into zinc selenide for example).[4] The mechanism for achieving modulation in these cases is principally the electro-optic effect.* An electric field in the waveguide produced by applying a voltage to appropriately positioned electrodes induces a change in the generally anisotropic permittivity components in the structure. This change in permittivity is translated into a variation in the propagation constants of two guided waves with orthogonal polarization. The polarization at the output depends on the relative phase of the two waves and by placing a polarizer there modulation is achieved.

Mode conversion methods have been demonstrated in a variety of ways. Surface acoustic waves have been employed, for example, to transfer energy from a wave in a planar guide into a diffracted guided wave, into unguided waves, and into a wave of different mode order.[5] The diffraction into another guided wave is achieved by launching an acoustic surface wave in a direction perpendicular to the direction of propagation of the optical wave. The optical wave passes through a grating at the acoustic wavelength and diffraction occurs. Energy conversion into unguided waves or into a guided wave of different mode order was achieved by launching an acoustic wave in a direction collinear with the direction of propagation of the optical wave. In accordance with momentum conservation concepts, energy from the optical wave (1) is transferred into optical wave (2) with propagation constant $k_{w2} = k_{w1} - \kappa$ where κ is the propagation constant of the surface acoustic wave.

---

*The reduction of free carriers in the junction and the change in mode structure that results from increasing the reverse bias can also contribute to the modulation. A reduction in free carriers increases the index of a region by a small amount $\Delta n \simeq 0(10^7\ N/f^2)$ where N is the carrier density/c. c. and f is the frequency.

When $k_{w1} - \kappa$ is less than $k_o n$ (n is the refractive index of the substrate) radiation occurs and when $k_{w1} - \kappa$ is equal to the propagation constant of another guided wave, mode conversion occurs.

Mode conversion to radiating fields using the electro-optic effect has been demonstrated in potassium dihydrogen phosphate (KDP)[6] and in a low carrier concentration gallium arsenide epitaxial layer.[7] In the former case the application of an electric field at the surface of a crystal generated a waveguide while in the latter case the application of an electric field changed the index of the film sufficiently to permit waveguiding.

Mode conversion using the electro-optic effect has also been demonstrated in zinc oxide and nitrobenzene films by placing an interdigital structure above a planar waveguide.[8] A voltage applied to the structure produces a spatially periodic electric field that results in a grating type refractive index variation in the planar guide. Energy from an optical wave propagating in the direction of the fingers of the interdigital structure is diffracted into other guided waves. Modulation is achieved as a result of the spatial separation of the waves.

Use of the magneto-optic effect to achieve mode conversion has been demonstrated in garnet films.[9] Energy from a plane guided wave of one polarization was converted into a guided wave of the same mode order but different polarization in a modified version of the Faraday effect. A magnetic field is applied to the guide in the direction of propagation by generating a current that flows perpendicular to the direction of propagation. Because the TE and TM modes have different propagation constants, the wire that carries the current is "snaked" around the propagation path to provide a magnetic field that is periodic in the direction of propagation. The wavenumber (2π/period) of the magnetic field is equal to the difference in propagation constants of the guided waves.

Perhaps the most useful modulation technique that can be expected from integrated optics is the linear waveguide switch.[10] By placing two linear waveguides close to one another so that their guided waves can interact, energy launched in one waveguide will pass periodically back and forth between the guides. By changing the index in the guides the rate of coupling can be varied. At the output the guides are separated and when the length of the structure is appropriate, energy appears in one guide or the other depending on the applied signal. The utility of such a device, which can also appear in the form of intersecting guides, lies in the useful electrical properties that accrue from small size as well as the geometrical compatibility with fibers.

The basis for modulation techniques lies in the refractive index change induced in the waveguide material or in that of the surrounding medium by the modulating field. These index changes are attributable to variations in the electric or magnetic polarizabilities that occur on the atomic or molecular level. Phenomenologically, the electro-optic effect is characterized by a change in the components of what is termed the optical indicatrix in the form [11]

$$\Delta_{ij} = \Delta\left(\frac{1}{n^2}\right)_{ij} = r_{ijq}E_q + R_{ijpq}E_pE_q \quad * \qquad (1)$$

where r_{ijq} is the linear (Pockels) and R_{ijpq} the quadratic (Kerr) electro optic coefficient and E_q is the component of the modulation field defined relative to the direction of the optic axis of the crystal. The permittivity matrix is related to the 3 x 3 indicatrix matrix as follows:

$$\underline{\underline{\epsilon}} = \epsilon_o\left(\frac{1}{\underline{\underline{n}}^2}\right)^{-1} \qquad (2)$$

When the coordinate system is aligned with the optic axes of a crystal, $\underline{\underline{\epsilon}}$ has the diagonal form

$$\underline{\underline{\epsilon}} = \epsilon_o \begin{bmatrix} 1/n_{11}^2 & 0 & 0 \\ 0 & 1/n_{22}^2 & 0 \\ 0 & 0 & 1/n_{33}^2 \end{bmatrix}^{-1} = \epsilon_o \begin{bmatrix} n_{11}^2 & 0 & 0 \\ 0 & n_{22}^2 & 0 \\ 0 & 0 & n_{33}^2 \end{bmatrix} \qquad (3)$$

*A summation over repeated indices is implied.

and the 1, 2, 3 directions determine which directions to take for the electric field. In uniaxial crystals $n_{11} = n_{22}$.

The changes in the ijth component of the indicatrix matrix expressed in (1) are placed in the middle term in (3) and the matrix is inverted to find $\underline{\underline{\epsilon}}$. Because $\Delta(1/n^2)_{ij}$ is small, the inversion of (3) may be carried out to first order in $\Delta(1/n^2)_{ij}$ and we find the result to be

$$\underline{\underline{\epsilon}} = \epsilon_o \begin{bmatrix} n_{11}^2(1 - n_{11}^2\Delta_{11}) & -\Delta_{21} n_{11}^2 n_{22}^2 & \Delta_{31} n_{11}^2 n_{33}^2 \\ -\Delta_{21} n_{11}^2 n_{22}^2 & n_{22}^2(1 - n_{22}^2\Delta_{22}) & -\Delta_{32} n_{22}^2 n_{33}^2 \\ \Delta_{31} n_{11}^2 n_{33}^2 & -\Delta_{32} n_{22}^2 n_{33}^2 & n_{33}^2(1\ n_{33}^2\Delta_{33}) \end{bmatrix} \quad (4)$$

Because r_{ijq} is symmetric in i and j ($r_{ijq} = r_{jiq}$) published values are expressed in the form r_{kq} and the correspondence between k and i, j is

k	=	1,	2,	3,	4,	5, 6
i, j	=	1, 1;	2, 2;	3, 3;	2, 3;	1, 3; 1, 2

The magneto-optic effect[11, 12] results in imaginary off diagonal terms in the permittivity matrix. For a uniaxial crystal a magnetic field B_3 applied in the optic axis direction can provide a permittivity matrix of the form

$$\underline{\underline{\epsilon}} = \epsilon_o \begin{bmatrix} 1 + \chi_{11} & i\,\chi_{12} & 0 \\ -i\,\chi_{12} & 1 + \chi_{11} & 0 \\ 0 & 0 & 1 + \chi_{33} \end{bmatrix} \quad (5)$$

where the susceptibility χ_{12} is expressed in terms of the Verdet constant V as

$$\chi_{12} = (1 - \chi_{11})^{1/2} \frac{\lambda}{\pi} V B_3 \quad (6)$$

Plane waves propagating in the (3) direction in a medium with a permittivity of the kind shown in (5) are circularly polarized with propagation constants

$$k_o n_R = (1 + \chi_{11} + \chi_{12})^{1/2} k_o \quad k_o n_L = (1 + \chi_{11} - \chi_{12})^{1/2} k_o \; . \tag{7}$$

If the waves are excited by a linearly polarized field, the direction of the electric field rotates by 90° when

$$k_o L(n_R - n_L) \simeq \frac{k_o L \chi_{12}}{(1 + \chi_{11})^{1/2}} = \pi \tag{8}$$

The angular rotation ($\pi/2$) divided by the distance L or rotation per unit length

$$\frac{\theta}{L} = \frac{\chi_{12}}{(1 + \chi_{11})^{1/2}} \frac{\pi}{\lambda} = B_3 V \tag{9}$$

is also used to express the relationship between the susceptibility χ_{12} and the applied magnetic field.

2. MODULATION ANALYSIS

To examine modulation in waveguides consider Maxwells's equations expressed as

$$\nabla \times \underline{E} = -\mu \frac{\partial}{\partial t} \underline{H} \qquad \nabla \cdot \mu_o \underline{H} = 0$$

$$\nabla \times \underline{H} = \frac{\partial}{\partial t} \underline{\underline{\epsilon}}^{\circ} \cdot \underline{E} + \frac{\partial}{\partial t} \underline{\underline{\epsilon}}^{(1)} \cdot \underline{E} = \frac{\partial}{\partial t} \underline{\underline{\epsilon}}^{\circ} \cdot \underline{E} + \frac{\partial}{\partial t} \underline{P}$$

$$\nabla \cdot (\underline{\underline{\epsilon}}^{\circ} \cdot \underline{E}) = -\nabla \cdot (\underline{\underline{\epsilon}}^{(1)} \cdot \underline{E}) = \rho \tag{10}$$

The equations are written for a generally anisotropic and inhomogeneous medium characterized by a permittivity matrix $\underline{\underline{\epsilon}}^{\circ} + \underline{\underline{\epsilon}}^{(1)}$. In anticipation of the perturbation analysis that will be employed, polarization and charge density $\underline{P}$ and ρ respectively are defined as functions of $\underline{\underline{\epsilon}}^{(1)} \cdot \underline{E}$.

The matrix $\underline{\underline{\epsilon}}^{(1)}$ is intended to principally represent the time varying permittivity components produced by electronic effects, but for purposes of simplifying the analysis we shall assume that $\underline{\underline{\epsilon}}^{\circ}$ represents an isotropic medium. Many materials useful for modulation purposes are uniaxial and have a permittivity referred to the optic axis of the form

$$\underline{\underline{\epsilon}} = \begin{bmatrix} \epsilon_{11} & 0 & 0 \\ 0 & \epsilon_{11} & 0 \\ 0 & 0 & \epsilon_{33} \end{bmatrix} \tag{11}$$

We shall assume that

$$\underline{\underline{\epsilon}} = \epsilon_{11} \begin{bmatrix} 1 & 0 & 0 \\ 0 & 1 & 0 \\ 0 & 0 & 1 \end{bmatrix} \tag{12}$$

and take the difference $(\underline{\underline{\epsilon}} - \underline{\underline{\epsilon}}^{\circ})$ as contributing to $\underline{\underline{\epsilon}}^{(1)}$. In general, the proper $\underline{\underline{\epsilon}}^{(1)}$ to employ in (10) requires a rotation to the coordinate system in which (10) is expressed. We will discuss this rotation later. In a similar manner, any spatially varying effects attributable to the presence of wires are lumped in $\underline{\underline{\epsilon}}^{(1)}$. The spatial variation in ϵ_{11} that results in waveguiding, the variation normal to the film in a planar waveguide, for example, is retained.

It is useful to examine equation (10) under the assumption that the modulating field is sinusoidal so that $\underline{\underline{\epsilon}}^{(1)}$ may be expressed in phasor notation as

$$\underline{\underline{\epsilon}}^{(1)} = \underline{\underline{\epsilon}}^{dc} + \underline{\underline{\epsilon}}^{M} e^{i\omega_M t} + \underline{\underline{\epsilon}}^{M*} e^{-i\omega_M t} \tag{13}$$

For traveling wave modulators, like acousto-optic devices, a subcarrier frequency appears in (13) in a form that depends on the particular method used to modulate the subcarrier. When a wave of a single optical carrier frequency is launched at the input to the modulator (even laser outputs have substantial bandwidths) the polarization density is of the form

$$\underline{P} = \underline{\underline{\epsilon}}^{(1)} \cdot \underline{E} = \left[\underline{\underline{\epsilon}}^{(1)} \cdot \left(\underline{E}^{(0)} e^{i\omega_c t} + \underline{E}^{(0)*} e^{-i\omega_c t}\right)\right] \quad (14)$$

where $\underline{E}^{(0)}$ is the phasor representation of the optical field. This polarization density generates fields at frequencies $\pm[\omega_c \pm (\omega_M)]$ that must also satisfy Maxwell's equations. In general, then, the field in the modulator is of the form

$$\underline{E}(t) = \sum_p \left(\underline{E}^{(p)} e^{j\omega_p t} + \underline{E}^{(p)*} e^{-j\omega_p t}\right) \quad (15)$$

where

$$\omega_p = \omega_c + p\omega_M \quad (16)$$

and p is an integer.

Equation (10) must be satisfied at each instant of time and this requires that the equations be satisfied separately for the fields at each frequency ω_p. We assume that $\underline{\underline{\epsilon}}^{\circ}$ is isotropic and locally homogeneous so that within each region $\underline{E}^{(p)}$ and $\underline{H}^{(p)}$ are found to satisfy the wave equation

$$\left(\nabla^2 + k^{(p)^2}\right) \underline{H}^{(p)} = -\nabla x \underline{J}^{(p)}$$

$$\left(\nabla^2 + k^{(p)^2}\right) \underline{E}^{(p)} = -\frac{k_o^{(p)^2}}{\epsilon_o} \underline{P}^{(p)} + \nabla \rho^{(p)} \quad (17)$$

where $k^{(p)^2} = \omega_p^{\,2} \mu \epsilon_{11}$, $k_o^{(p)^2} = \omega_p^{\,2} \mu \epsilon_o$, and $\underline{P}^{(p)}$, $\underline{J}^{(p)}$ and $\rho^{(p)}$ are the harmonic terms

$$\underline{P}^{(p)} = \underline{\underline{\epsilon}}^{dc} + \underline{\underline{\epsilon}}^{M} \cdot \underline{E}^{(p-1)} + \underline{\underline{\epsilon}}^{M*} \cdot \underline{E}^{(p+1)}$$

$$\rho^{(p)} = -\nabla \cdot \underline{P}^{(p)} \qquad \underline{J}^{(p)} = i\omega_p \underline{P}^{(p)} \quad . (18)$$

We now wish to reduce (17) to a set of coupled differential equations.[13] To do this in approximate fashion consider the components of $\underline{E}^{(p)}$ and $\underline{H}^{(p)}$ transverse to the direction of propagation expressed as a sum of modes in the form

$$\underline{E}_t^{(p)} = \sum_m \left(A_m^{(p)+} e^{ik_m^{(p)} x} + A_m^{(p)-} e^{-ik_m^{(p)} x} \right) \hat{\underline{E}}_m^{(p)} (y, z) / P_m^{1/2}$$

$$\underline{H}_t^{(p)} = \sum_m \left(A_m^{(p)+} e^{ik_m^{(p)} x} - A_m^{(p)-} e^{-ik_m^{(p)} x} \right) \hat{\underline{H}}_m^{(p)} (y, z) / P_m^{*\,1/2} \cdot \tag{19}$$

In (19) $\hat{\underline{E}}_m^{(p)} (y, z)$ and $\hat{\underline{H}}_m^{(p)} (y, z)$ represent the spatial variation of the transverse field components of the mth mode, $k_m^{(p)}$ is the propagation constant of the mode, and $P_m^{1/2}$ is the normalization

$$P_m = \left(\int \hat{\underline{E}}_m^{(p)} \times \hat{\underline{H}}_m^{(p)*} \cdot \hat{x}\, dydz \right) . \tag{20}$$

The summation in (19) is over all the waves that can propagate in the guide. For the continuous spectrum, e.g., the guided waves in a planar guide, the summation transforms to an integral. The normalization expressed in (20) should be considered at present as involving an integration over a finite area when the continuous spectrum is involved.

The fields $\hat{\underline{E}}_m^{(p)}$ and $\hat{\underline{H}}_m^{(p)}$ satisfy equations

$$\left(\frac{\partial^2}{\partial y^2} + \frac{\partial^2}{\partial z^2} + k^{(p)^2} \right) \left(\hat{\underline{E}}_m^{(p)}, \hat{\underline{H}}_m^{(p)} \right) = k_m^{(p)^2} \left(\hat{\underline{E}}_m^{(p)}, \hat{H}_m^{(p)} \right) \tag{21}$$

and can be expressed in general as a sum of TE_x and TM_x waves starting with solutions to the scalar wave equation.[14] The wave amplitudes $A_m^{(p)\pm}$ are constants in the absence of the terms on the right side of (17). More generally we assume $A_m^{(p)\pm}$ to be a slowly varying (with respect to the optical wavelength) function of x. This assumption permits us to suppress the term $d^2 A_m^{(p)\pm}/dx^2$ in the traditional manner of the coupled mode equations. Following the procedure: 1.) place (19) in (17) taking note of (21), 2.) cross multiply (17b)(17a) by $\hat{\underline{E}}_m^* x(\hat{\underline{H}}_m^* x)$, 3.) integrate both equations over the cross-section) and 4.) add both equations (subtract for dA_m^-/dx) we obtain

$$\frac{dA_m^{(p)+}}{dx} = i\,\frac{e^{-ik_m^{(p)}x}}{4k_m^{(p)}} \int \Bigg[\frac{\hat{\underline{E}}_m^{(p)*} \times (\nabla \times \underline{J}^{(p)})}{P_m^{*\,1/2}} + \frac{\left(k_o^{(p)2}/\epsilon_o\right)\hat{\underline{H}}_m^{(p)*} \times \underline{P}^{(p)} - \hat{\underline{H}}_m^{(p)*} \times \nabla\rho^{(p)}}{P_m^{1/2}} \cdot \hat{x}\, dydz \quad . \tag{22}$$

Equation (22) is a set of coupled equations that can be expressed in the general form

$$\frac{dA_m^{(p)+}}{dx} = i \sum_{n,\pm} \left(C_{mn}^{(p)\pm} A_n^{(p)\pm} + C_{mn}^{(p-1)\pm} A_n^{(p-1)\pm} + C_{mn}^{(p+1)\pm} A_n^{(p+1)\pm} \right) \tag{23}$$

where the summation is over all the forward and backward traveling modes. In addition to the omission of the second derivative term $d^2 A_m^{(p)}/dx^2$, eq. (22) is not in general exact because the coupling due to the longitudinal terms on the right side of (17) have not been included. To include this coupling we would be required to return to the TE_x and TM_x representations and deal with two generally hybrid modes at once.[14] We assume that this coupling can be neglected. As it is, to obtain a solution to (23) it is usually necessary to be fairly ruthless about eliminating terms. The procedure is based on the assumption that coupling into most modes is negligible.

The elimination of terms must be approached with caution, however, since it is an easy matter to arrive at erroneous results. As an example, in planar anisotropic guides it is possible to have waveguiding in certain angular directions and no guiding in others.[15] For isotropic guides one must have one case or the other. Although such a situation can be dealt with in principle using (22) or (23), in fact many terms must be retained and a specific problem can readily become intractable.

To examine the nature of the C_{mn}'s let us consider the coupling coefficient due to the term $\underline{\hat{H}}_m^{(p)*} \times \underline{P}$ in (22). The first term in (22) often contributes in almost equal magnitude to the coupling coefficient. For this reason we write below twice the coupling coefficient due to $\underline{\hat{H}}_m^{(p)*} \times \underline{P}$. From (13), (18), and (19) we find in this special case

$$C_{mn}^{(p)\pm} = \frac{k_o^{(p)^2}}{2\epsilon_o k_m^{(p)}} e^{i\left(\pm k_n^{(p)} - k_m^{(p)}\right)x} \frac{\int \underline{\hat{H}}_m^{(p)*} \times \left(\underline{\underline{\epsilon}}^{dc} \cdot \underline{\hat{E}}_n^{(p)}\right) \cdot \hat{x}\, dy\, dz}{P_n^{(p)^{1/2}} \; P_m^{(p)*^{1/2}}}$$

$$C_{mn}^{(p-1)\pm} = \frac{k_o^{(p)^2}}{2\epsilon_o k_m^{(p)}} e^{i\left(\pm k_n^{(p-1)} - k_m^{(p)}\right)x} \frac{\int \underline{\hat{H}}_m^{(p)*} \times \left(\underline{\underline{\epsilon}}^{M} \cdot \underline{\hat{E}}_n^{(p-1)}\right) \cdot \hat{x}\, dy\, dz}{P_n^{(p-1)^{1/2}} \; P_m^{(p-1)^{1/2}}}$$

$$C_{mn}^{(p+1)\pm} = \frac{k_o^{(p)^2}}{2\epsilon_o k_n^{(p)}} e^{i\left(\pm k_n^{(p+1)} - k_m^{(p)}\right)x} \frac{\int \underline{\hat{H}}_m^{(p)*} \times \left(\underline{\underline{\epsilon}}^{M*} \cdot \underline{\hat{E}}_n^{(p+1)}\right) \cdot \hat{x}\, dy\, dz}{P_n^{(p+1)^{1/2}} \; P_m^{(p-1)^{1/2}}} \tag{24}$$

In general $\underline{\underline{\epsilon}}^{(1)}$ may vary along the guide as well as in the transverse direction. Let us assume that the phasor permittivity is sinusoidal in x in the form

$$\underline{\underline{\epsilon}}^{(1)}(x, y, z) = \underline{\underline{\hat{\epsilon}}}^{(1)}_{(a)} e^{i\kappa_x x} + \underline{\underline{\hat{\epsilon}}}^{(1)}_{(-a)} e^{-i\kappa_x x} \tag{25}$$

The coupling coefficient can then be expressed in the form

$$C_{mn}^{(q)\pm} = \frac{k_o^{(p)^2}}{k_m^{(p)}} \left[\eta_{mn}^{(a)} e^{i(\pm k_n^q - k_m^p + \kappa_x)x} + \eta_{mn}^{(-a)} e^{i(\pm k_n^q - k_m^p - \kappa_x)x} \right] \tag{26}$$

where q is p, p+1, or p-1 and $\eta_{mn}^{(\pm a)}$ is the normalized overlap integral

$$\eta_{mn}^{(\pm a)} = \frac{1}{2\epsilon_o} \frac{\int \underline{\hat{H}}_m^{(p)*} \times \left(\underline{\underline{\epsilon}}_{\pm a}^{b_q} \cdot \underline{\hat{E}}_n^{(q)} \right) \cdot \hat{x} dydz}{P_n^{(q)^{1/2}} P_m^{(p)^{1/2}}} \; . \tag{27}$$

The symbol b_q in (27) represents dc, M^*, and M^{**} for q=p, p-1, and p+1, respectively.

Equation (23) combined with (26) and (27) represent an infinite set of coupled first order differential equations with non-constant coefficients. The functions $A_m^{(p)\pm}$ represent the wave amplitudes. The time average power density in the forward x direction at frequency ω_p is found as $\underline{E}^{(p)} \times \underline{H}^{(p)*} + \underline{E}^{(p)*} \times \underline{H}^{(p)}$. Using (19) and integrating over the cross section, the power in the mth mode is

$$\langle P \rangle_m = 2\left(A_m^{(p)+} A_M^{(p)*+} - A_m^{(p)-} A_m^{(p)*-} \right) \tag{28}$$

In (28) we assume that the wave propagation constants are real so that the waves are not attenuated. The proper treatment of unbounded waves such as the guided modes in planar structures require a modification of the present treatment. In the interest of avoiding a repetition of the equations such cases can be treated in approximate fashion by assuming that the fields involved in the various integrals that do not attenuate in the y and/or z directions as a result of the waveguiding properties of the structure are multiplied by some function g(y, z) whose square integral $\int |g|^2 dydz$ is unity. Derivatives of g(y, z) that would enter into Maxwell's equations are assumed negligible. Use of this method for planar guides is limited to permittivities $\underline{\underline{\epsilon}}^{(1)}$ that are uniform or periodic in the y-direction.

3. MODULATOR CHARACTERISTICS

The optical wave(s) launched in a modulator is at frequency ω_c. Within a short distance the wave and mode at this frequency contain most of the propagating power and coupling to other modes can be estimated by integrating equation (23) under the assumption that only $A_{m'}^{(0)}$ is non-zero on the right side. The first harmonic of the mth mode is approximately

$$A_m^{(1)+} \approx \frac{{k_o^{(1)}}^2}{k_m^{(1)}} \left[\eta_{mm'}^{a} \frac{\ell}{2} \frac{\sin\left(k_{m'}^{(0)} - k_m^{(1)} + \kappa_x\right)\frac{\ell}{2}}{\left(\kappa_m^{(0)} - \kappa_m^{(1)} + \kappa_x\right)\frac{\ell}{2}} e^{i\left(k_{m'}^{(0)} - k_m^{(1)} + \kappa_x\right)\frac{\ell}{2}} \right.$$

$$\left. + \eta_{mm'}^{-a} \frac{\ell}{2} \frac{\sin\left(k_{m'}^{(0)} - k_m^{(1)} - \kappa_x\right)\frac{\ell}{2}}{\left(k_{m'}^{(0)} - k_m^{(1)} - \kappa_x\right)\frac{\ell}{2}} e^{+i\left(k_m^{(0)} - k_m^{(1)} - \kappa_x\right)\frac{\ell}{2}} \right] A_{m'}^{(0)} \qquad (29)$$

The term $\eta_{mm'}^{(a)}$, expressed in (27) is of order $n\Delta n$ as may be gathered from equation (14) and Δn for electronically induced index changes is of order $10^{-4} - 10^{-3}$. The two terms in (29) are thus small until ℓ attains values of order $\ell/\pi\Delta n$. The sin x/x functions further reduce the amplitude of $A_m^{(1)}$ unless the argument $k_{m'}^{(0)} - k_m^{(1)} \pm \kappa_x / k_o \Delta n$ is small. By this reasoning we arrive at the concept that substantial coupling between modes occurs only under phase match (or momentum convervation) conditions. We can assume that substantial coupling occurs when

$$\left(k_{m'}^{(0)} - k_m^{(1)} \pm \kappa_x\right) < \Delta n\, k_o \qquad (30)$$

and use the inequality in (30) to eliminate terms in (23). This approximation improves as the inequality increases. A corresponding conservation of momentum argument is implicit in the integral in equation (27). One advantage of integrated optics lies in the fact that it is possible to achieve large values of κ_x and κ_y that are on the order of k_o. Coupling between a limited number of modes is thus facilitated.

When coupling is limited to two modes as, for example, in the Bragg angle diffraction or magneto-optic modulators described earlier, the coupled mode equations involve two A_m's, say m = 1 and m = 2 that are phase matched, e.g.,

$$k_2^{(p\pm1)} - k_1^{(p)} + \kappa_x = 0 \tag{31}$$

Equations (23) then reduce to a set of coupled equations involving the harmonic frequencies. In the absence of time independent dc coupling the nonzero terms in (23) are $A_1^{(0)}$, $A_1^{(\pm2)}$, $A_1^{(\pm4)}$ . . . and $A_2^{(\pm1)}$, $A_2^{(\pm3)}$, $A_2^{(5)}$ . . . . Coupling proceeds from the input mode to the first harmonic of the diffracted wave (and back) and on up through higher harmonics. When $\epsilon_a^M = \epsilon_a^{M*}$ the coupling coefficients are equal and $A_1^{(p)}$ and $A_2^{(p)}$ are the even and odd orders, respectively, of the Bessel functions $i^p J_p(C_{12}x/2)$. For low modulation frequencies the coupling can be treated by introducing the index change into the dc term and permitting C_{mn}^p to vary with time. With two modes present and the self coupling C_{mm}^p zero, the dc coupling is characterized by the equations

$$\frac{d\,A_1^{(0)}}{dx} = i\,C_{12}^{(0)}\,A_2^{(0)}$$

$$\frac{d\,A_2^{(0)}}{dx} = i\,C_{21}^{(0)}\,A_1^{(0)} = i\,C_{12}^{(0)}\,A_1^{(0)} \quad . \tag{32}$$

The solution to these equations when mode 1 is launched at x = 0 with amplitude $A_1^{(0)}(0)$ is

$$A_1^{(0)} = A_1^0(0) \cos C_{12}^{(0)} x$$

$$A_2^{(0)} = i\, A_1^{(0)}(0) \sin C_{12}^{(0)} x \quad . \tag{33}$$

The coefficient $C_{12}^{(0)}$ is proportional to the induced index variation Δn. When Δn varies sinusoidally the coefficients of the harmonic terms in (33) can be represented by Bessel functions.

We are now in a position to consider device modulation field and length requirements as well as optical transit time limitations. Let us assume that coupling is limited to two modes. Using (33) to describe the wave coupling, the device length required for full extinction of the excited mode is

$$L = \frac{\pi}{2C_{12}^{(0)}} \quad \text{or} \quad L = \frac{\pi}{2C_{12}^{(\pm 1)}} \quad . \tag{34}$$

The second equation holds for a traveling wave device in which the time varying permittivity has frequency components $\omega_M = \omega_{sc} \pm \omega_M$, where ω_{sc} is the subcarrier frequency and ω_M, is the modulation frequency. The coupling coefficients are related to the modulating fields through equations (1), (13) and (33). When the fields are expressed in terms of an applied voltage or current or some input wave power, equation (34) may be used to establish the signal magnitude required for input wave extinction in terms of the device length. In general $C_{1,2}^{(0)}$ and $C_{1,2}^{(\pm 1)}$ are proportional to $k_o \Delta n$ with the proportionality constant equal to or less than unity and Δn in the range 10^{-4} - 10^{-3}. Minimum device lengths of order 10^3 - $10^4 \lambda$ are therefore necessary to achieve full extinction.

There is a limitation to the rate at which the optical carrier can be efficiently modulated. This rate is essentially determined by the phase error expression (30). An alternative view may be obtained by considering two mode coupling in the traveling wave case in which k_x is substantial so that only $A_1^{(0)}$ and $A_2^{(-1)}$ couple, for example. The coupled equations are

$$\frac{d A_1^{(0)}}{dx} = i \frac{{k_o^{(0)}}^2}{k_1^{(0)}} \eta_{12}^{-a} e^{i\left(k_2^{(-1)} - k_1^{(0)} - \kappa_x\right)} A_2^{(1)}$$

$$\frac{d A_2^{(-1)}}{dx} = i \frac{{k_o^{(0)}}^2}{k_2^{(-1)}} \eta_{12}^{a} e^{-i\left(k_2^{(-1)} - k_1^{(0)} - \kappa_x\right)} A_1^{(0)} \tag{35}$$

and assuming both coefficients to be the same in (35), the solution for $A_2^{(-1)}$ is found to be

$$A_2^{(-1)} = \left(1 + \left[\left(k_2^{(-1)} - k_1^{(0)} - \kappa_x\right)/2\left({k_o^{(0)}}^2 k_1^{(0)}\right)\eta_{12}^{a}\right]^2\right)^{-1/2} e^{i\left(k_2^{(-1)} - k_1^{(0)} - \kappa_x\right)x/2}$$

$$\sin\left[\left[\left(\frac{{k_o^{(0)}}^2}{k_1^{(0)}} \eta_{12}^{a}\right)^2 + \left(\frac{k_2^{(-1)} - k_1^{(0)} - \kappa_x}{2}\right)^2\right]^{1/2} x\right] \tag{36}$$

The power coupled into $A_2^{(-1)}$ is reduced to $\frac{1}{2} \sin^2 (\pi/\sqrt{2})$ when the phase mismatch is

$$\left(k_2^{(-1)} - k_1^{(0)} - \kappa_x\right) = 2 \frac{{k_o^{(0)}}^2}{k_1^{(0)}} \eta_{12}^{a} = \pi/L \ . \tag{37}$$

The phase match requirement must be met over a band of frequencies corresponding to the range of modulation frequencies as well as the range of optical frequencies in the carrier. Neglecting the latter consideration for the moment, the propagation constant of the coupled mode at the frequency $\omega_c + \omega_M$ may be expressed

$$k_2^{(-1)} = k_2^{(0)} + \frac{\partial k_2^{(0)}}{\partial \omega} \omega_M + \dots \ . \tag{38}$$

For devices in which the modulation is applied with fixed fields so that κ_x is a constant the bandwidth limitation is found by inserting (38) into (37) to obtain

$$\omega_{M\,max} = \pi/L\left(\frac{\partial k_2^{(0)}}{\partial\omega}\right) = \pi v_{g_2}/L = \pi/\tau \qquad (39)$$

where v_{g_2} is the group velocity of the diffracted mode and τ is the transit time of the optical carrier through the device. The inverse of the right side of (39) τ/π may be interpreted as the minimum pulse width that can be generated by the structure.

If the device is a traveling wave modulator κ_x will also vary with frequency and the device length may be made long without sacrificing bandwidth. This is achieved by matching the group velocities. When the group velocity is not matched differentiation of (37) shows that the bandwidth is

$$\omega_{M_{max}} = \frac{\pi}{L\left(\frac{\partial k_2^{(0)}}{\partial\omega} - \frac{\partial\kappa_x}{\partial\omega}\right)} = \frac{\pi}{L\left(\frac{1}{v_{g_2}} - \frac{1}{v_{g_{sc}}}\right)} \qquad (40)$$

where $v_{g_{sc}}$ is the group velocity of the subcarrier. When the denominator of (40) is zero the maximum modulation frequency is expressed in terms of the difference in the dispersion or second derivative terms in (37), viz.,

$$\frac{1}{2}\,\omega^2_{M_{max}} = \frac{\pi}{L\left(\frac{\partial k_2^{(0)}}{\partial\omega^2} - \frac{\partial^2\kappa_x}{\partial\omega^2}\right)} \;. \qquad (41)$$

As for the bandwidth of the optical carrier, most lasers and injection lasers in particular have bandwidths that far exceed any modulation frequency that can be impressed. For two mode coupling (37) must be satisfied as both $k_2^{(-1)}$ and $k_1^{(0)}$ vary over the spectral range of the source. In devices in which coupling to many modes occurs and some degree of phase mismatch is tolerated, e.g., in Raman-Nath type diffraction, the usable bandwidth for both the carrier and subcarrier frequencies may be expected to be greater than in strongly resonant two mode devices.

4. CHARACTERISTICS OF OTHER MODULATION TECHNIQUES

The minimum length of any modulator, which determines the bandwidth, is of order $\lambda/\Delta n$ or 10^3 - $10^4 \lambda$. This characteristic may be deduced from the coupled mode equations, but for certain modulation techniques it is more readily deduced in a direct fashion. Consider, for example, modulation achieved in a structure in which the index of the guiding film differs by a small amount Δn from the surrounding medium so that on application of a field the change in index removes the waveguiding properties. As a model for investigating the distance required for modulation we may assume that the tangential electric field in the waveguide has an approximately Gaussian shape in the form

$$E_t^{(a)} \simeq \exp - z^2/2\sigma^2 \qquad \text{or} \qquad \exp - (y^2 + z^2)/2\sigma^2 \quad . \tag{42}$$

The latter expression applies to a linear guide and the former to a planar guide. The parameter σ may be taken as half the effective thickness of the guide

$$\sigma = \frac{t}{2} + \frac{1}{(k_w^2 - k^2)^{1/2}} \simeq \frac{t}{2} + \frac{1}{(2k\Delta n)^{1/2}} \tag{43}$$

where t is either the thickness or radius of the guides, k is the wavenumber of the surrounding material, and k_w is the guided wave propagation constant which differs from k by less than $k\Delta n$.

When the index of a portion of the waveguide of length L is lowered by Δn, the gaussian beam propagates as in free space. From the discussion in the chapter on periodic couplers the beam at a distance L is approximately

$$E_t^{(b)} \simeq \frac{1}{1 + i\, L/\sigma^2 k} \exp - \frac{z^2(1 - i\, L/\sigma^2 h)}{2\sigma^2\left[1+(L/\sigma^2 k)^2\right]} \quad . \tag{44}$$

The approximate relative power coupled back into the guide beyond the modulator region may be found by computing

$$P_c/P_{in} \simeq \left|\int E_t^{(a)} E_t^{(b)} \, dydz\right|^2 / \left|\int E_t^{(a)^2} \, dydz\right|^2 \qquad (45)$$

and this procedure yields the result

$$P_c/P_{in} = \frac{1}{\sqrt{1 + (L/2\sigma^2 k)}} \quad \text{or} \quad \frac{1}{1 + L/2\sigma^2 k} \qquad (46)$$

for a planar or linear guide respectively. Using (43) we note that to reduce the relative power in the continuing waveguide to $1/\sqrt{2}$ or $1/2$ a length

$$L \simeq 2\left(\frac{t}{2} + \frac{1}{(2k\Delta n)^{1/2}}\right)^2 k > \frac{1}{k\Delta n} \qquad (47)$$

is required.

The result expressed in (47) may be considered as representative of the device length required when the modulation couples to a broad region of the continuous spectrum. This in contrast to the behavior of a modulator that couples to a single radiating spectral component when, as the discussion of periodic couplers shows, the attenuation rate is proportional to $(\Delta n)^2$ rather than Δn. A structure that provides modulating fields with many spatial harmonics that couple to radiation fields produces a broad spectral radiation with the same modulating field amplitude required for coupling to a single radiating spectral component.

The length required for a mode interference device with stationary modulating fields may be found by determining the wave propagation constants $k_1^{(0)} + \Delta k_1^{(0)}$ and $k_2^{(0)} + \Delta k_2^{(0)}$ where the Δ terms represent electronically induced changes in propagation constant. $\Delta k_1^{(0)}$ and $\Delta k_2^{(0)}$ are approximately $C_{11}^{(0)}$ and $C_{22}^{(0)}$ when time varying components of $\underline{\epsilon}^{(1)}$ are included in $\underline{\epsilon}^{dc}$. The device length L required to rotate the polarization of the sum of a TE and a TM wave or to transfer energy from one side of a waveguide that can support two modes to the other is found as the solution of the phase difference equation

$$\left(\Delta k_1^{(0)} - \Delta k_2^{(0)}\right) L = \pi \quad . \tag{48}$$

Since the changes in propagation constant are proportional to $k_o \Delta n$, L is at least as great as $\pi / k_o \Delta n$. To avoid lengths greater than this, one must be certain that the wave propagation constants change differentially or that one propagation constant changes while the other remains constant. In junction modulators this is accomplished through the naturally occurring differences in induced index change in the direction of the applied field. In other configurations it may be accomplished by inducing an index change in the region in which the electric field of one optical mode is large while that of another mode is small.

As a final example we may consider the device length required to switch light between two parallel waveguides that are coupled together as a result of the fact that the fields in each guide overlap the other guide. The switching properties may be examined with aid of equations (23). We use the quasi-static approximation so that only $\underline{\underline{\epsilon}}^{dc}$ is nonzero. $A_1^{(0)}$ is the amplitude of the wave in one guide, $A_2^{(0)}$ the wave in the other, and C_{mn} are found using $\underline{P}$ taken as the sum

$$\underline{P} = \left(\underline{\underline{\epsilon}}^{(0)} - \underline{\underline{\epsilon}}^{(G_1)}\right) \cdot \underline{E}_1^{(0)} + \left(\underline{\underline{\epsilon}}^{(0)} - \underline{\underline{\epsilon}}^{(G_2)}\right) \cdot \underline{E}_2^{(0)} + \underline{\underline{\epsilon}}^{(1)} \cdot \left(\underline{E}_1^{(0)} + \underline{E}_2^{(0)}\right) \quad . \tag{49}$$

Here $\underline{\underline{\epsilon}}^{(0)}$ is the spatial permittivity distribution that assures that there are two guides, $\underline{\underline{\epsilon}}^{(G_1)}$ is the permittivity distribution of a single guide in the absence of the increased permittivity in the core of the other guide so that $\underline{\underline{\epsilon}}^{(0)} - \underline{\underline{\epsilon}}^{(G_1)}$ is only nonzero inside guide 2, $\underline{\underline{\epsilon}}^{(G_2)}$ is the same for guide 2, and $\underline{\underline{\epsilon}}^{(1)}$ is the electronic permittivity as before.

Although $\underline{E}_1$ and $\underline{E}_2$ are not orthogonal, consideration of the relative magnitudes of the integrals involved show that the same procedure as was used in going from (17) to (22) may be employed to arrive at coupled equations of the form

$$\frac{d\,A_1^{(0)}}{dx} = i\left(C_{11}^{(0)}\,A_1^{(0)} + C_{12}^{(0)}\,A_2^{(0)}\right) ,$$

$$\frac{d\,A_2^{(0)}}{dx} = i\left(C_{22}^{(0)}\,A_2^{(0)} + C_{21}^{(0)}\,A_1^{(0)}\right) . \qquad (50)$$

The coupling coefficient $C_{12}^{(0)}$ is the sum of a passive term $C_{12_p}^{(0)}$ due to $\underline{\underline{\epsilon}}^{(0)} - \underline{\underline{\epsilon}}^{(G_2)}$ and an active term $C_{12_a}^{(0)}$ due to $\underline{\underline{\epsilon}}^{(1)}$. The coefficients $C_{11}^{(0)}$ and $C_{22}^{(0)}$ contain only active parts due to $\underline{\underline{\epsilon}}^{(1)}$.

When the guides are identical, phase match is assured and the C's are constant. The passive parts are also equal. The solution for A_2 when energy is introduced into guide 1 ($A_1 = 1$ and $A_2 = 0$ at $x = 0$) is then found to be

$$A_2 = i\sqrt{\frac{C_{21}}{C_{12}}}\;\frac{1}{\left[1 + \left(\dfrac{C_{11} - C_{22}}{2\sqrt{C_{12}C_{21}}}\right)^2\right]^{1/2}}\; e^{i\left(\frac{C_{11} + C_{22}}{2}\right)x}$$

$$\cdot \sin\left[\left(\left(\frac{C_{11} - C_{22}}{2}\right)^2 + C_{12}C_{21}\right)^{1/2} x\right] . \qquad (51)$$

For passive effects C_{11} and C_{22} are zero and $C_{12} = C_{21}$. To make a switch we select the device length L so that the argument of the sine in (51) under passive coupling is

$$\left(C_{12_p} C_{21_p}\right)^{1/2} L = C_{12_p} L = \frac{\pi}{2} + m\pi \qquad (52)$$

where m is an integer. This assures complete coupling into guide 2 in the absence of applied fields. When modulating fields are applied, energy is removed from guide 2 so that the length L must also satisfy

$$\left[\left(\frac{C_{11_a} - C_{22_a}}{2}\right)^2 + \left(C_{12_p} + C_{12_a}\right)\left(C_{21_p} + C_{21_a}\right)\right]^{1/2} L = m'\pi \, . \qquad (53)$$

If the active parts of the coupling between guides is assumed small as occurs when the indices of the guide materials rather than the intervening regions are changed, C_{12_a} and C_{21_a} may be suppressed in (53). With this approximation the simultaneous equations (52) and (53) may be used to eliminate C_{12_p} and the device length is found to be

$$L = \frac{2\pi\left|m'^2 - \left(m + \frac{1}{2}\right)^2\right|^{1/2}}{\left(C_{11_a} - C_{22_a}\right)} \, . \qquad (54)$$

Minimum switching length occurs if $m' = 1$, $m = 0$ and the difference in the electronically induced propagation constants in each guide C_{11_a} and C_{22_a} is maximized. Since C_{11_a} and C_{22_a} are of order $k_o\Delta n$ the length required for coupling is again found to be of order $\lambda/\Delta n$. It should also be noted that the passive guide coupling is selected to be

$$C_{12_p} = \frac{\left(C_{11_a} - C_{22_a}\right)}{2\sqrt{3}} \qquad (55)$$

to achieve (54) with m' = 1, m = 0. To operate the switch, the guide separation is increased on the input and output ends of the coupling region.

The switch may also be viewed as a mode interference device in which the sum of normal modes of the form

$$\underline{E}_a = A_1^{(0)}\left(\hat{\underline{E}}_1 + \hat{\underline{E}}_2\right)e^{i\left(k_1 + C_{12_p}\right)x}$$

$$\underline{E}_b = A_1^{(0)}\left(\hat{\underline{E}}_1 - \hat{\underline{E}}_2\right)e^{i\left(k_1 - C_{12_p}\right)x} \tag{56}$$

propagate on the guide. These waves are obtained by collecting terms with the same propagation constant in (33) which is a solution of (50) when C_{11} and C_{22} are zero. Introduction of symmetric or antisymmetric electronic permittivity changes alter the relative phase of mode a or mode b, respectively and at a prescribed point determined by the phase difference energy appears in one guide or the other.

In the special case in which switching occurs between planar guides separated by a low index region, an exact solution for the propagation constants of the normal modes may be obtained when the index of each region changes uniformly. Using the notation introduced in the chapter on periodic couplers (Chapter 12), the eigenvalue equation of the 5-region structure is

$$D_{1,5}(k_x) = D_{1,3}D_{3,5} - D'_{1,3}\,{}'D_{3,5} = 0 \tag{57}$$

where layers 2 and 4 constitute a waveguide and layer 3 is an isolating region between the two guides. Values of k_x that satisfy (57) are the propagation constants of the normal modes. These values may be obtained by numerical means from (57) or, with generally small error, from a first order analysis. The factor $(-2i\,|k_1+k_2|\,|k_2+k_3|)(-2i\,|k_3+k_4|\,|k_4+k_5|)$ $\exp i(u_5 z_4 - u_1 z_1 - u_3 t_3)$ which appears in $D_{1,5}$ does not affect the solution for eigenvalues and may be removed from (57) to yield

$$\sin\psi_2^+ \sin\psi_4^+ - \exp -2|u_3|t_3 \sin\psi_2^- \psi_4^- = 0 \qquad (58)$$

where t_3 is the thickness of the coupling region and

$$\psi_2^\pm = u_2 t_2 - \tan^{-1}\frac{k_1}{ik_2} \pm \tan^{-1}\frac{k_3}{ik_2}$$

$$\psi_4^\pm = u_4 t_4 \pm \tan^{-1}\frac{k_3}{ik_4} - \tan^{-1}\frac{k_5}{ik_4} \quad . \qquad (59)$$

The eigenvalues for each guide acting alone satisfies $\psi_2^\pm = m\pi$ and $\psi_4^\pm = m'\pi$. So long as the coupling rate is small ($\exp -2|u_3|t_3 << 1$) the eigenvalues of the structure are approximately the eigenvalues of each guide acting alone. Taking m and m' to be zero for convenience, a first order expansion of $\psi_2^\pm$ and $\psi_4^\pm$ in (58) yields the change in eigenvalues of the coupled structure as solutions of

$$\left[\Delta k_x + \left(\frac{\partial\psi_2^+}{\partial n_i} \Big/ \frac{\partial\psi_2^+}{\partial k_x}\right)\Delta n_i\right]\left[\Delta k_x + \left(\frac{\partial\psi_4^+}{\partial n_i} \Big/ \frac{\partial\psi_4^+}{\partial k_x}\right)\Delta n_i\right] = \frac{(\exp - 2|u_3|t_3)\sin\psi_2^- \sin\psi_4^-}{\left(\frac{\partial\psi_2^+}{\partial k_x}\right)\left(\frac{\partial\psi_4^+}{\partial k_x}\right)} \qquad (60)$$

where n_i is the change in index of the ith region. In terms of the coupled wave parameters, (60) is

$$\left(\Delta k_x + C_{11_a}\right)\left(\Delta k_x + C_{22_a}\right) = \left(C_{12_p}\right)^2 \qquad (61)$$

The passive coupling coefficient C_{12_p} as presented on the right side of (60) is found to vary inversely with the effective guide thickness and to decay as $\exp -2|u_3|\, t_3$ where t_3 is the thickness of the layer between the guides.

5. CIRCUIT ASPECTS OF MODULATORS

The device length required for modulator operation is determined by the specific solution to the coupled mode equations. We will consider the analysis of a sample device in detail shortly, but as we have already observed, the required length is generally of order $\lambda_0/\Delta n$ where Δn is representative of the index change induced by the modulating fields. The significance of the device length lies in part on the limitation in modulator bandwidth that results from transit time or dispersive effects in lumped or traveling wave modulators, respectively, as expressed in (39) through (41). In addition to this limitation there are two other effects that serve to limit the bandwidth. These are related to the internal properties of the device and to the external circuit employed to supply the modulating fields.

Electrical properties of an electro-optic modulator can often be modeled by a single capacitor that represents the junction capacitance of a reverse biased junction device, the capacitance between the fingers of an interdigital structure, or the capacitance between electrodes used for modulation in a linear waveguide. Similarly, a magneto-optic device can be characterized by the inductance of the wire used to generate the magnetic field. These devices also have associated loss elements, however, that limit the bandwidth through reduction of the effective field in the modulation region or through excessive heating of the device. As an example of the bandwidth reduction due to the loss of effective field, consider the model shown in Fig. 2(a). Resistors R and R' represent the series resistance in a junction modulator or a diffused waveguide due to the finite conductivity of the substrate through which fields are applied to the modulator. The voltage across the capacitor is reduced to $1/\sqrt{2}$ the applied voltage at the frequency at which the impedance of the capacitor is equal to that of the resistors, i.e., at

$$\omega_M = 1/(R + R')C \quad . \tag{62}$$

This (angular) frequency represents a cutoff frequency for the device since the voltage across the capacitor is reduced and device heating increases at higher frequencies.

Figure 2(b) shows a model for an electro-optic device in which the loss properties of the material are included. If the complex permittivity of the material at some modulation frequency ω_M is $\epsilon' + i\epsilon''$, the capacitance of an electro-optic device is proportional to ϵ' and the model will include a parallel conductance

$$G = \left(\frac{\omega_M \epsilon''}{\epsilon'}\right) C \quad . \tag{63}$$

While the voltage across the capacitor is not reduced at high modulation frequencies, the current VG and associated heating of the device may become excessive as $\omega_M \epsilon''/\epsilon'$ approaches values of order unity. The imaginary part of the permittivity is itself a function of frequency described by characteristic absorption resonant frequencies or relaxation times. As an example, the Debye model for a polar liquid with relaxation time τ is

$$\epsilon = \epsilon' + i\epsilon'' = \epsilon_o \left(\epsilon_\infty + \frac{\epsilon_s - \epsilon_o}{1 - i\omega\tau}\right) \tag{64}$$

where ϵ_∞ is the high (optical) frequency and ϵ_s the low frequency relative permittivity of the medium. For nitrobenzene $\epsilon_s = 35$, $\epsilon_\infty = (1.55)^2$, and $\tau = 4.5 \times 10^{11}$ sec. Solid materials generally have their resonances at infrared frequencies. Without considering heating effects, the bandwidth limitation due to material absorption effects may be characterized by the frequency at which the current increases by a factor $\sqrt{2}$. This occurs when $G = \omega C$ or $\epsilon'' = \epsilon'$. For the medium described by (64), the bandwidth limitation is expressed as the solution of

$$\omega_M = \frac{1}{\tau}\left[1 + \frac{\epsilon_\infty}{\epsilon_s + \epsilon_\infty}\left(1 + \omega_M^2 \tau^2\right)\right] \simeq \frac{1}{\tau} \quad . \tag{65}$$

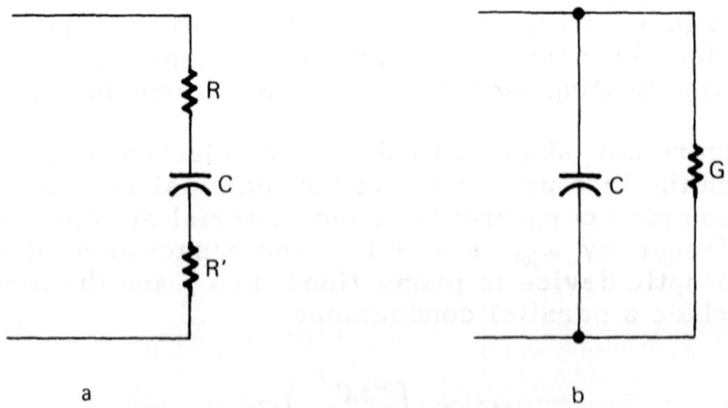

Figure 2. Examples of simple equivalent electrical circuits of electro-optic modulators.
a. Junction capacitance and resistance of p and n regions of a junction modulator.
b. Capacitance and associated conductance due to finite loss tangent of the electro-optic material.

The equivalent circuits of Fig. 2 indicate sample bandwidth limitations due to internal properties of the modulator. To apply efficient modulating signals to a lumped device, however, we require the addition of passive circuit elements that match the output impedance of the power supply or the characteristic impedance of an intervening transmission line to the device. Some examples of matching circuits are shown in Fig. 3. The circuits correspond to low pass or, with the addition of a matching inductance, to bandpass configurations. The lowpass circuits have a cutoff frequency (bandwidth) at which the voltage transfer decreases by a factor $1/\sqrt{2}$

$$\omega_M = \frac{1}{R_s C} \tag{66}$$

while the bandwidth of the tuned circuits resonant at ω_o is

$$2(\omega_M - \omega_o) = \omega_o^2 R_s C \quad . \tag{67}$$

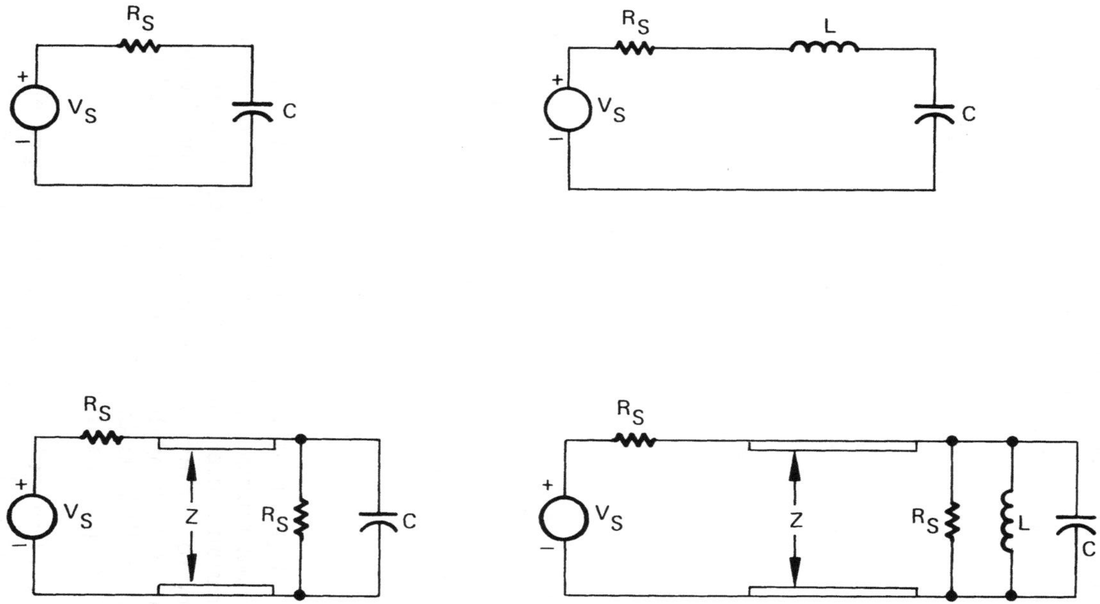

Figure 3. Examples of circuit configurations in which an electro-optic modulator modeled by capacitance C may be employed. R_s is the output resistance of the signal generator. Z is the characteristic impedance of a connecting cable and is taken to be R_s. L is the inductance added for bandpass operation.

It is customary to compare different modulators on the basis of the power expended in the resistor R at some modulation frequency in the middle of the modulation band when the rms signal voltage V is that required to achieve a prescribed degree of modulation. For parallel or series circuits of Fig. 3, this power is found to be, respectively,

$$P_R = \frac{1}{2}\frac{V^2}{R_s} \quad \text{or} \quad P_R = \frac{1}{2}\omega_o^2 C^2 R_s V^2 \tag{68}$$

The ratio of P_R to the cutoff frequency or the bandwidth as expressed in (66) or (67) is the same for all of the circuits, viz.,

$$\frac{P_R}{\omega_M} = \frac{P_R}{2(\omega_M - \omega_o)} = \frac{1}{2} C V^2 \quad . \tag{69}$$

This ratio is termed the figure of merit for the device. In analogy with (69), the figure of merit for a magneto-optic device would be $(1/2)\mathcal{L} I^2$ where $\mathcal{L}$ is the inductance of the wire and I is the required current. Traveling wave devices consisting, for example, of two wires placed parallel to a waveguide with the signal wave launched at one end and a matched termination at the other end require a power

$$P_L = \frac{1}{2} Y_o V^2 = \frac{1}{2c} C' V^2 \tag{70}$$

where Y_o is the characteristic admittance, C' is the capacitance per unit length, and $c = 1/\sqrt{L'C'}$ is the phase velocity of the transmission line. As before, V is the voltage necessary to achieve some degree of modulation. The bandwidth limitation of this device is determined by (40) or (41). For a TEM transmission line c is also the group velocity that appears in (40).

The objective of modulator design for integrated optics consists in providing as broad a frequency band as possible using as small a signal power as possible. As the figure of merit and traveling wave calculations show, this goal is achieved by reducing the device capacitance (or inductance) and signal voltage (or current). The capacitance of many two electrode configurations is proportional to the ratio of electrode width to electrode spacing and may therefore be the same for large bulk modulators as for small waveguides. The modulation process itself is dependent on the magnitude of the electric field in the modulator material, however, and since the electric field is proportional to the ratio of voltage to electrode spacing, a reduction in size of the device to the order of the optical wavelength can significantly reduce the figure of merit.

6. AN EXAMPLE

To demonstrate how some of the concepts presented in the preceding sections are used in the analysis and evaluation of a modulator, let us consider a planar diffraction modulator composed of interdigital fingers placed atop a waveguide (Fig. 4). We will assume that an isolation layer is present between the electrodes and the waveguide to reduce absorption losses and to minimize diffraction effects in the absence of modulating fields. When a voltage is applied to the electrodes, a periodic electric field is established on both sides of the electrode plane. Useful electro-optic materials often have a high permittivity at modulation frequencies relative to that of the substrate and an isolation layer, but we will assume that the permittivity of each region is the same. Under this assumption the electric field below the electrodes may be expressed

$$\underline{E}_M = \frac{V}{\Lambda}\sum_{n=0}^{\infty} B_{2n+1}\left[\hat{y}\cos(2n+1)\frac{\pi}{\Lambda}y + \hat{z}\sin(2n+1)\frac{\pi}{\Lambda}y\right]$$

$$e^{(2n+1)\frac{\pi}{\Lambda}(z-z_e)}\cos\omega_M t \qquad (71)$$

where Λ is the spacing between electrode centers, z_e is the electrode plane and n is an integer. Equation (71) is obtained as the gradient of the solution of Laplace's equation.

It has been shown[16] that for thin electrodes B_{2n+1} may be expressed

$$B_{2n+1} = 2\pi P_n(\cos\frac{\pi}{\Lambda}a)/K'(\sin\frac{\pi}{2\Lambda}a) \qquad (72)$$

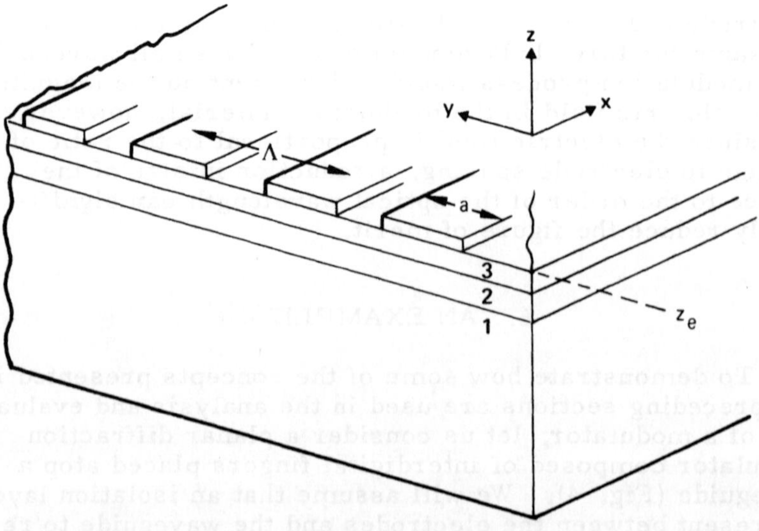

Figure 4. Cross section of a diffraction modulator showing electrodes of width a separated by a distance Λ. Region 3 is an isolation layer, region 2 is the waveguide, and region 1 is the substrate.

where P_n is the Legendre polynomial of order n, K' is the complete elliptic integral of the first kind, and a is the finger width. The numerical value of B_1 is found to range between 0.5 and 2 as a extends from small values to Λ (zero gap) and is approximately 1.5 when $a = \Lambda/2$.

Note that the electric field decays exponentially as z takes values less than z_e. The decay rate is determined by the finger spacing which we should like to make small in order to increase the resulting optical diffraction angle. If the finger spacing is less than the guide thickness, however, the modulating fields will not penetrate the guide. Let us assume that the penetrating electric field is limited to the first harmonic in (71) (n=0). The next step in the analysis is to compute the permittivity expressed in (4) using (71). This process will also generally involve coordinate rotations in which $\underline{E}_M$ is converted to the coordinates

$\hat{x}', \hat{y}', \hat{z}' = \hat{x}'_1, \hat{x}'_2, \hat{x}'_3$ of the medium using the transformation

$$\underline{E}'_M = \underline{\underline{T}}\, \underline{E}_M \tag{73}$$

where the elements of T are $T_{ij} = \hat{x}_i \cdot \hat{x}'_j$. Then (73) is rotated back into the coordinate system of the waveguide using

$$\underline{\underline{\epsilon}} = \underline{\underline{T}}^{-1}\, \underline{\underline{E}}'_M\, \underline{\underline{T}} \quad . \tag{74}$$

As an interesting variation on this computation we may consider the permittivity of a Kerr liquid for which, with reference to (1), only terms R_{iipp} are nonzero. The local optic axis in the otherwise isotropic liquid lies in the direction of the electric field. Taking this direction to be $\hat{y}'$, the permittivity of the Kerr liquid is expressed

$$\underline{\underline{\epsilon}}' = \epsilon_o \begin{bmatrix} \chi_{11} & 0 & 0 \\ 0 & \chi_{22} & 0 \\ 0 & 0 & \chi_{11} \end{bmatrix} \tag{75}$$

where

$$\chi_{11} = \left(n - \frac{1}{3}\lambda_o K\, E_{y'}^2\right)^2 \qquad \chi_{22} = \left(n + \frac{2}{3}\lambda_o K\, E_{y'}^2\right)^2 \tag{76}$$

and K is the Kerr constant ($K = 3.6 \times 10^{-12}\ m/V^2$ for nitrobenzene). Using the transformation for rotation about the x axis

$$\underline{\underline{T}} = \begin{pmatrix} 1 & 0 & 0 \\ 0 & \cos\theta & \sin\theta \\ 0 & -\sin\theta & \cos\theta \end{pmatrix} \tag{77}$$

where θ is the angle the modulating electric field makes with the y axis. The permittivity expressed in waveguide coordinates is

$$\underline{\underline{\epsilon}} = \epsilon_o \begin{pmatrix} \chi_{11} & 0 & 0 \\ 0 & \chi_{22}\cos^2\theta + \chi_{11}\sin^2\theta & (\chi_{22} - \chi_{11})\sin\theta\cos\theta \\ 0 & (\chi_{22} - \chi_{11})\sin\theta\cos\theta & \chi_{22}\sin^2\theta + \chi_{11}\cos^2\theta \end{pmatrix} \tag{78}$$

The angle θ is $\tan^{-1} E_{Mz}/E_{My}$ and for the first harmonic of the electric field in (71) may be expressed

$$\theta = \frac{\pi}{\Lambda} y \tag{79}$$

The value of $E_y'^2$ in (76) is $(\frac{V}{\Lambda} B_1)^2 \exp \frac{2\pi}{\Lambda}(z - z_e) \cos^2 \omega_M t$.

We assume that a plane guided TE optical wave

$$\underline{E}_1 = A_1 f(z) e^{ik_1 \hat{\underline{n}}_1 \cdot \underline{\rho}} \hat{n} \times \hat{z} \tag{80}$$

is incident on the modulator. The unit vector $\hat{n}_1$ is the direction of the incident wave, k_w is the guided wave propagation constant, and f(z) is the orthonormal field distribution as expressed, for example, in the chapter on periodic couplers. The question of which modes are excited is resolved by considering integrals such as appear in (22) and (24). Note that a sample diagonal term in (78) is

$$\epsilon_{22} = \epsilon_o \left[\frac{1}{2}(\chi_{22} - \chi_{11}) \cos \frac{2\pi}{\Lambda} y + \frac{1}{2}(\chi_{22} + \chi_{11})\right] \tag{81}$$

which exhibits a variation periodic in y. Nonzero values of the integrals over y in (22) and (24) are achieved when the coupled wave is

$$\underline{E}_2 = A_2 f(z) e^{ik_w \hat{\underline{n}}_2 \cdot \underline{\rho}} \hat{n}_2 \times \hat{z} \tag{82}$$

and

$$k_w n_{1y} - k_w n_{2y} = \frac{2\pi}{\Lambda} \equiv k_y \quad . \tag{83}$$

Furthermore, phase match is achieved when

$$k_w n_{1x} = k_w n_{2x} \quad . \tag{84}$$

Equations (83) and (84) are satisfied when $\underline{E}_1$ is incident at the Bragg angle ($n_{1y} = \kappa_y / 2k_w$).

The transverse components of the magnetic fields associated with (80) and (81) are

$$H_{z1,2} = A_{1,2}\left(\frac{k_w}{\omega\mu}\right) f(z) e^{ik_w \hat{\underline{n}}_{1,2} \cdot \underline{\rho}} \quad . \tag{85}$$

Limiting considerations to coupling coefficients expressed in (24), only $H_{z1,2}[1/4\, \epsilon_o(\chi_{22a} - \chi_{11a})\, E_{y1,2}]$ survive integration in y. Using the quasi-static approximation, the coupling coefficient obtained from (24) is

$$C_{12} = \frac{k_o^2}{\epsilon_o k_w} \frac{1}{8} \int_{z_1}^{z_2} (\chi_{22a} - \chi_{11a})\, f^2(z)\, dz \tag{86}$$

where

$$(\chi_{22a} - \chi_{11a}) = 2n\lambda_o k \left(\frac{V}{\Lambda} B_1\right)^2 \exp \frac{2\pi}{\Lambda}(z - z_e) \quad . \tag{87}$$

The integral in (86) serves to reduce the effective index change in a manner dependent on the ratio of finger width to guide thickness. When Λ is large compared to guide and isolation layer thickness and when the mode is well confined, we obtain

$$C_{12} \simeq \frac{k_o}{4} \lambda_o K \left(\frac{V}{\Lambda} B_1\right)^2 \quad . \tag{88}$$

The coupled mode equations and their solutions are expressed in (32) and (33). From (34) the device length for extinction of the undiffracted beam is required to be

$$L = \frac{1}{K\left(\frac{V}{\Lambda} B_1\right)^2} \tag{89}$$

and the figure of merit is

$$\frac{P_R}{BW} = \frac{1}{2}(C' L)V^2 = \frac{1}{2} C' \frac{\Lambda^2}{K B_1^2} \qquad (90)$$

where C' is the capacitance per unit length. For a linear electro-optic medium L would be proportional to $\Lambda / V B_1$ and the figure of merit would retain a factor V. The capacitance per unit length of an interdigital structure may be expressed[15]

$$C' = N\left(\frac{\epsilon_1 + \epsilon_2}{2}\right) \frac{K(\sin \frac{\pi}{2\Lambda} a)}{K'(\sin \frac{\pi}{2\Lambda} a)} \qquad (91)$$

where ϵ_1 and ϵ_2 are the permittivities on either side of the electrodes, K and K' are complete elliptic integrals of the first kind, and N is the number of fingers. Numerical values of C' are found to range from $6[N(\epsilon_1 + \epsilon_2)/2\epsilon_o]$ to $14[N(\epsilon_1 + \epsilon_2)/2\epsilon_o]$ pF/m as a/Λ ranges from 0.25 to 0.75. When $a = \Lambda/2$, C' is $8.85\ N(\epsilon_1 + \epsilon_2)/2\epsilon_o]$ pF/m. If the latter value is inserted in (20) and A_1 is taken as 1.5, the figure of merit for a nitrobenzene modulator is found to be

$$\frac{P_R}{BW} = 10.6\ N\Lambda^2 \quad . \qquad (92)$$

Expression (92) indicates that if a structure were 100 μm wide ($N\Lambda$) and had a finger spacing of 1 μm, the power required to modulate at a frequency of 1 GHz would be about 2π watts. This numerical result is representative of many potential waveguide materials. Furthermore, as the development of low loss linear waveguides proceeds, we can envision the power requirement dropping well below the 1 W/GHz level.

ACKNOWLEDGMENT

I am grateful for the help John Polky, Robert Smith and Clare Khasha gave me in preparing this article.

REFERENCES

1. D. J. Channin, Appl. Phys. Lett. 22, 364 (1973).

   J. P. Sheridan, J. M. Schnur, and T. G. Giallorenzi, Appl. Phys. Lett. 22, 560 (1973).

2. W. L. Bond, B. G. Cohen, R. C. C. Leite, and A. Yariv, Appl. Phys. Lett. 2, 57 (1963).

   A. Yariv and R. C. C. Leite, Appl. Phys. Lett. 2, 55 (1963).

   D. F. Nelson and F. K. Reinhart, Appl. Phys. Lett. 5, 148 (1964).

   F. K. Reinhart, D. F. Nelson, and J. McKenna, Phys. Rev. 177, 1208 (1969).

3. F. K. Reinhart and B. I. Miller, Appl. Phys. Lett. 20, 36 (1972).

4. W. E. Martin, J. Appl. Phys. 44, 3703 (1973).

5. L. Kuhn, M. L. Dakss, P. F. Heidrich, and B. A. Scott, Appl. Phys. Lett. 17, 265 (1970).

   L. Kuhn, P. F. Heidrich, and E. G. Lean, Appl. Phys. Lett. 19, 428 (1971).

   N. Chubachi, J. Kushibiki, and Y. Kikuchi, Electronics Lett. 9, 194 (1970).

   F. R. Gfeller and C. W. Pitt, Colinear Acousto-Optic Deflection in Thin Films, Dept. Electronic and Electrical Engrg. Rept., University College, London (1972).

   M. L. Shah, Appl. Phys. Lett. 23, 75 (1973).

6. D. J. Channin, Appl. Phys. Lett. 19, 128 (1971).

7. D. Hall, A. Yariv, and E. Garmire, Appl. Phys. Lett. 17, 127 (1970).

8. D. P. Giarusso and J. H. Harris, Appl. Opt. 10, 2786L (1971).

J. N. Polky and J. H. Harris, Appl. Phys. Lett. 21, 307 (1972).

M. T. V. Scibor-Tylski, Electronic Lett. 9, 309 (1973).

J. M. Hammer, D. J. Channin, and M. T. Duffy, Appl. Phys. Lett. 44, 3703 (1973).

D. P. Giarusso and C. S. Kumar, Appl. Phys. Lett. Lett. 23 (Sept. 1973).

9. P. K. Tisn, R. J. Martin, R. Wolfe, R. C. LeCraw, and S. L. Bank, Appl. Phys. Lett. 21, 394 (1972).

10. E. A. J. Marcatilli, Bell Syst. Tech. J. 48, 2103 (1969).

H. F. Taylor, J. Appl. Phys. 44, 3257 (1973).

11. W. R. Cook, Jr. and H. Jaffe, Chap. 6m, American Inst. of Phys. Handbook, 3rd ed., D. E. Gray, ed., McGraw-Hill (1972).

A. Yariv, Quantum Electronics, Chaps. 18, 19, John Wilson (1967).

12. P. S. Pershan, J. Appl. Phys. 38, 1482 (1967).

M. J. Freiser, IEEE Trans. MAG-4, 152 (1968).

13. W. H. Louisell, Coupled Mode and Parametric Electronics, John Wiley (1960).

D. A. Watkins, Topics in Electromagnetic Theory, Chap. 3, John Wiley (1958).

D. Marcuse, Light Transmission Optics, Chap. 10, Van Nostrant Reinhold (1972).

A. Yariv, IEEE Trans. QE-9, 919 (1973).

14. D. S. Jones, The Theory of Electromagnetism, p. 246, MacMillan (1964).

15. D. P. Giarusso and J. H. Harris, J. Opt. Soc. Am. 63, 139 (1973).

16. Helge Engen, IEEE Trans. ED-16, 1014 (1969).

Chapter 14 - ACOUSTO-OPTICAL INTERACTIONS IN GUIDED WAVE STRUCTURES

E. G. H. LEAN

IBM T. J. WATSON RESEARCH CENTER

YORKTOWN HEIGHTS, NEW YORK 10598

I. INTRODUCTION

Acousto-optic interactions using coherent light and microwave acoustics in solids have been extensively studied (for review see R. Damon, 1970; Quate, 1965). Many practical acousto-optic devices for laser beam modulation (Maydan, 1970), deflection (Gordon, 1966), filtering (Harris, 1969) and signal processing (Collins, 1967) have been realized. Recently, acousto-optic interactions in guided wave structures have been experimentally demonstrated (Kuhn et al, 1970; Kuhn et al, 1971). These thin film acousto-optical devices have significant advantages over their bulk wave counterparts because of the unique properties of guided waves. Guided optical waves propagate in a thin film with a thickness on the order of optical wavelength. The index of refraction of the film is required to be higher than that of the substrate material so that the light can be trapped in the film by total internal refractions between two interfaces. Depending on the thickness of the film, many discrete modes of either transverse electric (TE) or magnetic (TM) modes can propagate in the film. The phase velocities of the modes which depend on the thickness of the film and the refractive indices of the film and substrate can be readily calculated (Tien, 1970). The transverse field distribution of each mode, which is sinusoidal in the film and exponentially decaying outside the film, can also be calculated in a straightforward manner (Collins, 1960). The waveguide dispersion and the transverse field distribution are important properties of optical guided waves which give rise to significant advantages for thin film optical devices. The transverse distribution results in the concentration of the field in the thin film for high power density. The waveguide dispersion in thin films gives the possibility of maintaining the same phase velocity for a given mode over a long distance. Both high power density and long interaction length are essential conditions for efficient nonlinear devices which are based on parametric interaction processes. Efficient real time nonlinear devices for modulators, switches and deflectors are needed for any optical system.

Acoustic surface waves propagate on the surface of a substrate or in a thin film on a substrate with particle displacements decaying exponentially from the surface (White, 1970), similar to optical guided waves. The penetration depth of acoustic surface waves is of the order of an acoustic wavelength, which is typically 30 μm at a frequency of 100 MHz. The energy of acoustic surface waves is concentrated within this penetration depth, resulting in very high power density. The dispersion of acoustic surface waves can be controlled by providing a layer of different elastic properties on top of the substrate. These properties of high power density and controllable dispersion in both acoustic surface and guided optical propagation, when employed simultaneously, can provide efficient and controllable interactions.

Thin film acousto-optic devices, same as thin film electro-optic or magneto-optic devices have recently attracted considerable research attention and become practically important due to the advance of modern thin film technologies. These thin film active devices, plus other thin film optical devices for sources, waveguides and detectors can be called part of the new development in optics - integrated optics. It seems that optics has been following the same trend as electronics and acoustics, which have advanced from bulk solid state devices to surface and thin film devices to take advantage of miniaturization, planar technology and batch fabrication.

Integrated optics makes possible the fabrication of thin film optical devices within a single substrate, so as to perform optical signal processing functions without the usual problems of vibration and critical alignments. The integration of thin film devices has excited many researchers in the field of optics. Many devices and experiments concerning thin film lasers for sources, fast modulators for modulation and switching, passive elements such as waveguides, thin film lenses, beam splitters for channeling and waveguiding, couplers for coupling optical guided waves into and out of thin films, and detectors have been reported. Some of them are discussed elsewhere in this book. In this chapter, we shall discuss the unique properties, the device parameters and the applications of acousto-optic interactions in guided wave structures.

In Section II, a brief review on acousto-optic interaction mechanisms which include the basic photoelastic effect, the Raman-Nath diffraction, and the isotropic and anisotropic Bragg diffraction will be given. The calculation of the diffraction efficiency and the phase matching conditions will be included. In Section III, a detailed analysis of acousto-optic interaction in guided wave structures will be presented. Two unique features for guided wave structures, the overlap integral and the effect of waveguide dispersion for phase matching conditions, will be emphasized. In Section IV, theoretical calculations developed in Section III will be used to check the experimental results on the collinear interaction of optical guided waves by acoustic surface waves reported by Kuhn et al (1971), and the Bragg diffraction of optical guided waves by acoustic surface waves reported by Kuhn et al (1970) and by Chubachi (1973). In Section V, the device parameters for thin film acousto-optic devices will be discussed, with emphasis on comparisons between thin film and bulk devices. Some possible device applications and conclusions will be in Section VI.

II. REVIEWS OF ACOUSTO-OPTIC INTERACTIONS

Acousto-optic interactions in solids are basically due to the photoelastic effect, which causes a change in optical index of refraction in solids because of the presence of elastic waves. The change in the index of refraction due to the photoelastic effect, which is usually very small (typical value of $\Delta n/n$ is smaller than 10^{-4} under normal strain condition), can affect the phase of the light beam traversing the strained medium. The phase modulation of light by an acoustic wave, which provides a spatial and temperal variation of the index of refraction in the medium, has been extensively studied by Raman-Nath, (1935) and many others in the 1960's (for example, C. F. Quate, 1965). Under proper phase matching conditions, the interactions due to each small change in the index of refraction can be accumulated constructively, resulting in a substantial amplitude modulation of the incident light beam, as in the well known Bragg diffraction of light by acoustic waves (Willard, 1949). Bragg diffraction in crystalline media can be divided into isotropic (C. F. Quate, 1965) and anisotropic (Dixon, 1967) diffractions, depending on the presence of an optical birefringence in the diffraction process.

In this section, a brief review of the photoelastic effect, Raman-Nath diffraction and Bragg diffraction will be given. Phase matching conditions and diffraction efficiencies for isotropic and anisotropic Bragg diffractions will be included.

2.1 Photoelastic Effect

Photoelastic effect describes the change in optical properties in crystals due to the presence of elastic waves (Nye, 1957). Optical properties of a crystal are usually specified by the indicatrix. The indicatrix is an ellipsoid of wave normals whose coefficients are components of the relative dielectric impermeability tensor B_{ij} at optical frequencies. If coordinates are referred to principal dielectric axes of the crystal, the indicatrix is given by

$$B_{11}\, x_1^2 + B_{22}\, x_2^2 + B_{33}\, x_3^2 = 1 \qquad (2.1)$$

where

$$B_{ii} = \left(\frac{1}{n_{ii}}\right)^2 = \frac{1}{\varepsilon_{ii}} \qquad (i = 1, 2, 3) \tag{2.2}$$

n_{ii} and ε_{ii} are the optical index of refraction and the dielectric constant along principal crystal axes. Under elastic strains, the changes ΔB_{ij} in the impermeability is related to the strain tensor S_{kl} by the equation

$$\Delta B_{ij} = P_{ijkl} S_{kl} \tag{2.3}$$

where P_{ijkl} is the photoelastic tensor and

$$S_{kl} = \frac{1}{2}\left(\frac{\partial U_k}{\partial x_l} + \frac{\partial U_l}{\partial x_k}\right) \tag{2.4}$$

with U_k being the particle displacement component. The change in the index of refraction Δn_{ii} and the dielectric constant $\Delta\varepsilon_{ii}$ can be obtained from (2.2), (2.3) and (2.4):

$$\Delta n_{ij} = -\frac{1}{4} n_{ii}^{3} P_{ijkl}\left(\frac{\partial U_k}{\partial x_l} + \frac{\partial U_l}{\partial x_k}\right) \tag{2.5}$$

$$\Delta\varepsilon_{ij} = -\frac{1}{2} \varepsilon_{ii}^{2} P_{ijkl}\left(\frac{\partial U_k}{\partial x_l} + \frac{\partial U_l}{\partial x_k}\right) \tag{2.6}$$

The usual summation convention for tensor elements in (2.3), (2.5) and (2.6) is understood. It should be noted that subscripts i and j in Δn_{ij} indicate the polarization of the diffracted and the incident light respectively. The light diffraction involving Δn_{ij} with $i = j$ preserves the polarization of the incident light while in the cases of Δn_{ij} with $i \neq j$ the polarization of the diffracted light is different from that of the incident light.

Using standard matrix notations for symmetrical tensors in which the single subscript m ($m = 1, 2, . . . 6$) represents the double subscript ij with $i, j = 1, 2, 3$, Eq. (2.5) and (2.6) can be written as:

$$\Delta n_m = -\frac{1}{2} n^3 P_{mn} S_n \tag{2.7}$$

$$\Delta \varepsilon_m = -\varepsilon^2 P_{mn} S_n \tag{2.8}$$

where photoelastic constants P_{mn} form a 6x6 matrix and S_n constitutes a 1x6 matrix.

The indicatrix is a geometrical representation of the optical properties of a crystal. For a uniaxial crystal, the indicatrix is an ellipsoid with the major and minor axes being proportional to n_e and n_o. As illustrated in Fig. 2.1-a, a straight line parallel to the incident light propagation direction and a central section perpendicular to the straight line are drawn through the origin of the ellipsoid. The central section is an ellipse with two semi-axes which represent the indices of refraction for the ordinary (n_o) and the extraordinary (n_e) waves in the uniaxial crystal. The direction of the displacement vector D induced in the crystal by an incident E-field may also be found from the ellipse. As shown in Fig. 2.1-b, the induced D-vector is parallel to the normal to the ellipse at the intersect point of E-vector (the incident field) and the ellipse.

The photoelastic effect can be specified as a small change in the shape, size and orientation of the indicatrix. For example, an incident light polarized along the c-axis in a uniaxial crystal "sees" the n_e as shown in Fig. 2.2. A strain along the c-axis S_3 changes the shape of the ellipse seen by the incident light as indicated by the dotted line in Fig. 2.2. The change in Δn_3 or $\Delta\varepsilon_3$ gives rise to the induced displacement ΔD_3 by the strain.

2.2 Acousto-optic Interaction Mechanisms

In general, there are two types of light diffraction by acoustic waves. One is the Raman-Nath type of light diffraction which is essentially a phase grating phenomenon. The other is Bragg diffraction, which is more of a volume effect, in correspondence with Bragg diffraction of x-rays by crystal lattices. Bragg diffraction acquires somewhat different

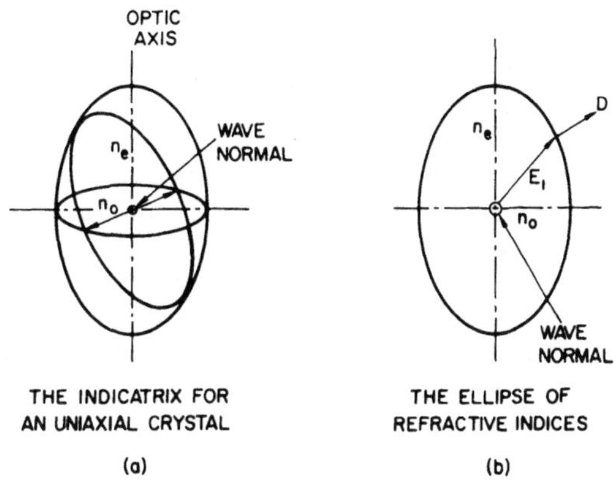

Fig. 2.1 - Geometrical representation of the indicatrix for a uniaxial crystal.

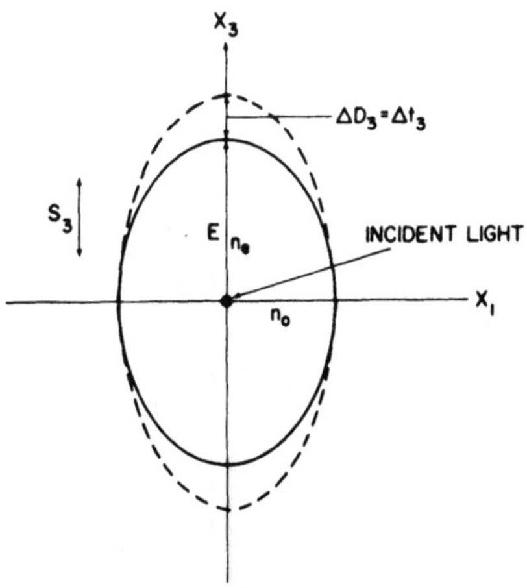

Fig. 2.2 - Geometrical representation of the photoelastic effect in a uniaxial crystal by an applied strain S_3.

properties in isotropic and anisotropic crystals. In this section, the basic acousto-optic interaction processes will be discussed.

Raman-Nath type of diffraction. This type of diffraction is of importance in interactions between acoustic surface waves and light or between bulk acoustic waves and light when bulk waves are confined to thin slabs. As shown in Fig. 2.3, the acoustic beam produces periodic change of the index of refraction in the crystal. As the incident plane wave passes through the crystal, the output wavefront is phase modulated by the periodic change in the index of refraction. The corrugated wavefront, in turn, results in many interference peaks in the far field region with the angle of the mth order given by

$$\sin\Theta_m = \sin\Theta_o + m\frac{\lambda}{\Lambda} \qquad m = \pm 1, \pm 2, \tag{2.9}$$

where Θ_o is the angle of incidence. For the case shown in Fig. 2.3, $\Theta_o = 0$. The intensity of the mth order interference peak is given by

$$\frac{I_m}{I_o} = J_m^2(\nu). \tag{2.10}$$

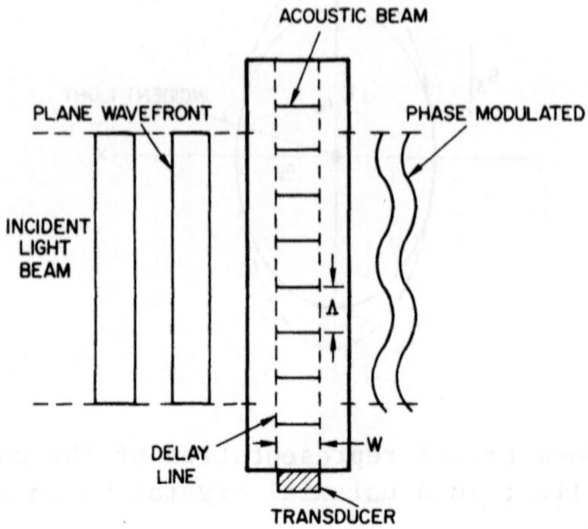

Fig. 2.3 - Configuration for a Raman-Nath type of light diffraction by acoustic waves.

where $J_m(\nu)$ is the Bessel function with the argument $\nu = (2\pi\Delta nw)/(\lambda_o n_o)$. Such diffraction of light was first studied by Raman and Nath thirty-eight years ago. Since then, many investigators have studied the problem and have come up with many practical devices (Berry, 1966).

The diffraction of light by acoustic surface waves is illustrated in Fig. 2.4. The propagating Rayleigh wave produces both a surface deformation and a periodic variation of the index of refraction within a surface layer about one wavelength in depth. The surface deformation and the periodic index variation provide a phase grating for the incident light and thus diffract the incident beam into many side orders. The mth order diffraction angle, Θ_m, is also given by Eq. (2.9). For the reflection case as shown in Fig. 2.4-a, only the surface deformation contributes to light diffraction. The intensity of the mth order is given by Eq. (2.10) with $\nu = (4\pi\delta_o/\lambda)$, where δ_o is the peak surface ripple. For the transmission case shown in Fig. 2.4-b, both the surface deformation and the periodic change in the index of refraction within the surface layer contribute to diffraction. The argument of the Bessel function is then

$$\nu = \frac{2\pi}{\lambda_o} [(n-1)\ \delta_o + |\int_0^\infty \overline{\Delta n}\ dx_2\ |] \tag{2.11}$$

The first term in Eq. (2.11) is due to the surface ripple while the second term is due to the photoelastic effect in the substrate. $\overline{\Delta n}$ is the effective change in the index of refraction by acoustic surface wave.

Light diffraction by acoustic surface waves has received much research attention. It provides a convenient tool for optical probing of propagation characteristics of acoustic surface waves (Lean, 1970).

Bragg diffraction of light by acoustic bulk waves. As the width of the acoustic beam increases and/or its frequency increases, the acoustically diffracted light begins to be dominated by an amplitude modulation instead of the phase modulation; the amplitude modulation arises from interferences in light reflected from the discontinuities in the refractive index induced by acoustic waves.

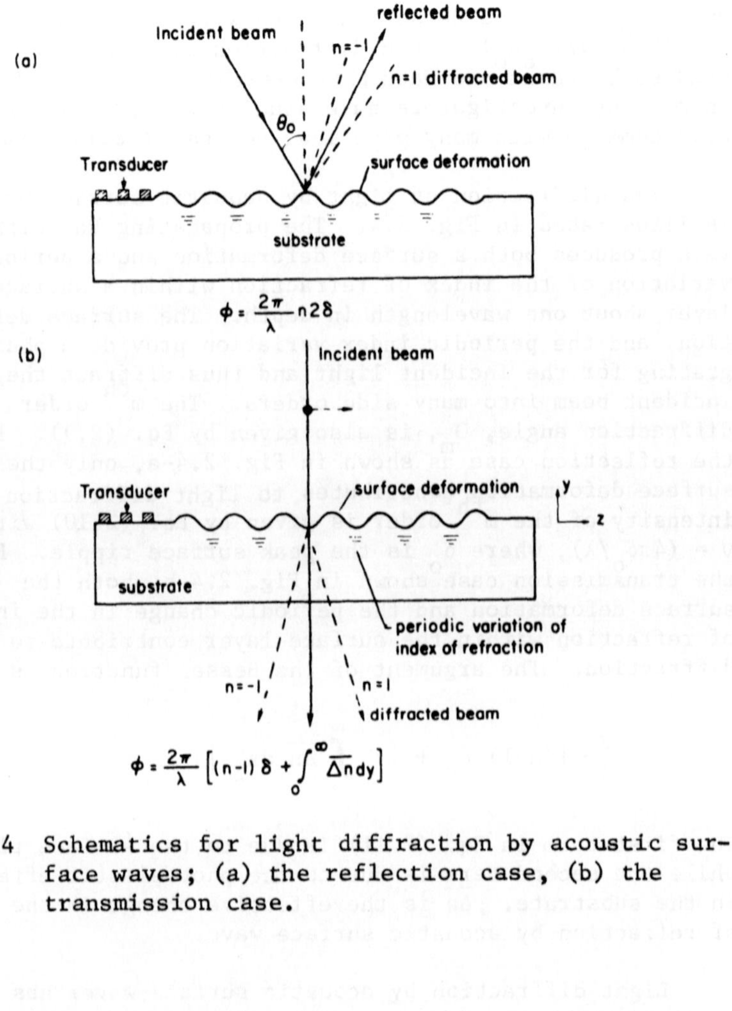

Fig. 2.4 Schematics for light diffraction by acoustic surface waves; (a) the reflection case, (b) the transmission case.

The condition for the Bragg diffraction region was discussed by Williard (1949) and is given by

$$Q = \frac{2\pi\lambda W}{\Lambda^2} > 1 \tag{2.12}$$

where λ and Λ are the optical and acoustical wavelength in the medium respectively; W is the acoustic beam width. Eq. (2.12) can be considered as the number of acoustic wavefronts in an acoustic beam having a width W transversed by the incident beam with an incident angle Θ

given by $\sin\Theta = \lambda/2\Lambda$ (the isotropic Bragg condition). For Bragg diffraction, the value of Q must be larger than one so that the reflection from each acoustic wavefront can be added constructively for the output of Bragg diffraction. The Raman-Nath region , where only the phase modulation of the incident beam is important, corresponds to $Q << 1$.

Acousto-optic Bragg Diffraction Efficiency. It has been shown that the Bragg diffraction of light by acoustic waves is a parametric interaction process (Quate, 1965, Lean, 1967). The parametric coupling results from a periodical change in the electrical susceptibility of a material in time and space due to acoustic waves acting through the photoelastic effect. The electric field $\vec{E}$, which exists in the strained crystal, can be obtained, in general, from the wave equation with proper boundary conditions. The wave equation is given by

$$\nabla^2 \vec{E} - \frac{[\varepsilon]}{C^2} \frac{\partial^2 \vec{E}}{\partial t^2} = \frac{1}{C^2} \frac{\partial^2}{\partial t^2} [\Delta\varepsilon] \vec{E} \tag{2.13}$$

where $[\varepsilon]$ is a diagonal 3x3 matrix consisting of dielectric constants of the medium; and $[\Delta\varepsilon]$ is a 3x3 matrix with the element $\Delta\varepsilon_{ij}$ given by Eq. (2.6) due to the photoelastic effect. With an acoustic wave at a frequency ω_s and a wavevector $\vec{k}_s$, $\Delta\varepsilon_{ij}$ can be written as

$$\Delta\varepsilon_{ij} = \frac{1}{2} G \exp [j (\omega_s t - \bar{k}_s . \bar{r})] + C.C. \tag{2.14}$$

where G is the amplitude of $\Delta\varepsilon_{ij}$. The right hand side of Eq. (2.13) is the coupling term for the propagating modes in the strained medium and the factor,

$$\Delta\vec{D} = [\Delta\varepsilon] \vec{E} \tag{2.15}$$

is the induced displacement for the parametric interaction process. Depending on the nonzero element $\Delta\varepsilon_{ij}$ in $[\Delta\varepsilon]$, the coupling between electric fields with the same polarization (when $i = j$) or with different polarizations (when $i \neq j$) results. The solutions for the electric fields of the wave equation in the strained medium may consist of an infinite number of plane waves (Slater, 1958). However, in the Bragg diffraction region, which requires a phase matching condition, only one diffracted light and the incident light are involved. Let the field of the incident light be

$$E_i = \frac{1}{2} A_i(r_i) \exp [j (\omega_i t - \bar{k}_i.\bar{r}] + C. C. \tag{2.16}$$

and the field of the diffracted light be

$$E_d = \frac{1}{2} A_d(r_d) \exp [j (\omega_d t - \bar{k}_d.\bar{r})] + C. C. \tag{2.17}$$

where $A_i(r_i)$ and $A_d(r_d)$ are the peak amplitudes of the incident and diffracted light respectively. ω_i, ω_d, $\bar{k}_i$ and $\bar{k}_d$ are the frequency and wavevectors of the incident and diffracted light respectively. $\underline{r}_i$ and $\underline{r}_d$ are the coordinates along the directions of $\bar{k}_i$ and $\bar{k}_d$. Because of the parametric coupling, $A_i(r_i)$ and $A_d(r_d)$ become functions of spatial coordinates. The required conditions for the Bragg diffraction of light by acoustic waves are given by

$$\vec{k}_d = \vec{k}_i \pm \vec{k}_s \tag{2.18}$$

and

$$\omega_d = \omega_i \pm \omega_s \tag{2.19}$$

Eqs. (2.18) and (2.19) are the conditions for conservation of momentum and energy in a parametric interaction process.

It can be shown that spatial variations of A_i and A_d can be obtained by solving the following pair of coupled amplitude equations (Yariv, 1971);

$$\frac{dA_i}{dx_2} = - j \frac{k_i \cos\Theta_i}{4} \frac{G}{\varepsilon} A_d \exp [-j (\Delta k_2 x_2 + \Delta k_3 x_3)] \tag{2.20}$$

$$\frac{dA_d}{dx_2} = -j\,\frac{k_d \cos\Theta_d}{4}\,\frac{G}{\varepsilon}\,A_i \exp\,[j\,(\Delta k_2\, x_2 + \Delta k_3\, x_3)] \qquad (2.21)$$

and

$$\vec{\Delta}\mathbf{k} = \vec{k}_d - \vec{k}_s - \vec{k}_i\,, \qquad (2.22)$$

where x_2 is the coordinate along the acoustic wavefronts as shown in Fig. 2.5 and $x_2 = r_i/\cos\Theta_d$ with Θ_i and Θ_d being the angles shown in Fig. 2.5. The solution for the coupled amplitude equations with boundary conditions such that at $x_2 = 0$; $A_i = A_i(0)$ and $A_d = 0$, is given by

$$A_d(x_2) = \frac{-j\,A_i(0)\,\cos\Theta_d}{T}\,\left(\frac{k_d}{4}\,\frac{G}{\varepsilon}\right)\,\sin(T\,x_2)$$

$$\exp\,[j\,(\tfrac{1}{2}\,\Delta k_2 x_2 + \Delta k_3 x_3)] \qquad (2.23)$$

where

$$T^2 = \frac{k_i k_d}{16}\,\left(\frac{G}{\varepsilon}\right)^2 \cos\Theta_i\,\cos\Theta_d + \frac{1}{4}\,\Delta k_2^2\,. \qquad (2.24)$$

The intensity of the diffracted light, which is proportional to $|A_d|^2$, at a distance $x_2 = L$ is given by

$$\frac{I_d(L)}{I_i(0)} = \left|\frac{k_d}{4}\,\frac{G}{\varepsilon}\,\cos\Theta_d\right|^2 L^2\,\frac{\sin^2(TL)}{(TL)^2}\,, \qquad (2.25)$$

where $I_i(0)$ is the incident light intensity.

At the phase matching condition where $\vec{\Delta}k = 0$, the maximum diffracted light intensity becomes

$$\frac{I_d(L)}{I_i(0)} = \left(\frac{k_d\,\cos\Theta_d}{k_i\,\cos\Theta_i}\right)\,\sin^2\left(\sqrt{\frac{k_i\,k_d\,\cos\Theta_i\,\cos\Theta_d}{16}}\,\left(\frac{G}{\varepsilon}\right)^2 L\right) \qquad (2.26)$$

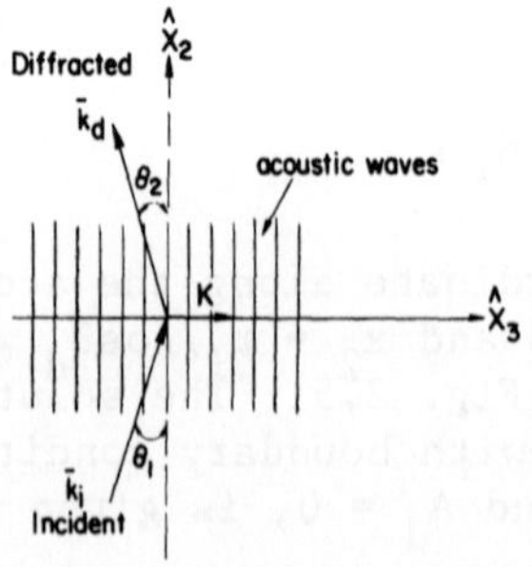

Fig. 2.5 - Schematic of Bragg diffraction of light by acoustic surface waves.

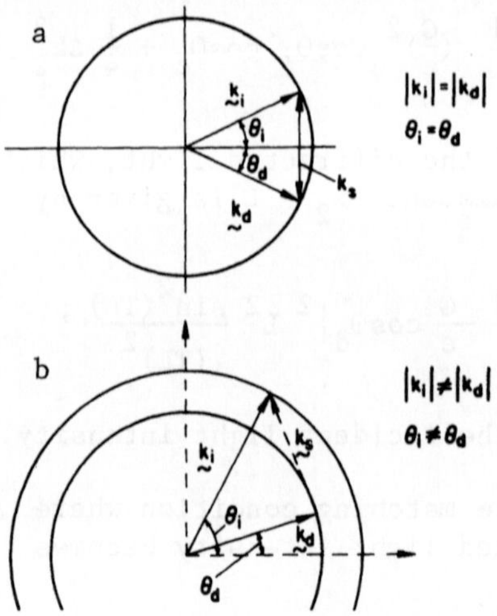

Fig. 2.6 Geometric representations of Bragg diffraction phase matching conditions; (a) isotropic diffraction case, (b) anisotropic diffraction case.

In the case where $k_i = k_d$, $\cos\Theta_i = \cos\Theta_d$, Eq. (2.26) reduces to the familiar equation for the acousto-optic diffraction efficiency

$$\frac{I_d(L)}{I_i(0)} = \sin^2 \left(\frac{\pi}{\lambda} L \sqrt{\left(\frac{n^6 p^2}{\rho v^3}\right) P_{ac}} \right), \qquad (2.27)$$

where ρ and v are the density and the acoustic velocity of the medium, p is the appropriate photoelastic constant and P_{ac} is the acoustic power density.

In obtaining (2.27), equation (2.26) and the power density with respect to the strain S, the density ρ and acoustic velocity, i.e.

$$P_{ac} = \frac{1}{2} \rho v^3 S^2, \qquad (2.28)$$

the magnitude of the photoelastic effect, $G = \varepsilon^2 P S$ in Eq. (2.8) were used.

It should be noted that the calculation leading to Eq. (2.26) is for the acousto-optic interaction in the Bragg region. A more detailed calculation which covers the Raman-Nath region, the Bragg region and in between the two regions, has been carried out by Klein and Cook, (1967).

Phase Matching Conditions - Isotropic and Anisotropic Diffractions. The diffraction efficiency is maximum at the phase matching condition,

$$\vec{k}_d - \vec{k}_s - \vec{k}_i = 0. \qquad (2.22)$$

Depending on the optical anisotropy of the diffraction medium, the diffraction can be either isotropic or anisotropic. If the medium is optically isotropic, the incident and the diffracted light "see" the same index of refraction so that $|k_i| = |k_d|$. This gives the well known isotropic Bragg diffraction with k-vectors as shown in Fig. 2.6-a. The incident angle Θ_i with respect to acoustic wavefronts, which is the same angle for the diffracted light , becomes

$$\sin\Theta_i = \frac{1}{2} \frac{k_s}{k} = \frac{1}{2} \frac{\lambda}{\Lambda} . \qquad (2.29)$$

If the medium is optically anisotropic and if the incident and the diffracted light "see" the different indices of refraction so that $|k_i| \neq |k_d|$, we have an anisotropic Bragg diffraction. The phase matching condition is shown schematically in Fig. 2.6-b, which gives the following relation,

$$k_s^2 = k_o^2 \, [\, n_i^2 + n_d^2 - 2n_i \, n_d \cos \, (\Theta_i - \Theta_d)]. \qquad (2.30)$$

For a given diffraction medium and a fixed incident angle Θ_i, the required acoustic frequency and the diffracted angle Θ_d can be calculated from (2.30). There are several special cases in the anisotropic diffraction, which are practically important:

i) when $\Theta_i = \Theta_d = 0$, we have a collinear interaction with k_i and k_d in the forward direction. Eq. (2.30) becomes

$$k_s = k_o \, (n_i - n_d). \qquad (2.31)$$

Such collinear forward interaction has been experimentally demonstrated and has become important for acousto-optic filter applications (Harris, 1969).

ii) When $\Theta_i - \Theta_d = 180°$, we have the condition for backward Brillouin scattering and Eq. (2.30) becomes

$$k_s = k_o \, (n_i + n_d). \qquad (2.32)$$

iii) When $\Theta_d = 0$, the diffracted light is propagating parallel to the direction of acoustic wavefronts. This direction gives a larger diffraction bandwidth for light deflection application (Lean, 1967). The detailed discussion of this special case will be in Section IV.

III. ACOUSTO-OPTIC INTERACTION IN GUIDED WAVE STRUCTURES - ANALYSIS

In this section, basic properties of acoustic surface waves and optical guided waves, which cause acousto-optic interactions in guided wave structures different from the bulk wave cases, will be briefly reviewed. The detailed analysis of acousto-optic interactions in guided wave structures will then be discussed with emphasis on their unique features.

3.1 Acoustic Surface Waves and Optical Guided Waves

Acoustic surface waves in solids, Rayleigh waves (Lord Rayleigh, 1885), Love waves (Love, 1911) and Stoneley waves (Stonely, 1966) have been well known for many years. There has been renewed interest in recent years in applying surface acoustic waves at microwave frequencies for signal processing (Stern, 1969) and nonlinear-interaction experiments (Lean, 1970). This interest arises from unique properties of surface acoustic waves, and due to the recent advance in photoresist and thin film technologies which are essential for fabricating surface acoustic wave devices at microwave frequencies.

Rayleigh waves, which are the most commonly used at the present time, are inhomogeneous waves propagating on the free surface of a solid substrate. The particle displacements are elliptically polarized and decay exponentially from the surface.

Consider a Rayleigh wave propagating along the x_3 axis on a surface whose normal is in the x_1 axis. The particle displacement of such a wave can be expressed by

$$V_i = B_i \sum_{n=1}^{3} a_{in} \exp\left[\alpha_n k_s x_1 + j\,(\omega_s t - k_s x_3)\right] \qquad i = 1, 2, 3 \tag{3.1}$$

where V_i are the particle displacement, k_s is the acoustic wave number, B_i is the peak amplitude, a_{in} and α_n are determined by the requirement that (3.1) be a solution of the wave equation in the medium subject to the stress free boundary condition. In isotropic substrates, solutions of the particle displacement for Rayleigh waves consist of only two components (Viktorov, 1967, White 1970) as shown in Fig. 3.1. These two components are 90° out of phase so that particle motions are ellipsoidal. The particle displacements decay exponentially from the surface with a penetration depth of the order of an acoustic wavelength. The energy of Rayleigh waves is concentrated within this penetration depth resulting in a very high power density. The acoustic energy is also accessible at any point along the propagation path so that Rayleigh waves can be easily influenced. The accessibility and the associated high power density render Rayleigh waves attractive both in device applications and as a research tool in thin film physics and surface physics.

The excitation of ASW on piezoelectric substrates such as $LiNbO_3$ or quartz, at lower frequencies is very simple (White, 1965); it is accomplished by means of interdigital metallic structures which can be fabricated

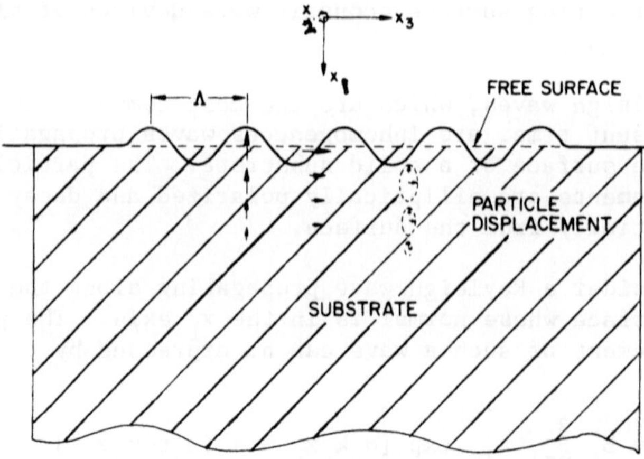

Fig. 3.1 Schematics for Rayleigh waves with elliptically polarized particle displacements propagating on a solid.

by standard photolithographic techniques. At frequencies higher than 1 GHz, the fabrication of interdigital transducers requires electron beam techniques (Lean, 1970).

Surface acoustic waves produce surface ripples as well as periodic changes in the index of refraction in the surface layer of about one wavelength in depth. They provide strong interaction with either a collimated light beam or a guided optical wave.

Guided optical waves can propagate in a thin film on top of a substrate; provided the thin film has a higher index of refraction than those of top and bottom media, i.e., $n_2 > n_1, n_3$, so that the light can be trapped in the film by total internal reflections and thus form guided optical modes. As shown in Fig. 3.2-a, a guided optical wave can be considered as a plane wave bouncing back and forth in the film due to total internal reflections from the interfaces. TE modes in a waveguide, as shown in Fig. 3.2, have field component E_2, H_1 and H_3 while TM modes have H_2, E_1 and E_3. The general forms for field components E_{m2} and H_{m2} of the m-th TE and TM modes can be respectively written as

$$\left.\begin{matrix} E_{m2} \\ \\ H_{m2} \end{matrix}\right\} = \frac{1}{2} A_{m2} U_{m,g} (P_g, x_1) \quad \exp [j (\omega t - \beta x_3)] + \text{C.C.} \qquad (3.2)$$

$$g = 1, 2, 3$$

where A_{m2} is the peak amplitude of the m-th mode; $U_{m,g}$ is the m-th mode transverse distribution function in the region g=1, 2, 3 (the air, film and substrate) respectively. P_g is the transverse wavevector in the g region, β is the propagation constant along the x_3-direction. The dispersion relation and all the characteristics of the propagating TE and TM modes can be obtained from the following eigenvalue equation (Nelson, 1967);

$$q_2 t_2 = \tan^{-1} \gamma_{21} \frac{p_1}{q_2} + \tan^{-1} \gamma_{23} \frac{p_3}{q_2} + m\pi \qquad (3.3)$$

where $\gamma_{gh} = 1$ for TE modes and $\gamma_{gh} = \varepsilon_g / \varepsilon_h$ for TM modes.

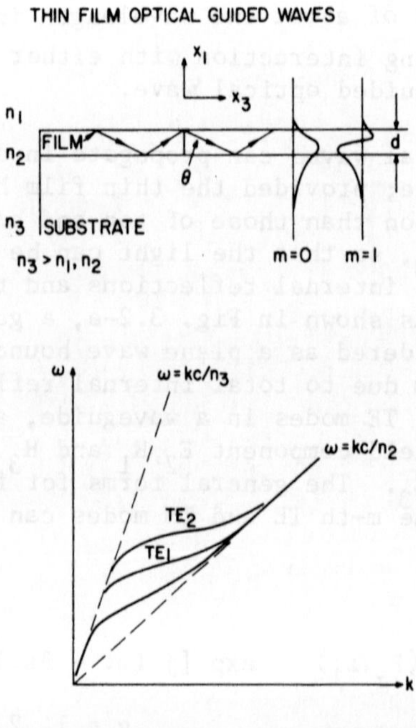

Fig. 3.2 (a) Thin film optical guided waves trapped by total internal reflection in a thin film with a thickness d. The conditions for propagating such optical guided waves are $n_2 > n_1$ and $n_2 > n_3$. The transverse field distribution for TE_o (m=0) and TE_1 (m=1) modes.

(b) The dispersion of optical guided waves for 3 TE modes.

ε_g = dielectric constant of medium g at vacuum optical wavelength λ_o.

$p_{1,3}$ = decay constant of evanescent tail in region.

$q_2 = p_2$ = transverse wavevector of guided mode in region 2.

m = mode order = 0, 1, 2, 3, . . .

p_1, q_2 and p_3 are given in terms of the guided wavevector β and the free-space wavevector $k_o = 2\pi/\lambda_o$:

$$p_1^2 = \beta^2 - n_1^2 k_o^2 \tag{3.4}$$

$$p_3^2 = \beta^2 - n_3^2 k_o^2$$

$$q_2^2 = n_2^2 k_o^2 - \beta^2$$

Eq. (3.4) is a transcendental equation in q_2, which can be solved for propagation constants of the guided modes.

The peak amplitude and the transverse distribution function can be obtained from the continuity of E and H fields at interfaces of the thin film. Inside the thin film, the field distribution is sinusoidal. Outside the film, fields decay exponentially with decaying constants of p_1 and p_3. These evanescent portions of the wave allow the coupling to and the interaction with other waves outside the film region. Fig. 3.2-a shows schematically the transverse field distribution of TE_o and TE_1 modes.

3.2 Collinear Interaction of Optical Guided Waves and Acoustic Surface Waves

In the presence of acoustic waves, either bulk waves or acoustic surface waves in the film, acousto-optic interaction occurs in guided wave structures with two important features. One is the transverse distribution of the amplitude in guided wave structures, and the other is the dispersion relation of the guided modes. Transverse distribution functions affect diffraction efficiencies.

The waveguide dispersion has the same effect as natural birefringence in crystals and provides the required condition for anisotropic Bragg diffraction even though the thin film and substrate materials are optically isotropic. In this section the analysis of acousto-optic interactions in guided wave structures will be discussed in detail.

As shown in Fig. 3.3 a guided (TE_m mode) optical wave is propagating along the x_3 axis with a field distribution E_i,

$$E_i = \frac{1}{2} A_i\ U_i(x_1) \exp\,[j(\omega_i t - k_i x_3)] + \text{C. C.} \qquad (3.5)$$

where A_i is the peak amplitude; $U_i(x_1)$ is the transverse distribution function; ω_i and k_i are the optical frequency and wavenumber of the TE_m mode. The subscript g for $U_{i,g}(x_1)$ as given in Eq. (3.2) is omitted in Eq. (3.5). An acoustic surface wave with strain components S_{q3} is propagating along the x_3 axis. The strain components S_{q3} are given by:

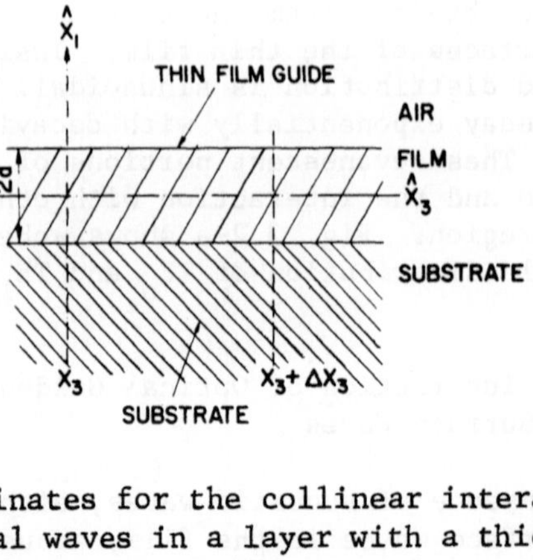

Fig. 3.3 Coordinates for the collinear interaction of optical waves in a layer with a thickness 2d on a substrate and acoustic surface waves.

$$S_{q3} = \frac{1}{2} B_{q3} V_{q3}(x_1) \exp [j (\omega_s t - k_s x_3)] + C.C. \quad (3.6)$$

$$q = 1, 2, 3$$

where B_{q3} is the peak amplitude and $V_{q3}(x_1)$ are transverse distribution functions in the film and in the substrate. Strain components are propagating along the x_3 axis with particle displacement directions indicated by q. ω_s and k_s are the acoustic frequency and wavenumber respectively. The change in the dielectric constant induced by the strain wave through the photoelastic effect has been discussed in Section 2.1. It can be shown from (2.8) that the induced dielectric constant is given by

$$\Delta\varepsilon_{ab} = - \varepsilon_{aa}^2 P_{abq3} S_{q3}, \quad a, b, q = 1, 2, 3 \quad (3.7)$$

where $\Delta\varepsilon_{ab}$ is the element of the change in the dielectric tensor $[\Delta\varepsilon]$, ε_{aa} and P_{abq3} are the dielectric and photoelastic constants in the film and in the substrate.

The interaction of the incident light $\vec{E}_i$ and $[\Delta\varepsilon]$ gives rise to an induced electric displacement $\vec{D}$ which has a component D_a given by

$$D_a = \Delta\varepsilon_{ab} E_{ib}$$

$$= - \frac{1}{4} \varepsilon_{aa}^2 P_{abq3} B_{q3} V_{q3} A_{ib} U_i \quad (3.8)$$

$$\times \{\exp j[(\omega_i + \omega_s)t - (k_i + k_s)x_3] + \exp j[\omega_i - \omega_s)t - (k_i - k_s)x_3]\}$$

where the index a and b run from 1 to 3. E_{ib} is the b component of $\vec{E}_i$ with an amplitude component A_{ib}.

The induced displacement current $\partial\vec{D}/\partial t$ can be considered as the driving source for the diffracted mode. Let the diffracted mode be a TE_n mode with m ≠ n with an electric field given by:

$$E_d = \frac{1}{2} A_d U_d(x_1) \exp [j(\omega_d t-k_d x_3)] + C. C. \qquad (3.9)$$

where A_d is the peak amplitude and $U_d(x_1)$ is the transverse distribution function for the TE_n mode. Assuming that the propagation losses are negligible, the power generated by $\partial D/\partial t$ in a volume should be equal to the power carried away by the mode from the same region. The power balance leads to the following equation:

$$\oint_a (\vec{E}_d^* \times \vec{H}_d)\, da = - \int_v \vec{E}_d^* \cdot \frac{\partial \vec{D}}{\partial t}\, dv \qquad (3.10)$$

where $\vec{E}_d$ and $\vec{H}_d$ are the electric and magnetic field associated with the generated TE_n mode. The surface "a" encloses the volume "v" in which the acousto-optic interaction takes place. In a waveguide mode, transverse field components of $\vec{E}_d$ and $\vec{H}_d$ are related by a wave impedance. For a TE mode, the wave impedance is given by

$\sqrt{\frac{\mu}{\varepsilon}}\, \frac{k}{k_d}$, where k is the wavenumber in the waveguide medium and k_d is the wavenumber for the TE_n mode. Kuhn (1969) has used the same argument to calculate the interaction of nonlinear optics in a finite geometry.

With the interaction region defined by two surfaces at x_3 and $x_3 + \Delta x$, for collinear interaction as shown in Fig. 3.3, the substitution of (3.8) and (3.9) into (3.10) results in the following equation:

$$\frac{1}{\Delta x_3} A_d^* A_d \Big]_{x_3}^{x_3+\Delta x_3} = \frac{j}{2}\sqrt{\frac{\mu}{\varepsilon}}\, \frac{k\omega_d}{k_d}\, \varepsilon_2^2\, P_{23}\, B_3\, A_i^*\, A_i\, \exp[(j\, \Delta k x_3)] \frac{\int U_d^* V_3 U_i\, dx_1}{\int U_d U_d^*\, dx_1} \qquad (3.11)$$

As a limit $\Delta x_3 \rightarrow 0$, we have

$$\frac{1}{\Delta x_3} A_d^* A_d \Bigg|_{x_3}^{x_3+\Delta x_3} = \frac{d\, A_d^* A_d}{dx_3}$$

$$= A_d^* \frac{d\, A_d}{dx_3} + A_d \frac{d\, A_d^*}{dx_3}$$

$$\sim 2A_d^* \frac{d\, A_d}{dx_3} \qquad (3.12)$$

In Eq. (3.12) it is assumed that A_d is a slow varying function with no rapidly changing phase. Eq. (3.11) then reduces to:

$$\frac{d\, A_d}{dx_3} = j\, \Gamma_{di}\, A_i \exp\,(j\, \Delta k_3 x_3) \qquad (3.13)$$

where

$$\Gamma_{di} = \frac{1}{4} \frac{v_d}{v} k\, \varepsilon_2\, P_{23}\, B_3 \frac{\int U_d^*\, V_3\, U_i\, dx_1}{\int U_d^*\, U_d\, dx_1} \qquad (3.14)$$

$$\Delta k_3 = k_i + k_s - k_d \qquad (3.15)$$

Eq. (3.13) is obtained under the condition that the output frequency is upshifted by ω_s, i.e.:

$$\omega_d = \omega_i + \omega_s. \qquad (3.16)$$

Eq. (3.13) describes the change in A_d as a function of x_3 due to the input A_i and the coupling constant Γ_{di}. The phase mismatch Δk is also included. If the input amplitude A_i is not depleted and can be assumed constant in the interaction region, the simple integration of (3.13) gives the result of A_d. However, in case A_i depends also on A_d, a similar derivation considering E_d as the incident wave while E_i as the diffracted wave leads to the following equation:

$$\frac{d\,A_i}{dx_3} = j\,\Gamma_{id}\,A_d \quad \exp\,(-j\,\Delta k_3 x_3) \tag{3.17}$$

where

$$\Gamma_{id} = \frac{1}{4}\,\frac{v_i}{v}\,k\,\varepsilon_2\,P_{23}\,B_3\,\frac{\int U_i^*\,V_3\,U_d\,dx_1}{\int U_i^*\,U_i\,dx_1} \tag{3.18}$$

The solution for the coupled amplitude equations (3.13) and (3.17) with the boundary conditions at $x_3 = 0$; $A_d = 0$, $A_i = A_i(0)$ can be readily obtained:

$$A_d(x_3) = j\,A_i(0)\,\frac{\Gamma_{di}}{T}\,\sin\,(Tx_3)\,\exp\,j\,\frac{\Delta k_3}{2}\,x_3) \tag{3.19}$$

where

$$T^2 = \Gamma_{di}\,\Gamma_{id} + \frac{\Delta k_3^2}{4}$$

The intensity of the diffracted mode at a distance $x_3 = L$ is given by:

$$\frac{I_d(L)}{I_i(0)} = \frac{|A_d(L)|^2}{|A_i(0)|^2} = (\Gamma_{di}L)^2\,\frac{\sin^2\,(TL)}{(TL)^2} \tag{3.20}$$

where

$$T^2 = \Gamma_{di}\,\Gamma_{id} + \frac{\Delta k_3^2}{4} \tag{3.21}$$

$$\Gamma_{di} = \frac{1}{4}\,\frac{v_d}{v}\,k\,\varepsilon_2\,P_{23}\,B_3\,F_{d3i};\quad F_{d3i} = \frac{\int U_d^*\,V_3\,U_i\,dx}{\int U_d^*\,U_d\,dx_1} \tag{3.22}$$

$$\Gamma_{id} = \frac{1}{4}\,\frac{v_i}{v}\,k\,\varepsilon_2\,P_{23}\,B_3\,F_{i3d};\quad F_{i3d} = \frac{\int U_i^*\,V_3\,U_d\,dx_1}{\int U_i\,U_i\,dx_1} \tag{3.23}$$

and $I_i(0)$ is the incident intensity of the input TE_m mode. At the phase matching condition where $\Delta k_3 = 0$, the diffraction efficiency of (3.21) becomes.

$$\frac{I_d(L)}{I_i(0)} = \left(\frac{\Gamma_{di}}{\Gamma_{id}}\right) \sin^2\left[\frac{\pi L}{\lambda_o}\sqrt{\frac{M}{2} P_{ac} \frac{v_d\ v_i}{v^2} F_{d3i}\ F_{i3d}}\right] \tag{3.24}$$

where

$$M = \frac{n^6\ p^2}{\rho\ v_3^3} \tag{3.25}$$

Eq. (3.24) reduces to the familiar equation of (2.32) in bulk wave cases where $v_d = v_i = v$ and $F_{d3i} = F_{i3d} = 1$.

Two unique features should be noted in the calculation for acousto-optic interactions in guided wave structures; one is the overlap integrals as shown in Eq. (3.22), (3.23) and the other is the different phase velocity v_i for each mode. Depending on transverse field distributions, $U_d^*(x_1)$, $U_i(x_1)$ and $V_{3q}(x_1)$, the overlap integrals, F_{i3d} and F_{d31}, have values from zero to unity. In cases of bulk acousto-optic interactions, transverse distribution functions become plane waves so that the values of overlap integrals reduce to unity. The velocity of each guided mode, which depends on the waveguide thickness and the refractive indices of the media, can be obtained from Eq. (3.3). With a waveguide which can propagate at least two modes, it is possible to have anisotropic interactions in this film. For example, let the phase velocity be v_o and v_1. The phase matching condition of Eq. (3.15) for a collinear acousto-optic interaction can be satisfied if the center frequency f_o of the acoustic wave is given by

$$f_o = \frac{v_s}{\lambda_o}\left(\frac{v_1}{v} - \frac{v_o}{v}\right) \tag{3.26}$$

3.3 Bragg Diffraction of Optical Guided Waves by Acoustic Surface Waves

The calculation of Bragg diffraction of optical guided waves by acoustic surface waves can be done using a similar approach as discussed in Section 3.2, and has been carried out elsewhere (Lean, 1973). For completeness, we shall repeat some of the derivations. Consider the interaction between two optical guided waves and one acoustic wave as shown in Fig. 2.5. Assume the input light having an electric field of a TE_m mode E_i:

$$E_i = \frac{1}{2} (\vec{x}_2 \sin \Theta_i - \vec{x}_3 \cos\Theta_i)\ A_i\ U_i(x_1)$$

$$\exp\ [j(\omega_i t - k_i \cos\Theta_i x_2 - k_i \sin\Theta_i x_3) + \text{C. C.} \qquad (3.27)$$

where $\vec{x}_2$ and $\vec{x}_3$ are unit vectors; A_i is the peak amplitude; $U_i(x_1)$ is the transverse distribution function for the TE_m mode; ω_i and k_i are the optical frequency and wavenumber of the mode.

Assuming that both the incident mode and the diffracted mode are TE modes, the incident mode has an electric field as given by Eq. (3.27) while the diffracted mode has an electric field of:

$$E_d = \frac{1}{2} (\vec{x}_2 \sin\Theta_d + \vec{x}_3 \cos\Theta_d)\ A_d\ U_d(x_1)$$

$$\exp\ [j(\omega_d t - k_d \cos\Theta_d x_2 + k_d \sin\Theta_d x_3)] + \text{C. C.} \quad (3.28)$$

where Θ_d is the diffracted angle with respect to the x_2 axis. The diffracted mode can be either the same mode as the incident wave (m = n), or a different mode (m ≠ n).

The acoustic surface wave with strain component S_{q3} propagates along the x_3 axis. The strain component S_{q3} is given by Eq. (3.7).

Following a similar calculation as shown in Section 3.2, it can be shown that the coupled amplitude equations for the acousto-optic Bragg diffraction in thin film are given by:

$$\frac{d\ A_d}{dx_2} = -\ j\ \Gamma_{di}\ A_i\ \cos\Theta_d\ \exp\ [j\ (\Delta k_3 x_3 + \Delta k_2 x_2)]\ (3.29)$$

$$\frac{d A_i}{dx_2} = - j \Gamma_{id} A_d \cos\Theta_i \exp [-j(\Delta k_3 x_3 + \Delta k_2 x_2)] \tag{3.30}$$

where

$$\Gamma_{di} = \frac{v_n k}{4v} \frac{(\sin\Theta_i \sin\Theta_d \varepsilon_{22} I_{22} + \cos\Theta_i \cos\Theta_d \varepsilon_{33} I_{33})}{\int_o^\infty U_d U_d^* dx_1} \tag{3.31}$$

$$\Gamma_{id} = \frac{v_m k}{4v} \frac{(\sin\Theta_i \sin\Theta_d \varepsilon_{22} I_{22} + \cos\Theta_i \cos\Theta_d \varepsilon_{33} I_{33})}{\int_o^\infty U_i U_i^* dx_1} \tag{3.32a}$$

$$I_{22} = \int_o^\infty P_{22q3} B_{q3} V_{q3} U_i U_i^* dx_1 \tag{3.32b}$$

$$I_{33} = \int_o^\infty P_{33q3} B_{q3} V_{q3} U_i U_d^* dx_1 \tag{3.32c}$$

$$\Delta k_2 = k_d \cos\Theta_d - k_i \cos\Theta_i \tag{3.33a}$$

$$\Delta k_3 = k_d \sin\Theta_d + k_i \sin\Theta_i - k_s \tag{3.33b}$$

The solution of the coupled amplitude equations (3.29) and (3.30) with boundary conditions that at $x_2 = 0$,

$$A_d = 0 \tag{3.34}$$

and

$$A_i = A_i(0) \tag{3.35}$$

can be readily obtained:

$$A_i(x_2) = \frac{-j A_i(0)\Gamma_{ki} \cos\Theta_d}{T} \sin (Tx_2) \tag{3.36}$$

$$\exp [j(\frac{1}{2} \Delta k_2 x_2 + \Delta k_3 x_3)]$$

where

$$T^2 = |\Gamma_{di}\Gamma_{id} (\cos\Theta_i \cos\Theta_d)| + \frac{1}{4}\Delta k_2^2 . \qquad (3.37)$$

The intensity of the diffracted light, which is proportional to $|A_d|^2$, at a distance $x_2 = L$ is given by:

$$\frac{I_d(L)}{I_i(0)} = |\Gamma_{di} \cos\Theta_d|^2 L^2 \frac{\sin^2(TL)}{(TL)^2} \qquad (3.38)$$

when $I_i(0)$ is the incident intensity of the input TE_m mode.

A similar solution of the interaction between two TM modes or between one TE and one TM mode by the acoustic surface wave can also be obtained. The conversion between one TE mode and one TM mode requires proper photoelastic constants which rotate the polarization of the incident mode to that of the diffracted mode. The results obtained in this section are valid in cases of small incident and diffracted angles so that only one direction dependence of the optical beams in x_2 direction is important. A more rigorous numerical calculation on the guided wave theory of light diffraction by acoustic microwaves has been reported by Chu and Tamir (1969).

If the medium is isotropic where $\varepsilon_{22} = \varepsilon_{33} = \varepsilon$, $\Theta_i = \Theta_d = \Theta$, $P_{22q3} = P_{33q3} = P_{33q3} = P$, and $v_m = v_n = v$, we have

$$\Gamma_{di} = \Gamma_{id} = \Gamma = \frac{k}{4} \varepsilon p B F \qquad (3.39)$$

where

$$F = \frac{\int_0^\infty U V U^2 dx_1}{\int_0^\infty |U| dx_1}$$

and the maximum diffraction efficiency as given in (3.38) at $\Delta k = 0$ can be reduced to:

$$\frac{I_d^{(d)}}{I_i(0)} = \sin^2 \left[\frac{\pi \cos\Theta}{\lambda_o} L\sqrt{\frac{M}{2} P_{ac} F^2}\right] \qquad (3.40)$$

where M is given by Eq. (3.25); P_{ac} is the acoustic power density, and is given by (2.28). It is noted that the Bragg diffraction efficiency for thin film cases as given by Eq. (3.40) is similar to Eq. (3.25) for the collinear interaction case and is different from the bulk wave case only by the overlap integral F.

IV. ACOUSTO-OPTIC INTERACTIONS IN GUIDED WAVE STRUCTURES EXPERIMENTAL RESULTS AND DISCUSSIONS

Interactions of optical guided waves in thin films and acoustic surface waves have been demonstrated. One is the mode conversion between two collinear TE modes or two TM modes by a collinear acoustic surface wave (Kuhn et al, 1971); one is the Bragg diffraction of a guided mode by an acoustic surface wave (Kuhn et al, 1970). Theoretical calculations of the collinear interaction use, which agree very well with measured values, have been reported (Lean, 1973). In this section, some of the experimental and theoretical results in the collinear interaction case will be repeated. In addition, calculations based on the results in Section 3.3 will be used to verify the experimental results of Bragg diffraction of optical guided waves by acoustic surface waves in a glass film on quartz (Kuhn, 1971) and in a ZnO film (Chubachi, 1973). Another experiment, which demonstrates the possibility of the anisotropic diffraction of a collimated light beam by acoustic surface waves on the substrate surface, will be included.

4.1 Collinear Interaction

The experimental setup for the collinear interaction is shown in Fig. 4.1a. Optical guided modes were propagating in a glass film on an aluminum coated lithium niobate ($LiNbO_3$) substrate (film thickness = 1.6 μm). Grating couplers (Dakss et al, 1970) were used to couple both TE_1 and TE_3 modes of a 6328 Å laser beam into and out of the thin film guide. An acoustic surface wave, launched on the substrate by interdigital transducers propagated across the surface forming a strain wave in the film. When phase matching conditions of Eq. (2.27) were satisfied as indicated in Fig. 4.1b, coupling was observed from one waveguide mode to the other. Fig. 4.2 shows the experimental results of the intensity depletion of the input TE_3 mode as the

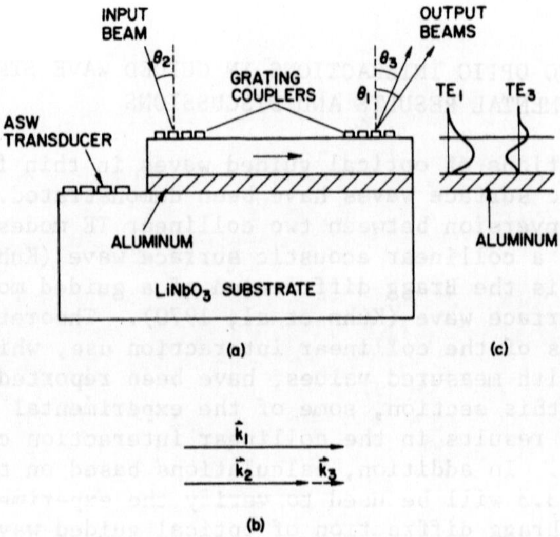

Fig. 4.1 (a) Experimental setup for the collinear acousto-optic interaction in a glass film on an aluminum coated $LiNbO_3$ substrate (reported by Kuhn, 1971).

(b) Phase matching condition for the collinear interaction.

(c) Field distributions of TE_1 and TE_3 in the glass film on top of the aluminum film.

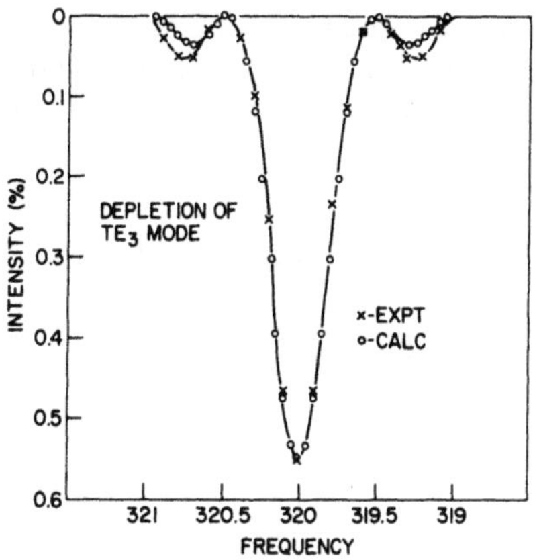

Fig. 4.2 Experimental and calculated results of the depletion of the TE_3 mode intensity as a function of acoustic frequency.

acoustic frequency was swept through the phase matching condition. The depletion of the TE_3 mode was about 55% at the center frequency of 320 MHz for an acoustic surface wave with a surface displacement of 7.5 Å. The surface displacement was measured by the optical probing technique described by Lean (1970). The augmentation of the output TE_1 mode which had a complementary shape of the experimental curve shown in Fig. 4.2 could also be observed.

To check the experimental results by the theoretical calculations in Section 3.2, the spatial dependence of the TE_1 and TE_3 modes, $U_1(x_1)$ and $U_3(x_1)$ as shown in Fig. 4.1c can be assumed as

$$U_m(x_1) = \begin{cases} \sin h_m d \exp\left[-p_m(x_1-d)\right] & x_1 \geq d \\ \sin h_m x_1 & 0 \leq x_1 < d \\ 0 & x_1 \leq 0 \end{cases}$$

$$m = 1,\ 3 \qquad (4.1)$$

where d is the thickness of the glass film. The parameters h_m and p_m are determined by the dispersion relation of the odd TE modes in a dielectric slab with a slab thickness being 2d (Collins 1965); i.e.,

$$p_m d = -\, h_m d \cot h_m d \qquad (4.2)$$

$$(p_m d)^2 + (h_m d)^2 = (n^2-1)\,(k_o d)^2 \qquad (4.3)$$

$$k_m^2 = k_o^2 + p_m^2 = n^2 k_o^2 - h_m^2, \quad m = 1,\ 3 \qquad (4.4)$$

where n is the index of refraction of the film and k_m is the mode propagation constant along the z-direction.

The acoustic wave can be assumed as having a constant amplitude over the film thickness because the film thickness is only a small fraction of the acoustic wavelength, i.e.,

$$V_{q3}(x_1) = \begin{cases} 0 & x_1 \geq d \\ 1 & 0 \leq x_1 > d \end{cases} \qquad (4.5)$$

Substituting (4.1) and (4.5) into (3.21) and using the measured values of the film thickness (d = 1.6 μm), the index of refraction (n = 1.57), the propagation constants for TE_1 and TE_3 modes ($k_1 = 1.56 \times 10^5$ cm^{-1}, $k_3 = 1.51 \times 10^5$ cm^{-1}) and the peak conversion efficiency, we have calculated the conversion efficiency as a function of acoustic frequency as shown in Fig. 4.2 together with the experimental values. The best fit was obtained with an interaction length L = 0.62 cm. The physical separation between input and output grating couplers of the sample was about 0.8 cm. Probably due to the glass film nonuniformity, not the total available interaction length was used in the interaction.

In order to match the peak conversion efficiency of 0.55, the product of the photoelastic constant P and the strain B in the film was calculated to be about 1.62×10^{-5}. Using the measured strain value $B = 4 \times 10^{-4}$, we have the photoelastic constant calculated to be P = 0.138 which is a reasonable value for a glass film. Such collinear acousto-optic interactions can be used to couple optical guided modes into radiation modes for electronically switchable input/output couplers. The calculation for such interaction has been carried out by Chang (1971). The experimental demonstration of the acousto-optic input/output coupler has been reported by Gfeller (1972).

4.2 Bragg Deflection of Optical Guided Waves by Acoustic Surface Waves

The experimental demonstration of Bragg deflection of optical guided waves by acoustic surface waves was reported by (Kuhn et al, 1970). The experimental configuration in their paper is shown in Fig. 4.3. Guided optical waves were propagating in a glass film which was sputtered on a y-cut α-quartz substrate (n=1.54). The glass waveguide had an index of refraction n=1.73 and thickness d=0.8 μm. A He Ne laser (λ_o = 0.63 μm) was coupled into the glass film by a grating coupler (Dakss et al (1970). Depending on the polarization, TE or TM modes can be excited in the film and propagated approximately parallel to the z-axis with a propagation constant k_3. A second grating coupler was used to couple the deflected and undeflected light beams out of the glass film.

The acoustic surface wave at f = 191 MHz was excited along the x-axis of quartz substrate by interdigital transducers which had a periodicity (acoustic wavelength) Λ = 16 μm, an aperture width W = 0.0889 cm and ten finger pairs. The amplitude and velocity of acoustic surface waves can be measured directly by an optical probe (Lean, 1970). In this technique, a laser beam reflected from the surface is diffracted by surface ripples associated with the acoustic surface wave. The probe efficiency, which is the ratio of the first order diffraction beam to that of the direct reflected beam, is a measure of normal components of the surface displacement. The strain amplitude B can then be determined.

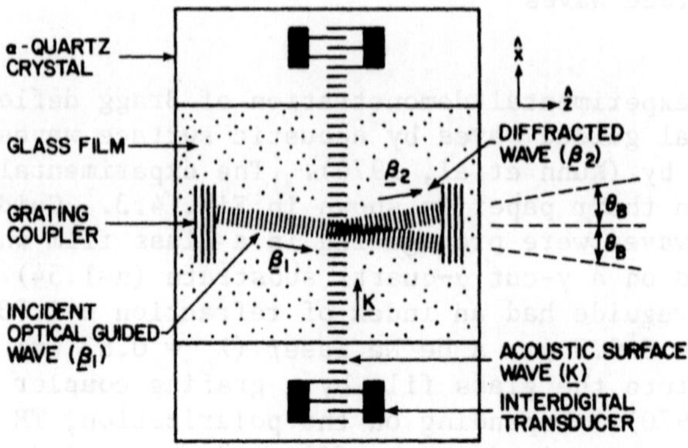

4.3 Experimental setup of Bragg deflection of optical guided waves by acoustic surface waves in a glass film on quartz (reported by Kuhn, 1970).

The diffraction efficiency was maximum when the incident angle was the Bragg angle. (Θ_B = arc sin ($\frac{1}{2}$ λ_g/Λ), where $\lambda_g = 2\pi/k_3$ is the wavelength of the guided mode). Using the experimental value of k_3 (= 1.67×10^5 cm^{-1}) and Λ = 10 μm, the Bragg angle was calculated to be 0.66°, while the experimentally observed value was 0.68°. Both experimental and theoretical results of the diffraction efficiency as a function of the amplitude of acoustic strain B is shown in Fig. 4.4. The calculated curve in Fig. 4.4 is based on Eq. (3.40) with F=1 (for isotropic diffraction and assuming V = 1), which can be reduced to

$$\frac{I_d}{I_i} = \sin^2 (\alpha B) \tag{4.6}$$

and

$$\alpha = \frac{1}{4} n^2 p \beta w \cos\Theta \tag{4.7}$$

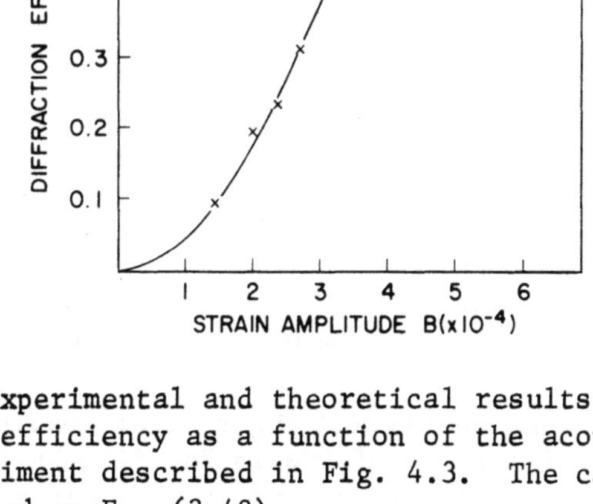

Fig. 4.4 - Experimental and theoretical results of the diffraction efficiency as a function of the acoustic strain in the experiment described in Fig. 4.3. The calculated curve is based on Eq. (3.40).

where p is the photoelastic constant of the glass film. The experimental points in Fig. 4.4 fit well to the calculated curve from (4.6) with $\alpha = 2.205\text{x}10^{3}$, which gives $p = 0.198$ from Eq. (4.7). The photoelastic constant is consistent with various measured values of glasses (Pinnow, 1970).

Recently a ZnO thin film acousto-optic deflector was demonstrated by N. Chubachi et al (1973). A ZnO film which was DC sputtered on a fused quartz substrate was used both to support the propagation of guided optical waves and to excite acoustic surface waves. More than 90% diffraction efficiency was observed in their experiment. Fig. 4.5 shows the experimental results reported by Chubachi together with the calculation based on Eq. (3.40) which can be written as

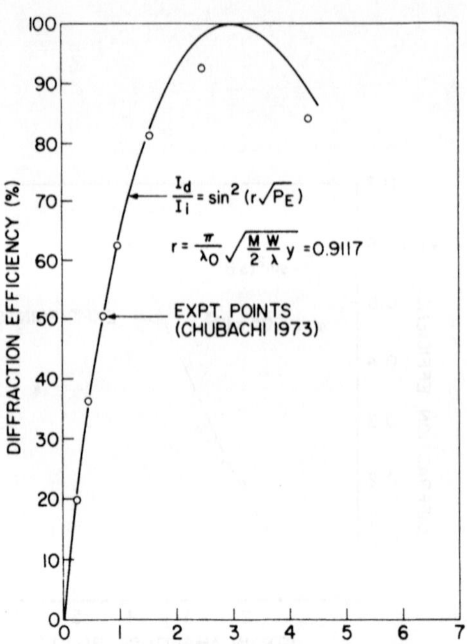

Fig. 4.5 - Experimental and theoretical results of Bragg diffraction efficiency of OGW in ZnO film by ASW as a function of acoustic frequency. Expt. points reported by Chubachi (1973) and the calculation is based on Eq. (4.8).

$$\frac{I_d}{I_i} = \sin^2 (\gamma \sqrt{P_E}) \qquad (4.8)$$

where

$$\gamma = \frac{\pi}{\lambda_o} \sqrt{\frac{M}{2} \frac{W}{\Lambda} \zeta}$$

and M is the M-factor (3.41a); W is the acoustic beam width, ζ is the conversion efficiency of the interdigital transducer. For the best fit between the experimental points and Eq. (4.8), γ is found to be 0.9117. Unfortunately W is unknown from Chubachi's paper. The value of M cannot be calculated. However, the agreement between the theory and experiment for the diffraction efficiency below 90% as shown in Fig. 4.5 is very good.

4.3 Anisotropic Light Diffraction by Acoustic Surface Waves

Light diffraction by acoustic waves in optically anisotropic crystals have been reported (Dixon, 1967) and its applications utilizing the 90° polarization rotation of the diffracted beam with respect to the incident beam polarization in high resolution deflection (Lean, 1967) and acousto-optic filters (Harris, 1969) have been demonstrated The anisotropic light diffraction requires an acoustic wave with a proper wavevector to match the difference of wavenumbers of the incident and the diffracted light due to the optical birefringence in the crystal. The anisotropic diffraction has also been reported in the thin guided optical wave structures (Kuhn, 1971). In thin film configurations, the required dispersion between the input and diffracted modes comes from the waveguide dispersion due to the film thickness instead of the crystal birefringence. In this section, experimental results of the anisotropic interaction of light and acoustic surface waves in yx quartz substrates will be reported.

The experimental configuration is shown schematically in Fig. 4.6. Acoustic surface waves with a wavelength Λ are propagating along the x_1-axis with strain components S_{q1}.

An incident light beam polarized along the x_3-axis

with a diameter D is incident from inside the substrate on the substrate surface with an incident angle Θ_i. If the incident angle Θ_i is greater than the critical angle, the light will be totally reflected from the substrate surface. The field of the incident beam is given by

$$E_i = A_i U_i(x_2) \exp [j(\omega_i t - k_i \sin\Theta_i x_i)] + C.C. \quad (4.9)$$

with

$$U_i(x_2) = \cos (k_i \cos\Theta_i x_2)$$

where A_i is the peak amplitude and $U_i(x_2)$ is the transverse field distribution.

Let the diffracted light have a field component E_d given by

$$E_d = A_d U_d(x_2) \exp [j(\omega_d t - k_d \sin\Theta_d x_1)] + C.C. \quad (4.10)$$

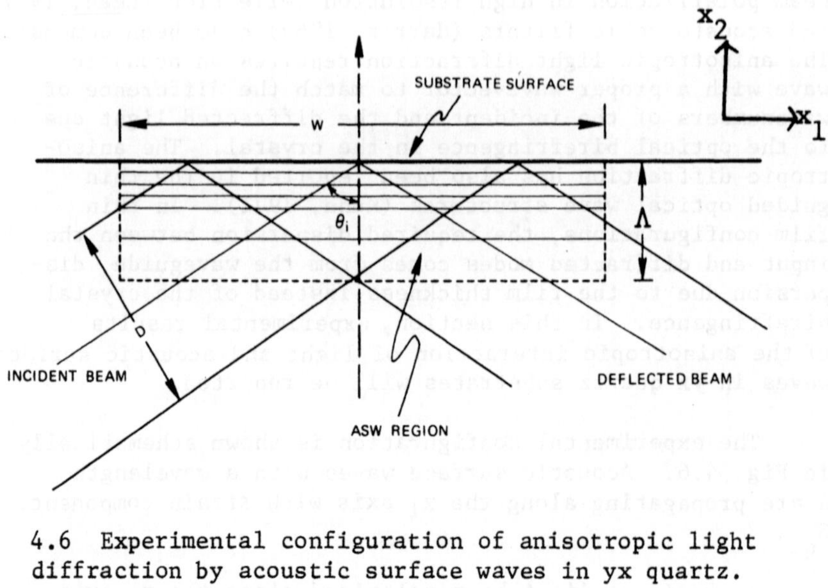

4.6 Experimental configuration of anisotropic light diffraction by acoustic surface waves in yx quartz.

and

$$U_d(x_2) = \cos (k_d \cos\Theta_d x_2)$$

where A_d is the peak amplitude of E_d.

In an anisotropic diffraction, the diffracted light must have its polarization rotated 90° with respect to the incident beam. It requires proper photoelastic constants and strain components. For the case having the acoustic surface wave propagating along the x_1-axis on a x_2-cut quartz substrate and the incident light propagating in the x_1 x_2 plane with the polarization along the x_3-axis, the proper photoelastic constant is P_{41} and the required strain component is S_{11}.

With the incident and diffracted fields E_i, E_d, and the strain component S_{11}, it can be shown that the diffraction efficiency is given by

$$\frac{I_d(L)}{I_i(0)} = |\Gamma_{di}|^2 L^2 \frac{\sin^2 TL}{(TL)^2} \tag{4.11}$$

with

$$T^2 = |\Gamma_{di} \Gamma_{id}| + \frac{1}{4} \Delta k_1^2 \tag{4.12}$$

$$\Gamma_{di} = \frac{1}{4} k_d \varepsilon P_{41} B_1 \frac{\int_o^\infty V_1 U_i U_d^* dx_2}{\int_o^\infty U_d U_d^* d_x^2} \tag{4.13}$$

$$\Gamma_{id} = \frac{1}{4} k_i \varepsilon P_{41} B_1 \frac{\int_o^\infty V_1 U_i U_d^* dx_L}{\int_o^\infty U_i U_i^* dx_2} \tag{4.14}$$

$$\Delta k_1 = k_d \sin\Theta_d - k_i \sin\Theta_i - k_s \tag{4.15}$$

where L is the interaction length and is determined by the diameter of the incident beam D, and the incident angle Θ_i, i.e., $L = D/\cos\Theta_i$.

The maximum efficiency of the diffracted light occurs at the phase matching condition where $\Delta k_1 = 0$, i.e.,

$$k_d \sin\Theta_d - k_i \sin\Theta_i = k_s . \tag{4.16}$$

Depending on the optical birefringence of the crystal and the incident angle Θ_i, Eq. (4.16) describes the relation for the acoustic k_s and the diffracted angle Θ_d. For example in a uniaxial crystal and with the incident beam in the plane normal to the c-axis, Eq. (4.16) reduces to

$$f_o = \frac{v}{\lambda_o} (n_e(\Theta_d) - n_o \sin\Theta_i) \tag{4.17}$$

where f_o is the center acoustic frequency, v is the acoustic velocity, n_e and n_o are the index of refraction for the uniaxial crystal.

If the incident beam is in the plane parallel to the c-axis, Eq. (4.16) can be reduced to

$$f_o = \frac{v}{\lambda_o} [n_e(\Theta_d) \sin\Theta_d - n_o \sin\Theta_i] \tag{4.18}$$

where $n_e(\Theta_d)$ is the index of refraction for the extra-ordinary wave seen by the light beam with a propagation angle $90-\Theta_d$ with respect to the c-axis. It can be shown in a uniaxial crystal that

$$n_e(\Theta_d) - n_o = \Delta n \cos^2\Theta_d \tag{4.19}$$

where $\Delta n = n_e - n_o$ and Eq. (4.18) becomes

$$f_o = \frac{v}{\lambda_o} [\Delta n \cos^2\Theta_d \sin\Theta_d + n_o (\sin\Theta_d - \sin\Theta_i)] \tag{4.20}$$

At the phase matching condition, the diffraction efficiency given by (4.11) can be reduced to Eq. (3.24) with $v_i = v_d = v$.

Experimentally we have shown the anisotropic light

diffraction by acoustic surface waves in a y-cut quartz crystal. A ten finger pair interdigital transducer was fabricated on the substrate surface to excite acoustic surface waves with a center frequency $f_o = 60.5$ MHz along the x-axis. The transducer has a 10% bandwidth. An argon laser was incident from the end of the quartz substrate with an incident angle Θ_i with respect to the substrate normal. The light propagated coplanarly with the acoustic surface waves. A polarizer and an analyzer were placed in front and after the crystal. The polarizer and analyzer were crossed so that the light was blocked by the analyzer. With the acoustic power on, the diffracted light with polarization rotated 90° could pass through the analyzer and was observed. As the acoustic frequency varied from 58 MHz to 64 MHz, two argon laser lines at $\lambda_o = 4765$ Å and $\lambda_o = 4880$ Å can be resolved as shown in Fig. 4.7. The center frequency for each line checks closely with the calculated value from Eq. (4.17) with $v = 3.26 \times 10^5$ cm/sec, $\Delta n = 0.0091$ and $\cos\Theta_i \cong 1$ because of the small Θ_i.

The observed diffraction efficiency with two watt input electric power was about one percent. With $\cos\Theta_i \cong 10^{-2}$, $W = 50\ \Lambda$, $M = 151 \times 10^{-15}$ (MKS) and $P_{ac} \cong 10$ mw (the transducer conversion loss was measured to be about 20 db loss). The calculated value of the diffraction efficiency from Eq. (4.11) is about 4 percent.

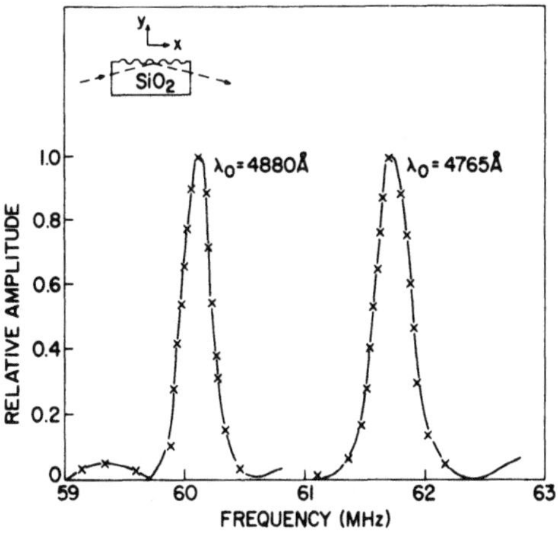

4.7 Experimental results of anisotropic diffraction of two argon laser lines at 4765 Å and 4880 Å as a function of acoustic surface wave frequency in yx SiO_2 substrate.

V. DEVICE PARAMETERS FOR ACOUSTO-OPTIC DEVICES

Acousto-optic devices have been shown to be important for laser beam control in such applications as beam deflection, modulation, filtering and optical signal processing. The essential parameters of acousto-optic devices are bandwidth, efficiency and speed. These device parameters for bulk acousto-optic devices as well as the material consideration have been reported (Gordon, 1967; Lean, 1969). In this section, similar device parameters for thin film acousto-optic devices will be discussed together with comparisons between bulk and thin film devices.

5.1 Efficiency

The diffraction efficiency is defined as the ratio of diffracted light intensity to that of the incident light. In isotropic media, the diffraction efficiency by acoustic surface waves is given by Eq. (3.40), i.e.

$$\frac{I_d(L)}{I_i(0)} = \sin^2 \xi \tag{3.40}$$

where

$$\xi^2 = \left(\frac{\pi \cos\Theta\, W}{\lambda_o}\right)^2 \frac{M}{2} P_{ac} F^2$$

and F is given by (3.39). If the acoustic power density can be approximated by

$$P_{ac} = \frac{P_T}{WH} \tag{5.1}$$

where P_T is the total power, W is the acoustic beam width and H is the height of acoustic beam, the diffraction efficiency becomes:

$$\frac{I_d(L)}{I_i(0)} = \sin^2 \left[\frac{\pi \cos\Theta}{\lambda_o} \sqrt{\frac{M}{2} P_T \frac{W}{H} F^2} \right] \tag{5.2}$$

A similar equation can be found for anisotropic Bragg diffraction cases. In a collinear interaction case, the diffraction efficiency is given by Eq. (3.24).

The overlap integral F in Eq. (5.2) is a unique feature for guided wave structures. Depending on transverse field distributions, F has a value between 0 to one. In addition, the diffraction efficiency as shown in (5.2) depends also on the M-factor, which is given by (3.25) and the aspect ratio W/H of the acoustic beam width to its height. In bulk wave cases where W/H typically is around 1 to 10, materials with a large M factor have to be used to increase the efficiency with a minimum acoustic power P_T. In guided wave structures, hwoever, because of the large aspect ratio (typically $W \sim 1$ cm and $H \sim 10$ μm and $W/H \sim 10^3$), the diffraction efficiency can be very high even if M is small. For example, if F = 1, it needs an acoustic power of

$$P_T = \frac{\lambda_o^2}{2M} \frac{H}{W} \tag{5.3}$$

for a theoretical 100 percent diffraction efficiency. Fig. (5.1) shows the required M-factor as a function of the aspect ratio W/H for several commonly used materials under the conditions that $P_T = 0.1$ watt and $\lambda_o = 0.63$μm. The new material TeO_2, which has a M-factor 650 larger than that of used quartz has been shown recently to be an excellent bulk acousto-optic deflector material (Warner, 1972). It takes only an aspect ratio of 4 to have 100% diffraction with 0.1 watt input acoustic power. Fused quartz, which has M = 1, can also be a good deflector material in thin film form (Kuhn, 1970). Also shown in Fig. 5.1, there are materials such as α-HIO_3, GaP and $PbMoO_4$, which can be made as in thin plates with W/H = 10 to 100. The thin plate acousto-optic devices in some particular applications can be very useful because of their good crystal quality and less problems in fabricating thin plates.

5.2 Bandwidths and Number of Resolvable Spots

As discussed in Section III, acousto-optic interactions in guided wave structures is a parametric interaction process in which phase matching is a basic requirement. The phase matching condition is given by Eq. (2.27), i.e.

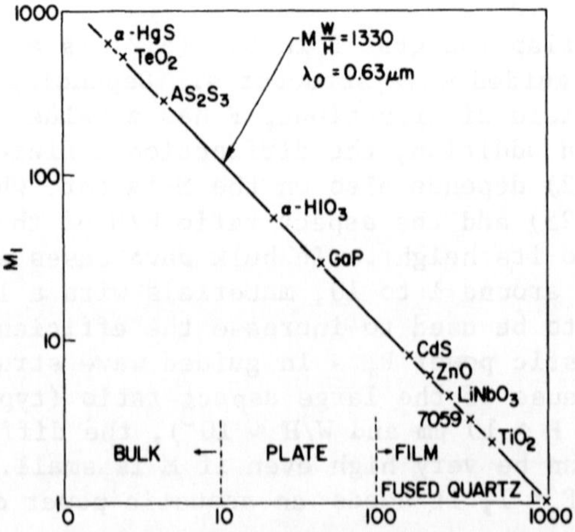

Fig. 5.1 The required M-value for several commonly used materials as a function of the aspect ratio W/H to obtain 100% diffraction efficiency using P_T = 0.1 watt at λ_o = 0.63 μm.

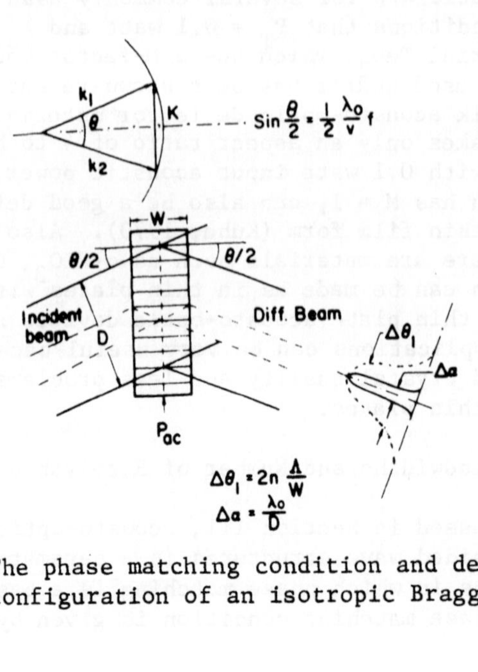

Fig. 5.2 The phase matching condition and device configuration of an isotropic Bragg deflector.

$$\vec{k}_d = \vec{k}_i \pm \vec{k}_s \tag{2.27}$$

If all three waves, the incident, the diffracted and the acoustic wave, are infinite plane waves so that $\vec{k}_i$, $\vec{k}_d$ and $\vec{k}_s$ are precisely defined in their respective directions, only one acoustic frequency as shown in Fig. 2.6 can satisfy the phase matching condition and diffracts the light. If the acoustic beam and the incident light have finite widths, there are uncertainties in the direction of $\vec{k}_i$ and $\vec{k}_s$. The diffracted light intensity which is a sinx/x function of $\vec{\Delta k}$ as given in (3.38) will not be zero even if there is a phase mismatch. Due to these finite beam widths which are usually the case in experimental and practical devices, finite bandwidths of these devices result. The bandwidths of isotropic and anisotropic diffraction as well as their resolutions and speeds in bulk acousto-optic devices have been discussed (Lean, 1969). Similar device parameters can be obtained for thin film devices.

5.2.1 Isotropic Diffraction

As shown in Fig. 5.2 , an incident light beam with a diameter D is incident on an acoustic column having a width W. The incident angle $\Theta/2$ is measured with respect to the acoustic wavefronts. The diffracted light which comes out with the same $\Theta/2$ angle has a maximum intensity under the Bragg condition, i.e.

$$\sin\frac{\Theta}{2} = \frac{1}{2}\lambda\frac{f}{v} \tag{5.4}$$

where $\lambda = \lambda_o/n$ is the optical wavelength of the incident light in the medium with an index of refraction n; v and f are acoustic velocity and frequency. Due to the finite acoustic beam width W, it is possible to deflect a collimated laser beam entering a Bragg cell at the Bragg angle over an angular bandwidth (measured outside the Bragg cell)

$$\Delta\Theta_1 = 2n\Lambda/W \tag{5.5}$$

by correspondingly adjusting the acoustic frequency over

a bandwidth

$$\Delta f_1 = 2n(v/\lambda_o)\ \Lambda/W. \tag{5.6}$$

For both equations (5.5) and (5.6), the bandwidth is measured between the -4 db points of the intensity distribution.

The bandwidth for the isotropic diffraction can be understood easily using simple diffraction theory (see Fig. 5.1). Each spatial period of the acoustic wave diffracts part of the incident light. An acoustic wavefront with a width W can be thought of as a "reflector" with an effective aperture,

$$d = W \sin \frac{1}{2}\,\Theta \tag{5.7}$$

e.g., that between the dotted lines in Fig. 5.2 intercepted by one such wavefront. Reflection from a single wavefront would produce a far-field diffraction pattern containing a central maximum of angular bandwidth

$$\Delta\Theta_1 = n\lambda/d = 2n\Lambda/W. \tag{5.8}$$

The incident light beam (of diameter D) covers many acoustic wave periods and the light waves diffracted from different periods interfere with each other. This interference leads to a series of diffraction maxima having a width

$$\Delta\alpha = \lambda_o/D \tag{5.9}$$

separated by the angle of λ_o/Λ. From Eq. (5.8) this interference spacing is much larger than $\Delta\Theta_1$ so that only one diffracted spot lies within the central peak of the diffraction envelope for a single acoustic wavefront. At the acoustic frequency for which the Bragg condition holds, the deflected beam is centered in this peak and has a maximum intensity. As the frequency is varied the peak is scanned within the envelope. The number N of resolvable spots, a primary parameter of acousto-optic devices used in deflection applications, is given by the ratio of the envelope width $\Delta\Theta_1$ to the spot width $\Delta\alpha$ or, using Eqs. (5.8) and (5.9)

$$N = 2n(D/\lambda_o)\lambda/W. \tag{5.10}$$

By using Eq. (5.6), Eq. (5.10) can be reduced to the well-known form

$$N = t\Delta f, \tag{5.11}$$

where

$$t = D/v \tag{5.12}$$

is the transit time of the acoustic wave across the laser beam of diameter D.

In the guided wave structure where one propagation mode exists in the waveguide, the bandwidth and N for the thin film devices are identical to Eq. (5.6) and (5.10). If there is more than one mode in the waveguide, there is additional limitation on the maximum available bandwidth (or the minimum W) without affecting the other guided modes.

5.2.2 Anisotropic Diffraction

For the case of an anisotropic Bragg diffraction, the incident and the diffracted light beams see different refractive indices. As shown in Fig. 2.6 and in Eq. (2.30) the center acoustic frequency f_o is given by

$$f_o = \frac{v}{\lambda_o}\sqrt{n_i^2 + n_d^2 - 2n_i\, n_d \cos(\Theta_i - \Theta_d)} \tag{5.13}$$

where n_i and n_d are the indices of refraction seen by the incident and the diffracted beams; Θ_i and Θ_d are the incident and diffracted angles with respect to a crystal axis (x_2-axis). If the acoustic wave is propagating along the x_3-axis, the Bragg condition for this anisotropic diffraction is then

$$k_s = k_i \sin\Theta_i - k_d \sin\Theta_d \tag{5.14}$$

which reduces to the familiar isotropic Bragg condition of (2.33) if $k_i = k_d$ and $\Theta_i = -\Theta_d$.

One interesting special case in anisotropic diffraction is at $\Theta_d = 0$ where the diffracted beam is parallel to the acoustic wavefront. In this case the center frequency becomes

$$f_o \cong \frac{v}{\lambda_o} \sqrt{2n_i B} \tag{5.15}$$

where $B = n_i - n_d$ and $\cos\Theta_i$ is approximately one, since Θ_i is usually small.

The angular bandwidth and the number of resolvable spots of the special anisotropic deflector ($\Theta_d = 0$) with an acoustic beam width W can be calculated easily. At the center frequency, the diffracted beam is traveling in the direction parallel to acoustic wavefronts. As the acoustic frequency is changed from f_o to $f_o + \Delta f$, the diffracted light has to leave the acoustic column at an angle Θ_d (not zero) with respect to the acoustic wavefronts to satisfy the phasematching condition. From Fig. 5.3 the condition for the maximum value of Θ_d before destructive interference from the acoustic grating reduces the diffracted light to an intensity 4 db below the intensity at $\Theta_d = 0$ is given by

$$k_i W \cos\Theta_i - k_d W \cos\Theta_d = 2\pi. \tag{5.16}$$

Expanding $\cos\Theta_d$ for small Θ_d and using the phasematching condition for the $\Theta_d = 0$ anisotropic Bragg diffraction, i.e.,

$$k_i \cos\Theta - k_d = 0, \tag{5.17}$$

which can be shown using Fig. 5.3 to be equivalent to Eq. (5.14) with $\Theta_d = 0$, we obtain the following relation for Θ_d for the diffracted intensity to reach -4db point;

$$\Theta_d = \sqrt{2\lambda/W}. \tag{5.18}$$

Because the variation in the diffraction intensity with Θ_d is approximately symmetric about the value $\Theta_d = 0$, the total angular bandwidth (measured outside the crystal) for the anisotropic deflector is

$$\Delta\Theta_2 = 2n\Theta_d = 2\sqrt{2n\lambda_o/W}. \tag{5.19}$$

The spot size $\Delta\alpha$ of the diffracted light for an incident beam having a diameter D that covers a large number of acoustic wavelengths is given by Eq. (5.9). The number of resolvable spots N, which is given by the ratio

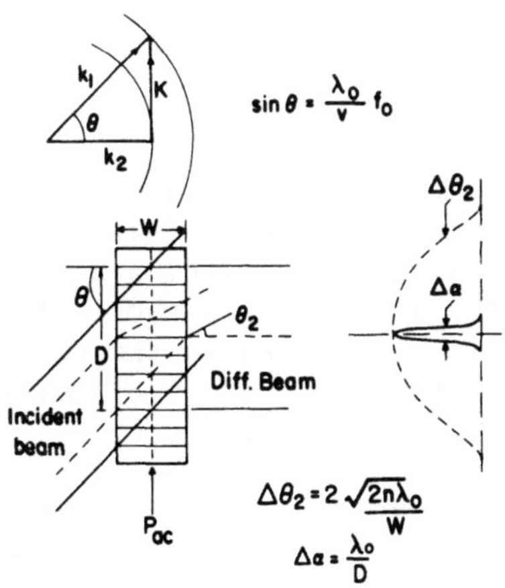

Fig. 5.3 - The phase matching condition and the device configuration of an anisotropic Bragg deflector in a uniaxial crystal.

of $\Delta\Theta_2$ to $\Delta\alpha$, can be written as

$$N = 2D\sqrt{\frac{2n}{\lambda_o W}} \qquad (5.20)$$

and can again be reduced to $N = t\Delta f$ if

$$\Delta f_2 = 2v\sqrt{\frac{2n}{\lambda_o W}} \qquad (5.21)$$

is the corresponding frequency bandwidth for the anisotropic deflector.

In guided wave structures, depending on the thickness of the film and the relative index difference between the film and the substrate, there may be more than one mode propagating in the waveguide. The multimode waveguides provide the required birefringence for the anisotropic diffraction. The birefringence is now due to the waveguide dispersion instead of the crystal birefringence. The waveguide birefringence is given by

$$B_{nm} = \frac{k_n - k_m}{k_o} = \frac{c}{v_n} - \frac{c}{v_m} \qquad (5.22)$$

where k_n and k_m are the wavevectors of the nth and m^{th} modes, and v_n and v_m are their respective phase velocities. Even if the film and the substrate are isotropic, the anisotropic diffraction can occur if two or more modes can propagate in the film. The bandwidths of anisotropic thin film devices are identical to those of bulk cases. However, if there are many modes in the waveguide, it has the possibility of more than one set of modes within the bandwidth of the Bragg diffraction process. Given a thin film thickness and the indices of the film and the substrate, the dispersion of all the propagating modes can be calculated according to section 3.1. With the known B_{nm}, the required center acoustic frequency is given by Eq. (5.15). The maximum available bandwidth can thus be calculated from Eq. (5.21).

As shown in Eqs. (5.11) and (5.21), the resolvable

spot N is determined by both Δf and $t = D/v$, the transit time across the laser beam diameter D. The available D is limited by the acoustic attenuation α in the medium, the physical size of the medium or the practical optical beam diameter. At frequencies higher than a few hundred megahertz, the acoustic attenuation may be the limiting factor. The number of resolvable spots become

$$N = \frac{\Delta f}{\alpha v} \tag{5.23}$$

where α (cm^{-1}) is the acoustic attenuation constant and Δf is given either in Eq. (5.6) or in Eq. (5.21) assuming that there is no bandwidth limitation from acoustic transducers.

5.3 Speed

The speed of acousto-optic devices are limited by the transit time across the laser beam. Approximately the speed τ of the acousto-optic device is given by

$$\tau \cong \frac{1}{\Delta f} + \frac{D}{v} \tag{5.24}$$

where the $1/\Delta f$ is determined by the acoustic transducer bandwidth and is the minimum pulse width which can be excited in the device; D is the optical beam diameter and v is the acoustic velocity. In a deflector where the number of resolvable spots is the major concern, τ can be made longer by increasing D for larger N.

In a modulator where the speed is the major concern, Δf has to be increased and D has to be decreased. With $\Delta f = 100$ MHz, the typical τ is about 15 nsec with $D = 10$ μm. The thin film guided wave structures are more suitable for such high speed modulator. This is because of the possibility of waveguiding the incident light beam to keep D small over a large interaction length.

5.4 Figures of Merit

To characterize the material dependences of bulk acousto-optic devices, a figure of merit given by Eq.

(3.25), i.e., $M = n^6 p^2/\rho\, v^3$ has been defined (Dixon, 1967). The larger the M value, the higher the diffraction efficiency as shown in Eq. (2.32). For thin film acousto-optic devices, the figure of merit can be defined as $M F^2$, where F is the overlap integral given in Eq. (5.2). In addition to the material dependence, the diffraction efficiency of the thin film devices also depend on the transverse distributions of the guided waves.

If both the diffraction efficiency and the bandwidth of the thin film acousto-optic devices are important, it can be shown from (5.6), (5.20) and (3.40) that

$$f_o\, \Delta f_1\, \xi^2 = \frac{\pi^2}{\lambda_o^3} \left(\frac{n^7 p^2}{\rho\, v}\right) \cos^2\Theta\, W\, P_{ac}\, F^2 \qquad (5.25)$$

for isotropic cases and

$$(\Delta f_2)^2\, \xi^2 = \frac{4\pi^2}{\lambda_o^3} \left(\frac{n^7 p^2}{\rho\, v}\right) \cos^2\Theta\, W\, P_{ac}\, F^2 \qquad (5.26)$$

for anisotropic cases. In comparison with the bulk wave case (Gordon, 1966), it is noted from both (5.25) and (5.26) that the material dependence of $n^7 p^2/\rho\, v$ and the overlap integral F are equally important for thin film devices. By comparing (5.25) and (5.26), we have

$$(\Delta f_2)^2 = 4\, f_o\, \Delta f_1. \qquad (5.27)$$

Thus the bandwidth of anisotropic deflector is much wider than that of the isotropic deflector. This makes anisotropic diffraction attractive for deflector applications. It should be noted that the bandwidth Δf can also be limited by the acoustic transducer bandwidth. This is particularly true for acoustic surface waves excited by interdigital transducers which usually have smaller bandwidths depending on the coupling constants in the substrate and the number of interdigital fingers (Smith, 1969).

VI. DEVICE APPLICATIONS AND CONCLUSIONS

Can thin film acousto-optic devices compete with their bulk wave counterparts? From the point of view of guided waves' unique properties, which provide higher diffraction efficiency and controllable dispersion for phase matching conditions, thin film acousto-optic devices clearly have significant advantages over the bulk wave counterparts. However, there are other practical considerations for thin film devices which cannot be neglected. For example 1) The coupling efficiency and the complexity of the input/output coupler to excite the optical guided waves will affect the overall efficiency of thin film devices; 2) It is more difficult to excite wideband acoustic surface waves, especially in non-piezoelectric substrates than bulk acoustic waves; 3) The bandwidths of the devices may not be limited by the bandwidths of diffraction mechanisms given by Section 5.2, but may be by bandwidths of acoustic wave transducers; 4) Thin film materials with good optical and acoustical properties, which can be either a single crystal layer on a passive substrate or a passive layer with a higher index of refraction on a piezoelectric substrate, are needed in thin film devices; 5) There may be power limitation due to nonlinear effects in guided wave structures.

The critical tolerances in thin film materials and fabrication techniques for thin film optical devices require pushing the available technologies developed for semiconductor integrated circuits to their limits. Good optical quality thin films for waveguiding by means of vacuum deposition, sputtering, epitaxial growth, ion implantation and thermal diffusion have been reported. However, good optical quality thin film waveguides with strong acousto-optic, electro-optic or magneto-optic constants either in the films or in the substrates still are rather difficult to attain.

The epitaxially grown ZnO films on sapphire substrates by chemical vapor deposition techniques (Hammer, 1973, Chubachi, 1973) have been demonstrated to be good electro-optic and acousto-optic thin film modulator materials. Ion-implanted GaAs waveguides may be suitable for modulator applications (Yariv, 1972). An approach to fabricate thin film optical waveguides in $LiNbO_3$ and $LiTO_3$ substrates by a novel out-diffusion technique reported by Kaminow (1973) is particularly interesting and promising. $LiNbO_3$ substrates have been among the most commonly used materials for acoustic surface waves and electro-optic devices due to the large piezoelectric and electro-optic constants in $LiNbO_3$ substrates. Efficient acoustic surface waves interdigital transducers with a center frequency above 1 GHz and a bandwidth of more than 25% can be fabricated on $LiNbO_3$ substrates. In conjunction with the novel out-diffusion technique for optical thin film waveguides, thin film acousto-optic deflectors in $LiNbO_3$ substrates with more than 1000 resolvable spots are possible. Fig. 6.1 shows schematically the thin film acousto-optic deflector with many interdigital transducers, which may be arranged to have different

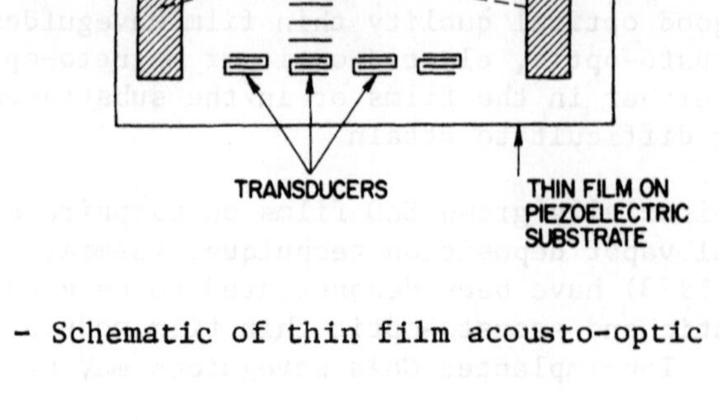

Fig. 6.1 - Schematic of thin film acousto-optic deflector.

center frequencies. The possibility of multiplexing in space and in frequency in thin film devices is an additional advantage.

The problem of efficiently coupling light beams into thin films has been partially overcome by the development of the prism coupler (Tien, 1969), grating coupler (Dakss, 1970) and tapered film coupler (Tien, 1970). More than 81% coupling efficiency by a prism coupler (Ulrich, 1970) has been reported. With the available input/output coupler techniques and good acousto-optic thin film materials, it is possible to replace some of the bulk acousto-optic devices such as the acousto-optic pulse compresser (Collins, 1967), the standing ultrasonic Bragg cell for optical switching and multiplexing (Tsai, 1972), etc. by thin film devices. The idea of integrating all the optical components in a single substrate to perform an optical signal processing function and thus avoid the usual problems of vibration and critical alignments will probably force us to consider thin film devices even more seriously in the future.

ACKNOWLEDGEMENTS

The author would like to thank Mr. P. F. Heidrich for his useful suggestions and his contributions to the experimental results in Section 4.3, and to thank Ms. L. Kastle for her skill and patience in preparing the manuscript.

REFERENCES

Berry, M. V., "The Diffraction of Light by Ultrasound" Academic Press, New York (1966).

Chang, W., IEEE J. Quantum Electronics, QE-7, No. 4, 167-170 (1971).

Chubachi, N., J. Kushibiki and Y. Kikuchi, Electronics Lett. Vol. 9, No. 9 193 (1973).

Chu, R. S. and T. Tamir, IEEE Trans. Microwave Theory & Tech. MTT-17, 1002 (1969).

Collins, J. H., E. G. H. Lean and H. J. Shaw, Appl. Phys. Lett. 11, 240 (1967).

Collins, R. E., Field Theory of Guided Waves (McGraw Hill, New York) (1960).

Dakss, M. L., L. Kuhn, P. F. Heidrich and B. A. Scott, Appl. Phys. Lett. 16, 523, (1970).

Damon, R. W., W. T. Maloney and D. H. McMahon, Physical Acoustics, Vol. VII, Chapter 5 (1970).

Dixon, R. W., J. of Appl. Phys., Vol. 38, 5149, 1967.

Gfeller, F. R., and C. W. Pitt, Electronics Letters (1972).

Gordon, E. I., Proc. IEEE 54 1391 (1966).

Gordon, E. I., IEEE J. Quant. Electr. QE-2, 104 (1966).

Harris, S., S. T. K. Nieh and D. K. Winslow, Appl. Phys. Lett. 15, 325 (1969).

Hickernell, F. S., 1971 Ultrasonics Symposium.

Kaminow, I. P., J. R. Carruthers, Appl. Phys. Lett. Vol. 22, No. 7, 326 (Apr. 1973).

Klein, W. R. and B. D. Cook, IEEE Trans. Sonics & Ultrasonics, Vol. SU-14, 123 (1967).

Kuhn, L., J. Quantum Electronics QE-5, 383 (1969).

Kuhn, L., M. L. Dakss, P. F. Heidrich and B. A. Scott, Appl. Phys. Lett. 17, 265, (1970).

Kuhn, L., P. F. Heidrich and E. G. H. Lean, Appl. Phys. Lett. 19, 428 (1971).

Lean, E. G. H., C. F. Quate, and H. J. Shaw - Appl. Phys. Lett. 10, 48 (1967).

Lean, E. G. H., "Studies of Microwave Shear Waves in Solids" PhD Dissertation, Stanford University (1967).

Lean, E. G. H. and C. G. Powell, Proc. IEEE 58 1939-1947 (1970).

Lean, E. G. H. and S. C.-C. Tseng, J. of Appl. Phys. 41, 3192 (1970).

Lean, E. G. H., Progress in Optics XI (Editor, E. Wolf) 125 (1973).

Lean, E. G. H. and A. N. Broers, Microwave Journal, (March 1970).

Lord Rayleigh, Proc. London Meth. Soc. 17 (1885); W. M. Ewing, W. S. Jardetsky and F. Press, Elastic Waves in Layered Media, 31 (1957).

Love, E. A. H. Some Problems in Geodynamics (Cambridge, 1911).

Maydan, D., J. Appl. Phys. 41, 1552 (1970).

Nelson, D. F. and J. McKenna, J. Appl. Phys. 38, 4057 (1967).

Nye, J. J., Physical Properties of Crystal (Oxford, Clarendon Press) 241 (1957).

Quate, C. F., C. D. W. Wilkinson and D. K. Winslow IEEE Proc. 53, 1604 (1965).

Ramon, C. V. and N. S. N. Nath, Proc. Ind. Acad. Sci. 2A, 413, (1935).

Slater, J. C., Rev. Mod. Phys. 30, 197 (1958).

Smith, W. R., H. M. Gerard, J. H. Collins, T. M. Reeder and H. J. Shaw, IEEE Trans. on Microwave Theory & Tech., vol. MTT-17, pp 865-873 (1969).

Stern, E., IEEE Trans. Microwave Theory & Tech. MTT-17, 835 (1969).

Stonely, R., Proc. Roy. Soc., A106, 416 (1964).

Tien, P. K., R. Ulrich and R. J. Martin, Appl. Phys. Lett. 14, 291 (1969).

Tien, P. K., Appl. Opt., Vol. 10, pp 2395-2413 (1971).

Tsai, C. S. and S. K. Yao, J. Appl. Phys. 43, 5081 (1972).

Ulrich, R., J. Opt. Soc. Am. 60, 1337 (1970).

Viktorov, I. A., Rayleigh and Lamb Waves, Plenum Press, (1967).

Warner, A. W., D. L. White and W. A. Bonner, J. Appl. Phys., Vol. 43, No. 11, 4489 (1972).

White, R. M. and F. M. Voltmer, Appl. Phys. Lett. 7, 314 (1965).

Willard, G. W., J. Acoust. Soc. Am. 21, 101 (1949).

Yariv, A., "Introduction to Optical Electronics," Holt, Rinehart & Winston, N. Y. 310 (1971).

Chapter 15 - LASER SOURCE CONSIDERATIONS IN INTEGRATED OPTICS*

C. L. Tang

Cornell University

Ithaca, New York

I. INTRODUCTION

With the recent development of low-loss optical fibers(1), there is now considerable interest in the development of integrated optical systems(2) consisting of miniaturized optical components. These include both passive elements such as input and output couplers, thin-film waveguides, etc., and active devices such as modulators, laser sources, etc. In this article, some of the basic considerations related to the development of miniaturized laser sources are discussed. For practical integrated optical systems, such sources will almost have to be solid state devices and most likely in thin-film forms. Our discussion will, therefore, primarily be limited to solid state thin-film lasers.

Although the basic principles underlying the devices are not significantly different from those for bulk lasers, especially those related to the active medium, there are nonetheless some important new problems to be considered. A different form of feedback structure is needed for thin-film lasers because Fabry-Perot cavities made up of mirrors as used commonly in bulk lasers are difficult to construct in thin-film form. In addition, because of the desired small size of the device, the gain in the active medium must generally be much higher than those required for bulk lasers. In Section II, we consider some of the new concepts related to the resonant feedback structure in thin-film forms; and in Section III, we discuss the basic numerical considerations involved in the stimulated emission process in various active media in order to see which ones can be reasonably expected to be useful for laser sources in integrated optics.

It should be pointed out at the outset, however, that at the present time the major problems involved in developing such lasers are primarily technological, not in just understanding the physics or in knowing the orders-of-magnitude of things. In other words, the hard part of the problem is in making the devices work; the technological problems related to making the devices work are, however, only very briefly touched upon in some cases and in most cases not at all in this article. In this sense, the title of this article is not really appropriate; it is only used here for want of a better choice.

II. RESONANT FEEDBACK STRUCTURES

In thin-film lasers, the active medium may be either the thin-film waveguide or the substrate. In either case, a resonant structure analogous to the Fabry-Perot cavity of the bulk lasers is still required. Since mirror reflectors in bulk optical devices are generally replaced in thin-film devices by periodic structures with a spatial period on the order of the half-wavelength of the light, resonant feedback structures for thin-film lasers are generally of the form of periodic structures either deposited onto or etched into the waveguides.

Periodic structures or gratings in various forms can be used as the resonant feedback structure for lasers in two basic ways: In the distributed feedback structure (or DFB)(3), the grating spans the whole length of the active medium; feedback is achieved through Bragg reflection distributed throughout the medium. DFB lasers have been analyzed in detail on the basis of the coupled-mode equations by Kogelnik and Shank (KS)(3). The discussion presented here is based upon an analysis of the dispersion relation (or ω-k curve) for the periodic structure. The final results are of course the same. Hopefully, a different approach to the same problem will help to clarify the nature of the approximations in the analysis and the physics involved, which is really the aim of the present discussion. It turns out that the qualitative features of all the important properties of the DFB laser can be explained very simply in terms of the general behavior of the dispersion curve of the structure.

Periodic structures can also be used simply as resonant reflectors placed at the two ends of the active region as in conventional Fabry-Perot type of laser cavities. Although DFB structures have been used most extensively in experiments on thin-film lasers, there are important advantages in using the gratings or periodic structures as simple reflectors in the Fabry-Perot type of configuration even for thin-film lasers as will be shown later.

It should be mentioned that high quality optical gratings with lattice spacings down to the 1000-2000 Å range can now be fabricated and are being used in experimental devices. These gratings are fabricated either in photoresist by exposure to laser interference fringe patterns(4) or in electron-resist(5) by exposure to an electronically controlled scanning electron microscope. Such gratings deposited on thin-film waveguides can be used either directly or as masks for etching by various means the structure into the thin-films. Technically, much work remains to be done to perfect these devices for practical use, but the feasibility and importance of these techniques have now been unequivocally demonstrated; such techniques are being used routinely in laboratory experiments.

A. Dispersion Relations for Periodic Structures

To avoid overly complicating the picture, the discussion will be based upon a simple one-dimensional picture and will begin with a consideration of the dispersion relation (or ω-k diagram) of the electromagnetic wave in the spatially modulated medium. Furthermore, a simple sinusoidal modulation of the refractive index will be assumed. The medium is, therefore, characterized by a unmodulated gain constant α and a modulated index of refraction of the form:

$$n(\omega) = n_o(\omega) + 1/2[n_1 \exp iGz + n_{-1}\exp(-iGz)] \qquad (1)$$

where $G = 2\pi/d$ is the fundamental reciprocal lattice constant of the periodic structure and d is the spatial period of the index modulation. $n_{\pm 1}$ become complex if there is also gain modulation. A plane monochromatic electromagnetic wave $E_o \exp(ikz - i\omega t)$ propagating in the medium will generate first order polarization side-band waves with propagation constants $k \pm G$, which will in turn re-radiate electromagnetic waves 'forced' to propagate with the same phase velocity or propagation constant. The amplitude of a particular re-radiated or forced side-band wave will be large only when its propagation constant is also near that of a free propagating wave; for first order side-bands, this would be the case if

$$k + G \cong -k \quad \text{for } k < 0, \quad \text{and } k-G \cong -k \quad \text{for } k > 0,$$

which are the conditions for first order Bragg diffraction:

$$k \cong \pm G/2. \tag{2}$$

Thus, for $k \cong G/2 > 0$, we can assume the field in the medium to have the form:

$$E(z,t) = E(z)e^{-i\omega t} \cong E_o e^{ikz-i\omega t} + E_- e^{i(k-G)z-i\omega t} \quad ; \tag{3}$$

for $k \cong -G/2 < 0$, it has the form:

$$E(z,t) \cong E_o e^{ikz-i\omega t} + E_+ e^{i(k+G)z-i\omega t} \quad . \tag{4}$$

Substituting either Eq. (3) or Eq. (4) into the wave equation and following the usual procedure (see appendix), one obtains the dispersion relation for waves near where the Bragg condition is satisfied:

$$[\bar{k}^2-k^2][\bar{k}^2-(k\pm G)^2] - 4K^2\bar{k}^2 = 0 \tag{5}$$

where

$$\bar{k} = n_o(\omega)\omega/c \tag{6}$$

$$K = \left(\frac{n_1}{n_o}\right)\left(\frac{\bar{k}}{2}\right) \tag{7}$$

K is a very important parameter that characterizes the strength of coupling between the waves propagating in opposite directions. For the one-dimensional structure considered here, it is directly proportional to the fractional index modulation and the wave number. For a more complicated structure such as a grating deposited on or etched into a thin-film waveguide, one can still characterize the coupling between a wave and its side bands in terms of a phenomenological parameter K, even though it will now be more difficult to calculate or estimate and may have to be determined experimentally.

The properties of the periodic structure as a resonant feedback structure and the mode structures of lasers using such structures depend very much on the dispersion relations of the waves in the medium. The dispersion relation, Eq. (5), is shown schematically in Figure 1. With the periodic modulation, for each frequency ω(or $\bar{k}$ as defined in Eq. (6)) there are now four possible waves, corresponding to a forward (k_o>0) and a backward ($-k_o$<0) propagating wave and their respective lower (k_{-1}) and upper (k_{+1}) side bands, which are forced waves. Furthermore, the shape of the dispersion curve

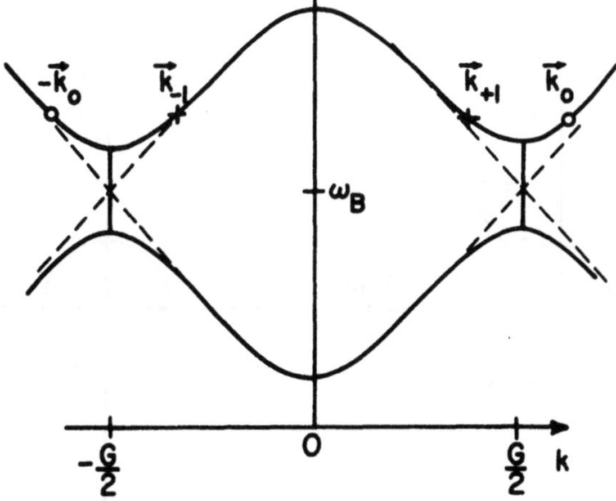

Figure 1 - Dispersion characteristics of a spatially periodically modulated medium.

is strongly modified, with the appearance of gaps, near where the Bragg conditions are satisfied due to the strong coupling of the waves. Near and within each of the gap regions, the dispersion relation Eq. (5) can be simplified:

$$(k \pm \frac{G}{2})^2 = (\bar{k} - \frac{G}{2})^2 - K^2, \quad \text{for } k \lessgtr 0, \tag{8}$$

by making use of the approximations $\bar{k}^2 - k^2 \cong 2\bar{k}(\bar{k} \pm k)$ and $\bar{k}^2 - (k \pm G)^2 \cong 2\bar{k}[\bar{k} \mp (k \pm G)]$ for $k \lessgtr 0$. This immediately gives the propagation constants $(k \pm G/2)$ for each frequency ω or $(\bar{k} - G/2)$. The results are shown schematically in Figure 2, for the gap region near $+G/2$; except for a shift in origin, the behavior of the dispersion curve near $-G/2$ is exactly the same. Outside the gap $|\bar{k} - G/2| > K$, there are two purely real propagation constants for each ω; inside the gap, the two propagation constants become complex with the same real

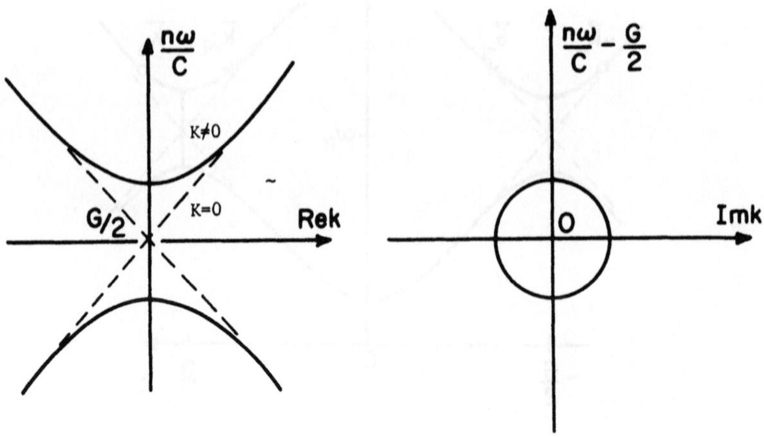

Figure 2 - Real and imaginary parts of the propagation constant near and in the gap region.

part equal to G/2 but imaginary parts of opposite signs. Physically, it means that for each frequency outside the gap region there are two undamped waves of different phase velocities traveling in each direction. Inside the gap region, the two waves travel with the same phase velocity equal to $2\omega/G$ but have different damping constants, one with increasing amplitude and the other with decreasing amplitude in the direction of propagation. In the absence of any gain in the medium, the wave propagating with a decreasing amplitude corresponds to a freely propagating wave; the wave which propagates with an increasing amplitude is a side-band wave of a damped freely propagating wave in the opposite direction, which means that it is a forced wave and does not exist by itself. A wave incident upon a finite periodic structure will, therefore, always be damped in going through the structure if it falls within the gap region.

Suppose now the medium has gain and the spatial gain constant is α, the dispersion relations Eqs. (5) and (8) are modified:

$$[\bar{k}^2 - 2i\alpha\bar{k} - k^2][\bar{k}^2 - 2i\alpha\bar{k} - (k\pm G)^2] - 4K^2\bar{k}^2 = 0 \quad (9)$$

and

$$(k \pm \frac{G}{2})^2 \cong [(\bar{k} - \frac{G}{2}) - i\alpha]^2 - K^2. \quad (10)$$

Figure 3 shows the corresponding modifications in the behavior of the dispersion curves near the gap region. Note that the wave with $\bar{k}>G/2$ and $Rek>G/2$ or $\bar{k}<\frac{G}{2}$ and $Rek<\frac{G}{2}$ also has an $Imk<0$; thus, the freely propagating wave is a growing wave due to the gain in the medium. The wave with $\bar{k}>G/2$ but $Rek<G/2$ or $\bar{k}<G/2$ but $Rek>G/2$, which corresponds to the upper side-band of a freely propagating wave in the opposite direction, has an $Imk>0$ and is a decaying wave in the direction of propagation. Furthermore, in the high gain limit, or $\alpha>>K$:

$$Re(k - \frac{G}{2}) \equiv k' - \frac{G}{2} \cong \pm(\bar{k} - \frac{G}{2}), \quad (11)$$

$$-Imk \equiv k'' \cong \pm(\alpha^2 + K^2)^{1/2} \quad (12)$$

for $Rek \cong G/2$.

Thus, in the high gain limit, the waves propagate as if

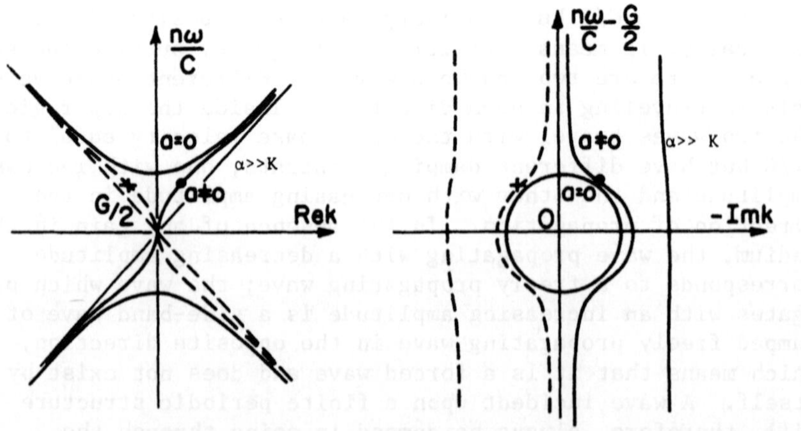

Figure 3 - Effect of gain in the medium on the dispersion characteristics of the modulated medium near and in the gap region. Note that the solid curves give the real and imaginary parts of one root and the dashed curves give the other of the dispersion equation for each frequency ω.

it is in an unmodulated amplifying medium even in the gap region, but there are in addition strong side-bands if the medium is modulated. These characteristics will have important implications when the periodic structure is used as the resonant feedback structure for lasers.

If the gain of the medium is modulated periodically spatially instead of the index of refraction, the corresponding dispersion relations can be obtained and understood in the same way. Without entering into further detailed discussion, we show in Figure 4 the corresponding dispersion curves. The only major difference is that now the frequency or $\bar{k}$ becomes complex.

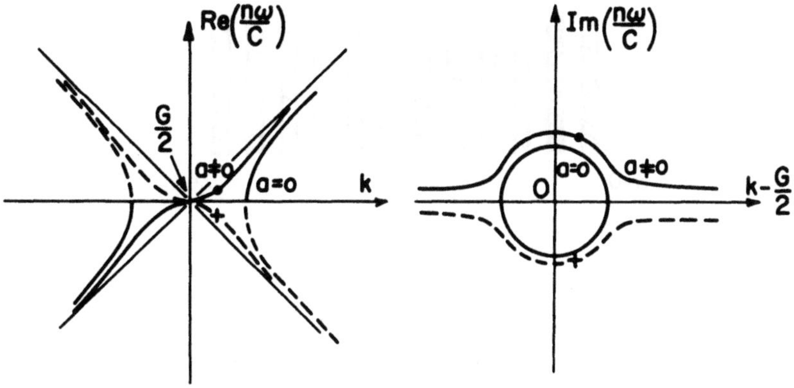

Figure 4 - Dispersion characteristics of a gain modulated medium near and in the gap region.

B. Lasers Using Periodic Structures as the the Distributed Resonant Feedback Structure (DFB Lasers)

a. <u>General conditions for self-sustained oscillations.</u> Suppose a finite periodic structure located from z=0 to L as shown schematically in Figure 5 is used as the resonant feedback structure for a laser. It turns out that the more important properties of such a laser can be explained simply in terms of the behavior of the dispersion relation of the periodic structure in the presence of gain in the medium.

Consider first the feedback mechanism. Figure 1 shows that a wave at k_o near the gap region G/2 will be scattered

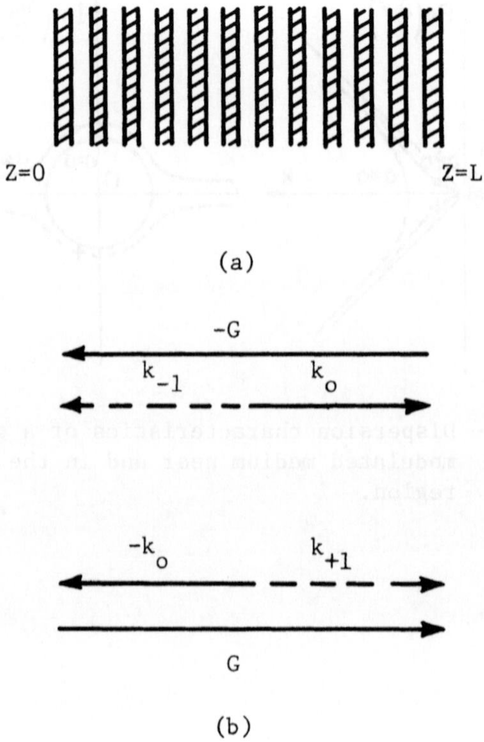

Figure 5 - (a) A periodic structure used as the distributed feedback structure of a laser (DFB Laser). (b) Wave-vector picture demonstrating the feedback mechanism based upon Bragg-reflections in a DFB structure.

by the periodic structure and create a side-band wave in the opposite direction at $k_{-1}= k_o-G$, which is a forced wave but is near $-k_o$. To satisfy the boundary condition at z=L, however, a free wave at $-k_o$ must also be exicted. Repeating the same process for this wave in the opposite direction, a wave at $-k_o$ will in turn excite or feedback to the original

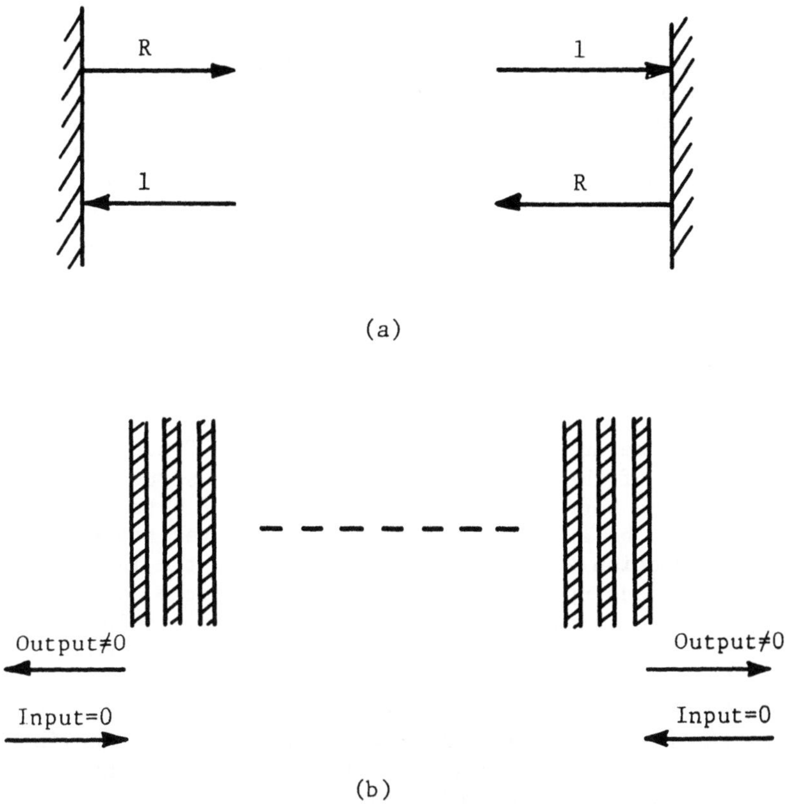

Figure 6 - Boundary conditions for a Fabry-Perot laser cavity (a) and a distributed feedback laser (b).

$+k_o$ wave, thus completing the feedback loop. This is shown schematically in Figure 5(b). Without any gain in the medium each scattering process leads to a loss in the amplitude of the wave and no oscillation can be started or sustained. In the presence of gain in the medium, this loss in amplitude can be compensated by the gain, leading to the build-up or maintaining of self-sustained oscillation.

The general condition for the initiation and maintaining of such oscillations can be determined from the wave equation subject to the appropriate boundary conditions. The boundary conditions for the distributed feedback structure are fundamentally different from those for the Fabry-Perot type of laser structure; this is illustrated qualitatively in Figure 6. The key point is that in the Fabry-Perot structure, the amplitudes of the waves propagating in different directions at the boundaries or the mirrors are related by the mirror reflectivities. In the distributed feedback structure, at the boundaries z=0 and L, the total amplitude of the wave traveling into the structure is always equal to zero and is independent of the wave traveling away from the structure which can be anything not equal to zero. Since the total amplitude of the waves traveling into the structure must be zero at the boundary but finite elsewhere in the structure there must be at least two <u>exponential</u> waves traveling in each direction so that they can cancel at the input end and still add up to a finite amplitude at the output end of the structure. This is precisely the situation in the periodic structure as the previous discussions showed. Furthermore, with gain in the medium, the spatial variations of the four exponential waves traveling in the laser structure are shown qualitatively in Figure 7. Suppose we begin with the freely propagating wave traveling to the right at z=0 and take its complex amplitude there to be 1. At z=L, it becomes $\exp(ik_oL) \equiv \exp(ik_o'+k_o'')L$; it is a growing wave to the right because $k''>0$ as shown in Figure 3(b). It can be shown (see appendix) on substituting Eqs. (3) or (4) into the wave equation that the complex amplitude of the lower side-band of this wave is

$$\frac{k_o-\bar{k}+i\alpha}{K}\, e^{(ik_o'+k_o'')L} \qquad (13)$$

at this point; this is a forced wave traveling in the opposite direction with the wave vector $k_{-1}=k_o-G$. Since the boundary condition at z=L is that the total amplitude of the waves

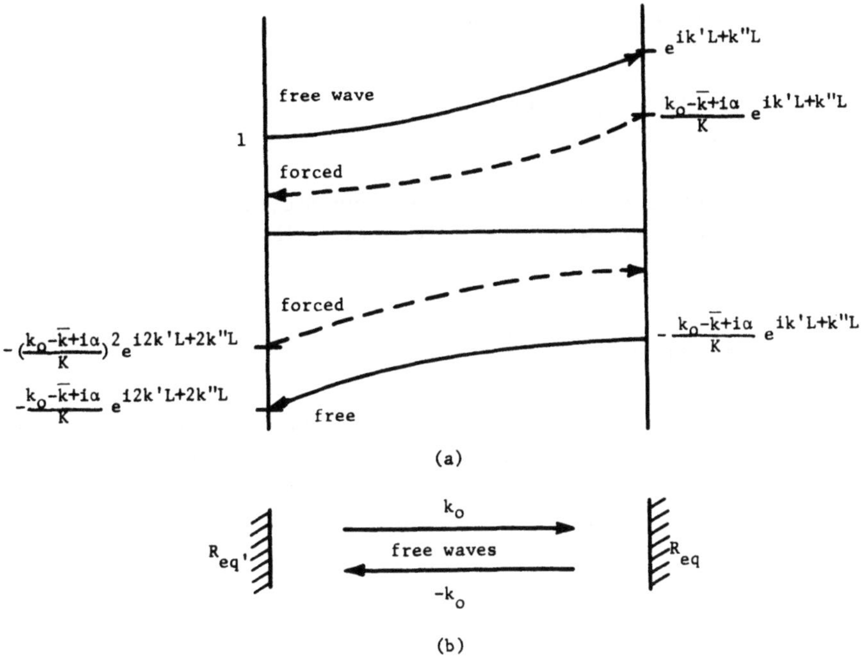

Figure 7 - (a) Spatial variations of the complex amplitudes of the free waves (solid curves) and the forced side-band waves (dashed) in the DFB laser.
(b) The DFB laser shown in (a) can be visualized as a Fabry-Perot type of laser where the mirror reflectivity R_{eq} is equal to $-(k_o-\bar{k}+i\alpha)/K$.

traveling to the left must be zero, a free wave with the wave vector $-k_o$ must be launched also at this point to cancel the forced wave here. This free wave must, therefore, have the complex amplitude

$$-\frac{k_o-\bar{k}+i\alpha}{K}\, e^{(ik_o'+k_o'')L}$$

On reaching the opposite boundary at z=0, the complex

amplitude of this $-k_o$ wave becomes

$$-\frac{k_o-\bar{k}+i\alpha}{K}\, e^{2(ik_o'+k_o'')L}$$

At this point following the same argument, it will in turn launch a free wave with the wave vector $+k_o$ and complex amplitude

$$\left[\frac{k_o-\bar{k}+i\alpha}{K}\right]^2 e^{2(ik_o'+ k_o'')L} \tag{14}$$

For self-sustained oscillation, this wave must be that which we assumed started here in the first place. Thus, the all-important general condition for self-sustained oscillation in this laser is:

$$\left[\frac{k_o-\bar{k}+i\alpha}{K}\right]^2 e^{2(ik_o'+k_o'')L} = 1 \tag{15}$$

The phasor part of this condition determines the frequencies of oscillation and, hence, the spectral and mode properties of the distributed feedback laser; the amplitude part determines the threshold condition for oscillation and, hence, the requirements on the pumping rate and coupling parameter. For the purposes of deducing and elucidating these properties and requirements however, it may be helpful to note that Eq. (15) can be cast in the same form as that for the Fabry-Perot type of laser:

$$R_{eq}^2\, e^{2(ik_o'+k_o'')L} = 1\ , \tag{16}$$

if an equivalent mirror reflectivity

$$R_{eq} = -\frac{k_o-\bar{k}+i\alpha}{K} \tag{17}$$

is defined for the distributed feedback structure. Thus, as shown schematically in Figure 7(b), the periodic structure in the distributed laser can be visualized as being replaced by a set of mirrors as in the Fabry-Perot cavity with an equivalent mirror reflectivity given by Eq. (17). For given coupling parameter K, gain constant α, and frequency ω or $\bar{k}$, the wave number k_o follows from the dispersion relation Eq. (9) or the approximate Eq. (10); R_{eq} is therefore always known. The characteristics of the distributed feedback laser can, therefore, be readily understood following the usual procedures for analyzing conventional Fabry-Perot lasers. We now examine briefly some of the more important ones.

b. Spectral and mode properties. To determine exactly the laser oscillation frequencies ω or the corresponding $\bar{k}$, one must solve Eq. (15) simultaneously with the dispersion relation Eq. (9) or Eq. (10), which can always be done at least numerically. The qualitative features are relatively easy to understand, however. Let us define the phase shift of the equivalent mirror reflectivity defined in Eq. (17) to be $\emptyset$:

$$R_{eq} \equiv |R_{eq}| \exp(i\emptyset) \quad . \tag{18}$$

Eq. (15) therefore requires that

$$\emptyset + k_o'L = N\pi \quad \text{, where N is an integer.} \tag{19}$$

$k_o'L/\pi$ is the number of half-wavelengths in L; it is in general huge (at least 100 or more) compared with $\emptyset/\pi$ (<2) even for miniaturized lasers. Neglecting the frequency dependence of $\emptyset$, the spacing between the modes of the laser modes are, therefore, determined by the condition $\Delta k_o'L \cong \pi$. From the discussions on the dispersion relations, in particular Figure 3, we know that, in the limit of high gain $\alpha >> K$, k' becomes approximately equal to $\bar{k} = 2\pi n_o \nu/C$; therefore, in this region, the frequency spacing of the laser modes $\Delta\nu$ becomes the same as that of conventional Fabry-Perot cavities:

$$\Delta\nu \cong C/2n_oL \quad ; \tag{20}$$

this is shown schematically in Figure 8. At where the Bragg condition $\bar{k} \cong G/2$ is satisfied exactly, Figure 3 shows that $k_o \equiv k'-ik'' \cong \bar{k} - i(\alpha^2+K^2)^{1/2}$; the phase factor ϕ of R_{eq} is therefore $\pi/2$ according to Eq. (17). On the other hand, $k'L = GL/2$ is equal to integral number of π's. Therefore,

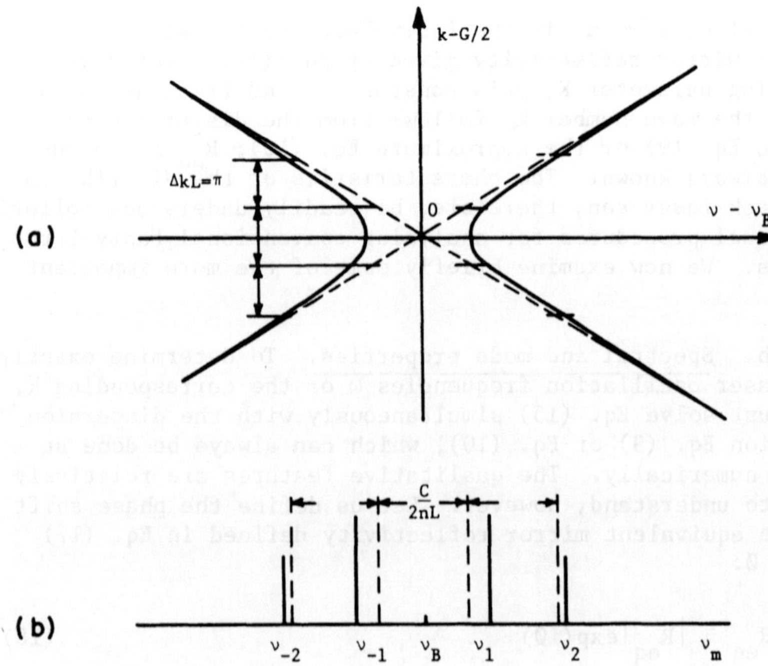

Figure 8 - (a) Influence of the dispersion characteristics on the frequencies of the laser modes; the separations between the laser mode frequencies are determined by the condition $\Delta k'L \cong \pi$.
(b) The corresponding mode spectrum of the DFB laser in the high gain limit (dashed lines) and in the low gain limit (solid lines).

from Eq. (19), exactly at the Bragg frequency $\nu_B = C/2n_O d$ (d is the period of the periodic structure), there can be no laser oscillation. Furthermore, since the dispersion curve is more or less symmetric about the Bragg frequency (see

Fig. 8), the laser modes should be symmetrically located about ν_B as shown in Figure 8(b).

If the coupling parameter is not small compared with the gain constant, then the dispersion curve is more severely distorted near the gap region as shown in Figure 3 or 8; the condition $\Delta k_o' L \cong \pi$ will lead to certian amount of frequency-pushing of the modes from those separated by $C/2n_oL$. Furthermore, the modes closer to the Bragg frequency will be pushed more relative to those further away. The actual laser oscillation frequencies in this case are shown schematically in Figure 8 also. Note that because the dispersion curve is distorted in such a way that the magnitude of its slope $\left|\frac{\partial\nu}{\partial k_o'}\right|$ becomes smaller than $|\partial\nu/\partial\bar{k}|$ and the modes are determined by $\Delta k_o' L \cong \pi$, the modes are pushed further away from ν_B than the Fabry-Perot modes or the modes in the high gain limit. Furthermore, it is also clear from Fig. 8 that, because of the shape of the dispersion curve and $|k'-G|L \sim \pi/2$ for the lowest order modes, these modes must be pushed outside the region $|\nu-\nu_B| \le \frac{CK}{2\pi n_o}$ around ν_B.

Finally, the qualitative features of the intensity distributions of the laser modes can also be understood simply based on considerations of the dispersion relations. Interference of the waves traveling in opposite directions inside the cavity leads to a rapid variation on the scale of the optical wavelength; the slow spatial variation of the intensity of the laser light inside the cavity is primarily due to the interference of the two waves, one free and one forced, traveling in the same direction. In the high gain limit, as shown in Figures 7(a) and 3(a) the freely propagating wave in each direction grows rapidly but the forced waves decays rapidly. As a result, there is little interference between the waves in the same direction and the spatial variation of the intensity is dominated by the strong exponential character of the free waves. The intensity distribution must have, therefore, the form $\exp(k''|z-L/2|) \sim \exp(\alpha|z-L/2|)$ approximately and is relatively independent of the mode numbers. As the coupling parameters increases relative to the gain constant, the amplitudes of the free wave and the forced wave in the same direction remain closer to being equal to each other in the structure. As a result, the interference effect becomes more pronounced. Since the difference between the k-vectors of the two waves, $(k_o'-k_{+1}'')$, are approximately integral multiples of $2\pi/L$ as shown in Figure 8, the intensity distribution in the cavity should have very approximately the forms $\sin^2\pi z/L$, $\sin^2 2\pi z/L$, $\sin^2 3\pi z/L$, etc., in the limit of $K>>\alpha$ for successive modes symmetrically located from the Bragg frequency ν_B. Examples of such intensity distributions are shown qualitatively in Figure 9 for both the high and low gain limits and the intermediate cases.

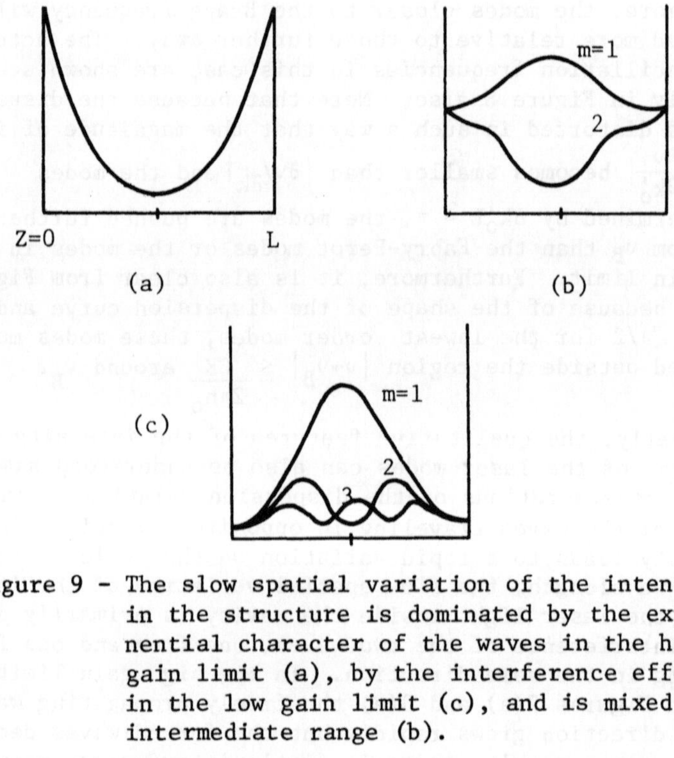

Figure 9 - The slow spatial variation of the intensity in the structure is dominated by the exponential character of the waves in the high gain limit (a), by the interference effect in the low gain limit (c), and is mixed in the intermediate range (b).

c. Threshold conditions. The exact threshold condition for oscillation follows of course from the amplitude part of the general condition for sustained oscillation, Eq. (15), and the dispersion relation Eq. (9) or Eq. (10); from these one can always determine the requirements on the pumping rate or the gain constant α and the coupling parameter K. Again, this may require numerical solutions in general. Simple approximate results can be obtained from these equations in the limits of high or low gain, however.

In the high gain limit, $k_o \cong \bar{k} - i(K^2+\alpha^2)^{1/2} \cong \bar{k} - i\alpha(1+K^2/2\alpha^2)$ from Eqs. (11) and (12). Substituting this into Eq. (15) and assuming that the phasor part of the equation is automatically satisfied on resonance, one obtains immediately the threshold condition:

$$\left(\frac{K}{2\alpha_{threshold}L}\right)^2 e^{2\alpha_{th}L} \cong 1 \quad \text{for } K << \alpha, \tag{21}$$

for all modes not too far away from the Bragg frequency or $|k'-G/2| \lesssim \alpha$.

In the low gain limit $K>>\alpha$ and for the lower order modes near $G/2$, $|k_o'-G/2| \cong (m\pi/L) << K$ (where m=1,2,3... refers to the order of the successive modes counting from ν_B); it follows from Eq. (10) and Figure 3 that

$$\bar{k} - G/2 \cong K$$

$$k_o'' \cong \frac{K\alpha}{k_o'-G/2} \cong \left(\frac{KL}{m\pi}\right)\alpha \; .$$

Substituting these into Eq. (15) and expanding the resulting equation, one finds the threshold condition

$$\alpha_{threshold}L \cong \left(\frac{m\pi}{KL}\right)^2 \tag{22}$$

assuming in addition $\alpha_{th}L << m\pi$.

These approximate threshold conditions are particularly useful for gaining a qualitative understanding of the orders of magnitude of the pumping rates and coupling parameters required in practical lasers. Physical means for meeting such requirements are discussed in Section III.

C. Lasers Using Periodic Structures as the Resonant Reflectors in the Fabry-Perot Type of Cavities

Despite the fact that the distributed feedback structure is almost always used in thin-film type of lasers, there are some important advantages in using the Fabry-Perot type of laser cavity even for thin-film lasers. To form a Fabry-Perot cavity, one needs high-reflectivity mirrors. In bulk lasers, one uses multilayered dielectric-coated mirrors. In thin-films, one can use periodic structures deposited onto or etched into the thin-film waveguide as shown schematically in Figure 10. In contrast to the distributed feedback structure shown in Figure 5(a), there is no periodic perturbation in the active region. This type of laser structure would have lower threshold and possibly narrower output spectrum than a comparable distributed feedback laser.

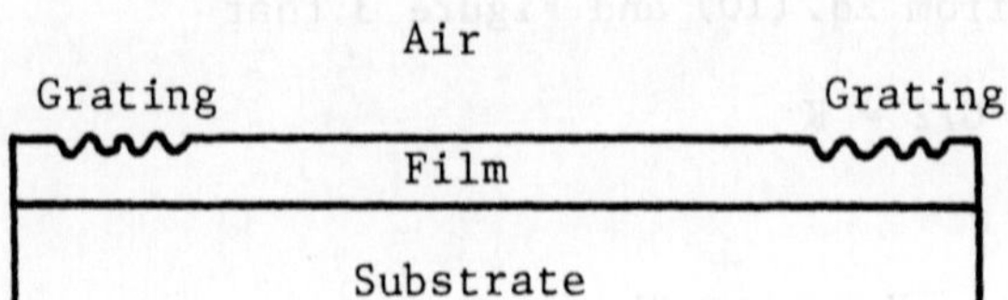

Figure 10 - Periodic structures etched into the thin-film waveguide used as the resonance reflectors of a Fabry-Perot type of thin-film laser. The active region is between the reflectors.

Consider first the threshold requirement. When used as a resonance reflector, near where the reflectivity is the highest, the threshold condition in this case is approximately:

$$\alpha_{th} L \cong \ell n\left(\frac{4}{KL}\right) \qquad \text{for } KL \lesssim 1 \tag{23}$$

if the length of the active region is L and the periodic structure of the total length L is split into two and used as resonance-reflectors at the ends of the active region. α_{th} given by Eq. (23) is generally much smaller than that given by Eq. (22) for the lowest order (m=1) mode.

Figure 11 shows the change in the reflectivity as a function of the wave number $|k-G/2|L$ for an M-element periodic structure when used as a resonance reflector. Also shown are the rough locations of the oscillating modes if the same periodic structure is used as a laser distributed feedback structures. In the distributed feedback configuration, the laser modes are always located at where the feedback or the reflectivity is relatively large and discrimination against the higher order modes by the periodic structure is relatively small and does not vary much with the period or the length of the structure. On the other hand, in the Fabry-Perot configuration, because the lengths of the active region and the periodic structure and its period can all be varied separately, it would be possible to design the structure for maximum discrimination against the unwanted laser modes in order to obtain a narrower laser spectral width.

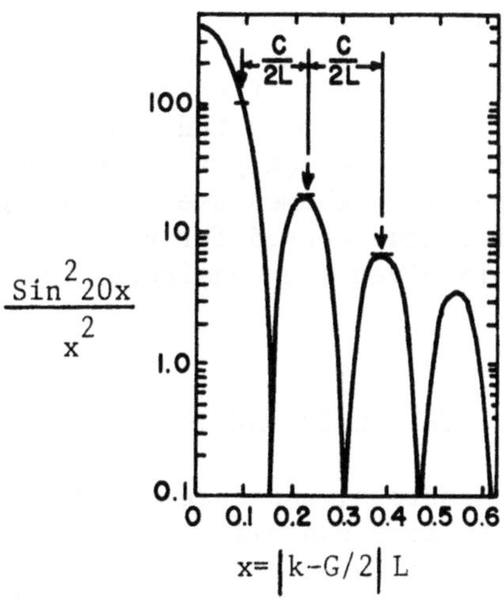

Figure 11 - Reflectivity of a 20-element periodic structure as a function of $|k-G/2|L$.

D. Order-of-Magnitude Estimates

In laser considerations, numerically the most important parameters are the coupling parameter K and the threshold value of the gain constant α_{th}.

For thin-film devices, because the periodic structure is often deposited onto or etched into the thin-film waveguide, to estimate accurately the coupling parameter K is a much more difficult problem than the one-dimensional situation considered so far. The difficulty is often compounded by the fact that the profile of the structure is not known well enough. As a rough estimate, however, we can take the coupling parameter K^2 estimated on the basis of Eq. (7) for the corresponding bulk structure reduced by a factor roughly equal to the fraction of the energy of a given guided mode that is strongly affected by the surface periodic structure. For the purpose of our discussions here, we take this reduction factor to be say 10 to 100. For bulk structures, coupling parameters on the order of 100 to 1000 cm^{-1} are achievable. Therefore, K values on the order of 10 to 100 cm^{-1} for thin-film devices are not unreasonable.

From Eq. (22), the gain coefficient at the threshold α_{th} should thus be in the range of $\sim(10 \text{ to } 1000)L^{-3}$. For an active region L on the order of a mm, α_{th} would be on the order of 0.1 to 1 cm^{-1} which is also in the range of typical losses for thin-film waveguides. Thus, for thin-film lasers, the spatial gain coefficient should be on the order of 1 cm^{-1} or more, depending of course on the actual coupling parameter and the length of the active region. Gain constants of this order of magnitude is considered high for bulk lasers; therefore, only a few of the many types of active medium and pumping schemes useful for bulk lasers can be suitably modified and adopted for use in integrated optics. The problems involved are considered briefly in the following section.

III. THE ACTIVE MEDIUM

Dye films, semiconductors, and insulating solids suitably doped with impurity ions can all be used as the active medium for laser sources in integrated optics. Although thin-film dye lasers appear to be the most popular with the experimentalists in the field at the present time, in the long run such lasers will probably be of limited use in practical optical systems, because these lasers are limited fundamentally to pulsed operations. The immediate advantage of the dye system is that such lasers are simple to construct and can, therefore, be used to test and try out various ideas on thin-film lasers proposed for integrated optical systems.

The main consideration is to achieve a large enough (amplitude) gain constant α for each medium. The net gain constant for any laser transition would depend upon the population inversion n, the stimulated emission cross section σ_e, and whatever absorption α_a there is in the medium at the laser wavelength:

$$2\alpha = n\sigma_e - 2\alpha_a \; . \tag{24}$$

The emission cross section depends (6) in turn upon the Einstein A-coefficient (or the spontaneous lifetime $\tau_{sp}=A^{-1}$ for model two-level systems), the total linewidth $\Delta\nu$, the relative dielectric constant of the host material ε_r, etc., for a Gaussian line:

$$\sigma_e \cong \sqrt{\frac{\ell n 2}{\pi}} \cdot \frac{A}{8\pi} \cdot \frac{\lambda^2}{\Delta\nu\varepsilon_r} \tag{25}$$

for the particular transition under consideration. The population inversion depends upon the pumping scheme and the relaxation processes of all the levels involved; they vary with different systems obviously and would have to be considered separately for each case.

A. Dye Lasers (7)

Figure 12 shows schematically a simplified energy diagram for the most widely used laser dye RH6G and the corresponding measured absorption and fluorescence curves. The laser transitions in the 5600-6300 Å range are those between a vibrational state of the first excited electronic singlet level S_{1v}, and an excited vibrational state of the ground electronic level S_{0v}. Population inversion is achieved or maintained through optical pumping from the ground to the excited electronic level followed by non-radiative relaxations within the manifold of vibrational states of each electronic level that are extremely fast and can be considered almost as instantaneous for the purpose of the present discussion. The

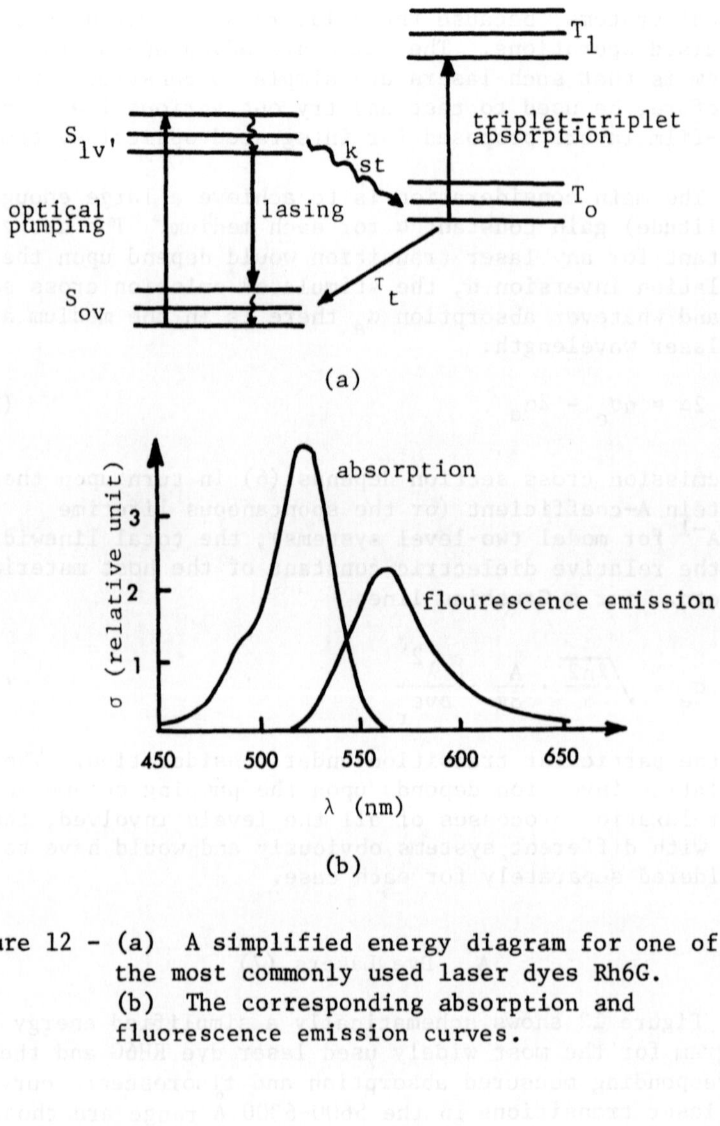

Figure 12 - (a) A simplified energy diagram for one of the most commonly used laser dyes Rh6G. (b) The corresponding absorption and fluorescence emission curves.

spontaneous radiative lifetime from $S_{1v'}$ to S_{ov} is, however, relatively long ($\tau_{sp} \sim 5\times10^{-9}$sec). In addition to the singlet levels, the triplet (T) levels also play a significant role in the operation of the dye laser. In fact, it is the triplet-triplet absorption due to the long lived ($\tau_t \sim 10^{-3}$ sec) lowest triplet level (T_o) of the dye molecules in solid hosts that is mainly responsible for the fact that such lasers can only be operated on pulsed basis.

The singlet-triplet relaxation rate k_{st} is on the order of 10^7 sec^{-1}. Thus, for any given steady-state population density n_s for the upper laser state S_{1v}, the corresponding population density n_t of the triplet level T_o would be:

$$n_t = n_s k_{st} \tau_t \quad ; \qquad (26)$$

because $\tau_t k_{st} >> 1$, there will be an accumulation of population in the level T_o. Since the triplet-triplet transition frequency is close to that of singlet-singlet transitions, the net laser gain coefficient of Eq. (24) becomes

$$2\alpha = n\sigma_e - n_s k_{st} \tau_t \sigma_t \quad ,$$

where σ_t is the triplet-triplet absorption cross section. The population of the lower laser level can usually be considered negligibly small or $n \cong n_s$; therefore,

$$2\alpha \cong n_s(\sigma_e - k_{st}\tau_t\sigma_t) \quad . \qquad (27)$$

Numerically, $\sigma_e \cong 10^{-16}$ cm^2 while $\sigma_t \cong 5x10^{-17}$ cm^2; it is clear that unless τ_t can be reduced to the order of k_{st}^{-1} or 10^{-7} sec or shorter, there cannot be any net gain on a steady-state basis. In the bulk dye lasers, the dye molecules are in liquid solvents that can be saturated with, for example, oxygen; collisions between the oxygen and dye molecules can lead to quenching of the triplet states. Furthermore, the dye solution can be flown rapidly across the laser region to physically remove the triplet state molecules also. In a solid matrix, there is no comparably effective means for quenching the triplet states. Therefore, population inversion can be maintained or achieved only on the order of 10^{-7} sec or less, before an appreciable number of molecules have relaxed into the T_o level via the S_1 level. Another equally fundamental limitation is that the dye molecules will eventually bleach or chemically decompose under prolonged illumination by light. In the bulk lasers, the molecules are replenished in the flowing system, which again cannot be done in the solid dye film. The laser is therefore limited to pulsed operation. For a pulse duration shorter than the singlet-triplet relaxation time 10^{-7} sec, the triplet state population can be neglected and Eq. (27) reduces to

$$2\alpha \cong n_s \sigma_e . \tag{28}$$

The excited singlet state population n_s is determined by the molecular concentration N (number per cm^3), the $S_o \to S_1$ optical pumping rate $\sigma_p I_p / h\nu_p$, and the $S_1 \to S_o$ spontaneous rate τ_s^{-1};

$$n_s = N \frac{\sigma_p \tau_s I_p / L\nu_p}{1+\sigma_p \tau_s I_p / h\nu_p} , \tag{29}$$

where I_p is the pumping light intensity and ν_p is the corresponding frequency. Numerically, $\sigma_p \tau_s / h\nu_p$ is approximately 10^{-13} esu (using $\sigma_p \approx 2 \times 10^{-16} cm^2$); therefore, most of the dye molecules will be inverted and in the S_1 level if I_p is approximately 1 MW/cm^2 or more. Much higher pumping intensity will not lead to much higher α in a pulsed Rh6G dye laser; this is, therefore, the pumping intensity to aim for. In this range, we have the extremely simple result

$$2\alpha \approx N\sigma_e . \tag{30}$$

Since $\sigma_e \cong 10^{-16}$ cm^2, one needs a dye concentration of at least 2×10^{-5} M in the solid film to achieve an α of 1 cm^{-1} at a minimum pumping intensity of $I_p \cong 1$ MW/cm^2. Since the gain scales with the dye concentration upto may be 10^{-2}M, above which it may become difficult to get good quality doped films, one can readily achieve a net gain on the order of 10 db/cm or more for Rh6G dye films pumped by, for example, a pulsed N_2 laser at 3370 Å. This is indeed the most popular system with the experimentalists and the observed gain is consistent with our estimates here. The most commonly used host substance appears to be polyurethane.

Such dye lasers are reported widely in the literature. Not only the gain, but also the observed general spectral or mode properties are in agreement with the picture developed in the previous section.

B. Semiconductor Lasers (8)

Semiconductor lasers are already in miniaturized form

and have been intensively developed long before the advent of integrated optics. Recent efforts in the area are mainly directed towards developing semiconductor lasers using artificial periodic structures as the resonant feedback structure. Since an extensive literature on the basic mechanisms taking place in semiconductor lasers already exists, we will confine our discussions to quoting a few numbers in order to compare these with the other lasers potentially useful for integrated optical systems.

In semiconductors, the laser transitions take place between the conduction and valence bands or between these bands and the impurity levels. For rough numerical estimates, the laser can be analyzed using a two-level model and on the basis of Eq. (24) and Eq. (25). GaAs is perhaps the most important laser material at the present time; we will use it as an example for the purpose of our numerical discussions here. In comparison with Rh6G, for GaAs at 77° K the spontaneous lifetime or the recombination time τ_s is about a factor of 2 smaller, the linewidth a factor of ~ 5 smaller, the relative dielectric constant ~ 10 larger, and λ^2 is ~ 2 larger; the emission cross section is, therefore, approximately 2×10^{-16} cm^2 on the basis of Eq. (25). Assuming α_a to be negligibly small or the internal quantum efficiency to be nearly 1, an electron-hole pair concentration n_c of 2×10^{17} cm^3 would then lead to a gain constant of $2\alpha \simeq 40$ cm^{-1}, which is enough for laser oscillation in a 100μ long GaAs sample with cleaved ends. With a periodic structure as the resonant feedback structure, one should be able to reduce the required α for oscillation.

Electron-hole pairs can be created near the polished surface of bulk GaAs crystals or platelets by optical or electron-beam pumping or by injection through a p-n junction. In the case of optical pumping, since τ_s for GaAs at 77° K is about 3 nsec, to achieve an electron-hole pair concentration of 2×10^{17} cm^{-3} for laser oscillation, an optical pumping rate $2\alpha_p I_p/h\nu_p$ of 4×10^{26} per sec-cm^3 would be needed. Since α_p corresponds to band-to-band transitions, it can be very large and depends upon the frequency ν_p of the pumping light relative to the fundamental absorption-band edge. If ν_p is such that α_p becomes too large (say $\gg 10^4$ cm^{-1}) then the depth of the active region near the surface of the crystal would be too small leading to a large diffraction loss for the laser mode; this is not desirable. Take for example $\alpha_p \gtrsim 10^3$ cm^{-1}; the required pumping intensity for oscillation is approximately 50 KW/cm^2, which is achievable using laser pumping on a pulsed basis. Experimentally, the reported threshold pumping intensity is of this order of magnitude.

Pumping can also be achieved using an electron-beam (50 KV or more) on pulsed basis. The required deposition rate of pumping energy density is comparable to the optical-pumping case. One advantage of electron-beam pumping is that the penetration depth of the beam can be more easily adjusted by changing the beam voltage, while in the optical case a tunable source would be needed.

The most important pumping scheme for semiconductor lasers is still injection through a junction. The requried threshold current density J_{th} for a junction of width d is approximately:

$$J_{th} = \frac{n_c ed}{\tau_s} , \qquad (31)$$

assuming again unit quantum efficiency. Eq. (31) follows from the simple fact that the total electric power dissipated per unit volume $J_{th}V/d$ goes into exciting the electrons across the band-gap of V; thus, the rate of pumping (number of electrons per cm^3) is J_{th}/ed. The lifetime of the electrons in the upper laser level is τ_s; the corresponding population density is, therefore, $J_{th}\tau_s/ed$, which must be equal to n_c at the threshold for oscillation. For $d \sim 10^{-4}$ cm, J_{th} is on the order of 1000 Amp/cm^2, which is again achievable in cooled diodes.

Junction diodes using periodic structures as the feedback structure have not yet been achieved. The technical task of growing a junction with a periodic structure imbedded in it is not trivial but appears solvable.

C. Doped Insulating Solids (9,10,11)

Solid-state lasers using insulating crystals or glass hosts doped with suitable impurity ions are beginning to receive increased attention as possible laser sources for integrated optical systems. These lasers can be optically pumped by light emitting diodes and as such would be useful for practical systems. We consider several possible materials:

ruby (9) and Nd^{3+} doped YAG (10) or glass (11).

Ruby or Cr^{3+} doped Al_2O_3 crystal is a three-level system as shown schematically in Figure 13. In laser operation, Cr^{3+} ions in the ground state 1 is optically pumped (rate τ_p^{-1}) to level 3 followed by extremely fast nonradiative decay

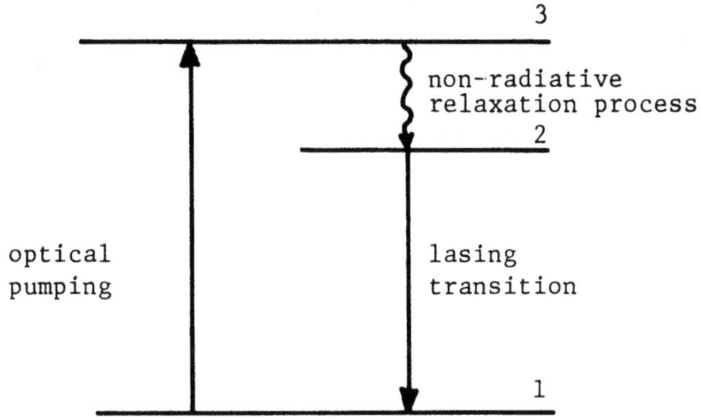

Figure 13 - Schematic energy diagram of an optically pumped three-level laser system.

to the upper laser level 2, which has a relatively long radiative lifetime of $\tau_s \sim 3\times10^{-3}$ sec. The linewidth $\Delta\nu_{21}$ for the radiative transition 2→1 at 6943 Å is, however, much wider than τ_s^{-1} due to the two-phonon Raman process in the solid; it is at least 10^{11} sec^{-1} or wider at room temperature. Scaling again from the dye results on the basis of Eq. (25), one obtains an emission cross section of $\sim 2\times10^{-20}$ cm^2, which is close to the measured value. To obtain a gain constant $\alpha \sim 1$ cm^{-1} would require according to eq. (24) a population inversion of $n \gtrsim 2\alpha/\sigma_e \sim 10^{20}$ cm^{-3}, which is not easy but still possible to achieve; it would require rather heavy doping compared with the ruby crystals typically used for bulk lasers. The actual laser will probably take the form of a high quality or low-loss thin-film waveguide deposited on a polished ruby crystal. In this case, because of the fact that only the fringe field of a guided mode will "see" the active medium, the effective gain constant of the waveguide mode will be proportionally reduced or the required population inversion would be still higher. If the film loss is substantially lower than 1 cm^{-1}, then it would be possible by making the laser longer, say 1 cm or longer, to reduce the required α and hence the doping concentration by a factor of 10 or more. If the film loss is already 1 cm^{-1} or more, then increasing the laser length would not help much since one must still overcome the film loss first, which may already be too high.

For the three-level laser shown in Figure 13, there is also a minimum required optical pumping rate in order to establish a population inversion:

$$1/\tau_p \equiv \sigma_p I_p / h\nu_p > 1/\tau_s , \tag{32}$$

or

$$I_p > h\nu_p / \sigma_p \tau_s . \tag{33}$$

For ruby, σ_p for the blue-green pumping band is approximately 10^{-19} cm^2. Thus, a minimum pumping intensity on the order of $\sim$2KW/cm^2 is required to establish a population inversion in the ruby crystal independent of the Cr^{3+} ion impurity concentration. This is needed just to make $\alpha = 0$. To make α large enough to overcome the losses in the laser in order to reach the threshold for oscillation would of course require more and is subject to the considerations discussed in the previous paragraph.

Nd^{3+} ions in either a crystalline host such as YAG (yittrium aluminum garnet) or glass hosts may also be a useful system. It is a four-level system with the lower laser level somewhat above the ground and hence essentially empty; population inversion is, therefore, easier to achieve than the three-level system. For bulk lasers where the loss in the medium is small and the laser can be made quite long, this is a very important advantage. For thin-film lasers, a substantial population inversion is in any case required to overcome the relatively high losses in the waveguide and the feedback structure, this is a less important advantage. In addition, it is partly offset by the fact that absorption cross section for the orange pumping band is smaller by almost an order of mangitude for Nd^{3+} in YAG or glasses than for ruby. The emission cross section at 1.06μ for Nd^{3+} in YAG is, however, larger $\sim 10^{-19}$ cm^2; thus, this system is comparable to ruby if the required α is on the order of 1 cm^{-1} or more. Otherwise, the Nd-YAG may require less pumping intensity. The emission cross section for Nd^{3+} in glass at 1.06μ is about 10^{-20} cm^2 or about a factor of 10 smaller because of the large linewidth due to inhomogeneous broadening in the amorphous glass matrix. Therefore, Nd-glass would require a larger population inversion or pumping intensity to reach oscillation. On the other hand, because of the larger linewidth, the requirement on the accuracy of the lattice spacing of the periodic structure used for the resonant feedback structure becomes less stringent. This is an important practical consideration in device fabrication.

D. Gaseous Lasers

Finally, there has also been a great deal of interest in developing miniaturized gaseous lasers using capillary waveguides as the discharge tube. The emission cross sections of the gaseous laser transitions are generally much larger: ranging from 10^{-17} cm^2 for the 10.6μ line in CO_2 (12), to 10^{-13} cm^2 for the 1.15μ line of Ne (13) and 10^{-12} cm^2 for the 4880 Å line (14) of Ar^+; but, because of the low volume densities of the molecules, atoms, or ions in the gaseous media, the net gain for these systems is relatively low. However, by making use of the fact that the plasma parameters of the laser discharges tend to scale roughly as (pressure X tube diameter $\approx$ constant), it is possible to operate at a higher pressure and hence higher active particle density or gain using a smaller diameter tube. Short capillary gaseous lasers (15) on the order of inches have recently been developed and could have some specialized applications in integrated optics.

Appendix - Derivation of Eqs. (5) and (13).

The fields in the spatially modulated medium must satisfy the wave equation

$$\frac{\partial^2}{\partial z^2} E(z) + \frac{n^2(\omega)\omega^2}{c^2} E(z) = 0. \qquad \text{(A1)}$$

Substituting Eqs. (1), (3), and (4) in Eq. (A1) and equating the coefficients of terms with the same exponential dependence gives (16):

$$[\bar{k}^2 - k^2]E_o + 2K\bar{k}E_\pm = 0 \qquad \text{(A2)}$$

$$[\bar{k}^2 - (k\pm G)^2]E_\pm + 2K\bar{k}E_o = 0 \qquad \text{(A3)}$$

where $\bar{k}$ and K are defined in Eqs. (6) and (7). For nontrivial solutions to exist, the secular determinant of Eqs. (A2) and (A3) must be zero, which yields the dispersion relation Eq. (5).

To obtain Eq. (13), one notes that the amplitude E_- of the lower side band is related to the amplitude of the forward-traveling free wave E_0 as follows:

$$E_- = -\frac{\bar{k}^2-k^2}{2K} E_o \cong \frac{k-\bar{k}}{K} E_o \qquad \text{(A4)}$$

from Eq. (A2). In the presence of gain, $\bar{k}^2-k^2$ in all the previous equations is replaced by $\bar{k}^2-2i\alpha\bar{k}-k^2$; therefore,

$$E_- = \frac{k-\bar{k}+i\alpha}{K} E_o$$

which is Eq. (13).

REFERENCES

* Work supported in part by the National Science Foundation through Grant No. GK-33848.

1. F. P. Kapron, D. B. Keck, and R. D. Maurer, Appl. Phys. Letters 17, 423 (1970).

2. S. E. Miller, Bell System Tech. J., 48, 2059 (1969).

3. H. Kogelnik and C. V. Shank, J. Appl. Phys., 43, 2327 (1972).

4. M. L. Dakss, L. Kuhn, P. F. Heidrich, and B. A. Scott, Appl. Phys. Letters, 16, 523 (1970).

5. J. Turner, B. Chen, L. Yang, J. M. Ballantyne, and C. L. Tang, Appl. Phys. Letters, 23, 333 (1973).

6. W. V. Smith and P. P. Sorokin, The Laser, McGraw Hill Co. (New York, 1966), Eqs. (2-12) and (7-1).

7. B. B. Snavely, Proc. IEEE, 57, 1374, (1969).

8. G. Burns and M. I. Nathan, Proc. IEEE, 52, 770 (1964).

9. T. H. Maiman, Phys. Rev. Letters, 4, 564 (1960); T. H. Maiman, R. Hoskins, I. D'Haenens, C. Asawa, and V. Evtuhov, Phys. Rev. 123, 1151 (1961).

10. J. E. Geusic, H. Marcos, and L. G. Van Uitert, Appl. Phys. Letters, 3, 183 (1964).

11. E. Snitzer and C. C. Young, in Lasers, Vol. 2, edited by A. K. Levine, M. Dekker, Inc. (New York, 1971).

12. H. Statz, C. L. Tang and G. F. Koster, J. App. Phys., 37, 4278 (1966).

13. H. Statz, C. L. Tang, and G. F. Koster, J. App. Phys., 34, 2625 (1963).

14. R. I. Rudko and C. L. Tang, J. App. Phys., 38, 4731 (1967).

15. P. W. Smith, Appl. Phys. Letters, 19, 132 (1971); R. L. Abrams and W. B. Bridges, IEEE J. of Quant. Elect., QE-9, 694 (1973); D. T. Hodges and T. S. Hartwick, Appl. Phys. Letters 23, 252 (1973).

16. C. L. Tang, IEEE J. of Quant. Elect., QE-9, 9 (1973), Eq. (7) and Eq. (8).

INDEX

MIX
Papier aus verantwortungsvollen Quellen
Paper from responsible sources
FSC® C105338

If you have any concerns about our products,
you can contact us on
ProductSafety@springernature.com

In case Publisher is established outside the EU,
the EU authorized representative is:
Springer Nature Customer Service Center GmbH
Europaplatz 3, 69115 Heidelberg, Germany

Printed by Libri Plureos GmbH
in Hamburg, Germany